W0049851

TOWARDS UNDERSTANDING GALAXIES AT LARGE REDSHIFT

ASTROPHYSICS AND SPACE SCIENCE LIBRARY

A SERIES OF BOOKS ON THE RECENT DEVELOPMENTS
OF SPACE SCIENCE AND OF GENERAL GEOPHYSICS AND ASTROPHYSICS
PUBLISHED IN CONNECTION WITH THE JOURNAL
SPACE SCIENCE REVIEWS

Editorial Board

R.L.F. BOYD, *University College, London, England*

W. B. BURTON, *Sterrewacht, Leiden, The Netherlands*

L. GOLDBERG †, *Kitt Peak National Observatory, Tucson, Ariz., U.S.A.*

C. DE JAGER, *University of Utrecht, The Netherlands*

J. KLECZEK, *Czechoslovak Academy of Sciences, Ondřejov, Czechoslovakia*

Z. KOPAL, *University of Manchester, England*

R. LÜST, *European Space Agency, Paris, France*

L. I. SEDOV, *Academy of Sciences of the U.S.S.R., Moscow, U.S.S.R.*

Z. ŠVESTKA, *Laboratory for Space Research, Utrecht, The Netherlands*

VOLUME 141
PROCEEDINGS

TOWARDS UNDERSTANDING GALAXIES AT LARGE REDSHIFT

PROCEEDINGS OF THE FIFTH WORKSHOP OF THE
ADVANCED SCHOOL OF ASTRONOMY OF THE
ETTORE MAJORANA CENTRE FOR SCIENTIFIC CULTURE,
ERICE, ITALY, JUNI 1-10, 1987

Edited by

RICHARD G. KRON

Department of Astronomy and Astrophysics,
University of Chicago, Chicago, Illinois

and

ALVIO RENZINI

Dipartimento di Astronomia,
Università di Bologna, Bologna, Italy

Springer-Science+Business Media, B.V.

Library of Congress Cataloging in Publication Data

CIP

Ettore Majorana International Centre for Scientific Culture. Advanced School of Astronomy.
    Workshop (5ht: 1987: Erice, Sicily)
    Towards understanding galaxies at large redshift: proceedings of the Fifth Workshop of
the Advanced School of Astronomy of the Ettore Majorana Centre for Scientific Culture,
Erice, Sicily, Italy, June 1–10, 1987 / edited by Richard G. Kron and Alvio Renzini.
        p.   cm. — (Astrophysics and space science library; v. 141)
    Includes index.
    ISBN 978-94-010-7814-6   ISBN 978-94-009-2919-7
    DOI 10.1007/978-94-009-2919-7

    1. Galaxies—Congresses. 2. Red shift—Congresses. I. Kron, Richard G.
II. Renzini, Alvio. III. Title. IV. Series.
QB856.E77   1987
523.1'12—dc 19

87–36756
CIP

Published by Kluwer Academic Publishers,
P.O. Box 17, 3300 AA Dordrecht, Holland.

Kluwer Academic Publishers incorporates
the publishing programmes of
D. Reidel, Martinus Nijhoff, Dr W. Junk and MTP Press.

Sold and distributed in the U.S.A. and Canada
by Kluwer Academic Publishers,
101 Philip Drive, Norwell, MA 02061, U.S.A.

In all other countries, sold and distributed
by Kluwer Academic Publishers Group,
P.O. Box 322, 3300 AH Dordrecht, Holland.

All Rights Reserved
© 1988 Springer Science+Business Media Dordrecht
Originally published by Kluwer Academic Publishers 1988
Softcover reprint of the hardcover 1st edition 1988
No part of the material protected by this copyright notice may be reproduced or
utilized in any form or by any means, electronic or mechanical
including photocopying, recording or by any information storage and
retrieval system, without written permission from the copyright owner

TABLE OF CONTENTS

PART 2. STAR-FORMING DISK SYSTEMS

PART 3. **FIELD AND CLUSTER GALAXIES AT LARGE REDSHIFT**

PREFACE

This workshop was intended as an update and an extension of the workshop on the "Spectral Evolution of Galaxies" that was held in Erice two years ago. It concentrates on new developments concerning galaxies seen at large lookback times. This seemed also a good opportunity to look ahead to the next generation of ground- and space-based instrumentation, and to consider various future strategies for collecting information concerning the edge of the observable universe.

The main idea was to bring together people with specialities in modelling galaxy components (such as stars, clusters, gas, and dust) as well as whole stellar systems (stellar populations, star formation rates, chemical enrichment), and people specialized in making direct measurements of galaxies and clusters at large lookback times.

The confrontation of expectations and observations was planned to be the central theme of the conference, which explains the title "Towards Understanding Galaxies at Large Redshift". The first part of the workshop focussed on the physical processes that operate in galaxies, and that would likely have some observable manifestation at large redshifts. In the second part the most recent observational work was reported, and we were pleased to have the participation of most of the groups active in this field. The last part was directed towards new approaches to be made possible by the next generation of instrumentation, although in general all the contributions were indeed in this spirit of setting more ambitious goals. The papers are correspondingly organized in this sequence, even though for practical reasons the actual program did not follow this plan exactly.

The format of the meeting and the ambiance of Erice naturally provided for an excellent interaction among the participants, and in many ways most of the *real work* was perhaps accomplished outside of the meeting room, as usual in these cases! To promote a sense of informality and stimulate discussion we did not pay much attention to the time schedule of the meeting, and this often resulted in very long, spirited discussions, that made hopeless any attempt of record. This is certainly a limitation for the present proceedings, although we urged the authors to try incorporating in their written contribution the flavor of the discussion and the resulting new insights.

We would like to thank the "Ettore Majorana Center for Scientific Culture" for the ideal handling of all the logistics of the meeting, and in particular Ms. Pinola Savalli for her invaluable contribution through all the phases of the organization. We are also grateful to the Italian National Research Council (CNR) for a small grant that allowed the participation of a few young researchers, and we are most grateful to all participants for having used their own travel grants in order to attend the meeting.

Richard Kron & Alvio Renzini

PARTICIPANTS

A. Aragon: Departamento de Astrofisica, Universidad de Madrid, Madrid, Spain
Nobuo Arimoto: Tokyo Astronomical Observatory, Mitaka, Japan
Eulalia Athanassoula: Observatoire de Marseille, Marseille, France
Jacques M. Beckers: Advanced Development Program, NOAO, Tucson, AZ, USA
Matthew Bershady: University of Chicago, Chicago, IL, USA

Eduardo Bica: Observatoire de Paris-Meudon, Meudon, France
Albert Bosma: Observatoire de Marseille, Marseille, France
Richard Bower: Department of Physics, University of Durham, Durham, UK
Gustavo Bruzual: Centro de Investigaciones de Astronomia, Merida, Venezuela
David Burstein: Physics Department, Arizona State University, Tempe, AZ, USA

Lucio M. Buson: Osservatorio Astrofisico, Asiago, Italy
Alberto Buzzoni: Osservatorio Astronomico, Merate, Italy
Alfonso Cavaliere: Dipartimento di Fisica, II Università di Roma, Roma, Italy
Veronique Cayatte: Observatoire de Paris-Meudon, Meudon, France
Guido Chincarini: Osservatorio Astronomico, Merate, Italy

Gisella Clementini: Space Telescope Science Institute, Baltimore, MD , USA
Paolo Conconi: Osservatorio Astronomico, Merate, Italy
Arlin P. S. Crotts: McDonald Observatory, Austin, TX, USA
A. Diaz: Departamento de Fisica Teorica, Universidad de Madrid, Madrid, Spain
Sperello di Serego: ESO, Garching b. München, FRG

S. George Djorgovski, Center for Astrophysics, Cambridge, MA, USA
Sandro D'Odorico: ESO, Garching b. München, FRG
Richard Ellis: Department of Physics, University of Durham, Durham, UK
S. Michael Fall: Space Telescope Science Institute, Baltimore, MD, USA
Klaus J. Fricke: Universitäts-Sternwarte Göttingen, Göttingen, FRG

Jay A. Frogel: NOAO, Tucson, AZ, USA
Riccardo Giovanelli: NAIC, Arecibo Observatory, Arecibo, Puerto Rico
Laura Greggio: Dipartimento di Astronomia, Università di Bologna, Bologna, Italy
Preben J. Grosbøl: ESO, Garching b. München, FRG
Puragra Guhathakurta, Princeton University Observatory, Princeton, NJ, USA

Bruno Guiderdoni: Institut D'Astrophysique, Paris, France
James E. Gunn: Princeton University Observatory, Princeton, NJ, USA
J. Patrick Henry: Institute for Astronomy, University of Hawaii, Honolulu, HI, USA
Charles R. Jenkins: Royal Greenwich Observatory, Hailsham, UK
Wolfram Kollatschny: Universitäts-Sternwarte Göttingen, Göttingen, FRG

David C. Koo: Space Telescope Science Institute, Baltimore, MD, USA
Gerald E. Kron: Pinecrest Observatory, Honolulu, HI, USA
Katherine G. Kron: Pinecrest Observatory, Honolulu, HI, USA
Richard G. Kron: University of Chicago, Chicago, IL, USA
Edwin D. Loh: Department of Physics, Princeton University, Princeton, NJ, USA

Steven Majewski: Yerkes Observatory, Williams Bay, WI , USA
Bruno Marano: Osservatorio Astrofisico, Catania, Italy
Francesca Matteucci: ESO, Garching b. München, FRG
Y. Mellier: Observatoire de Toulouse, Toulouse, France
Jorge Melnick: ESO, Garching b. München, FRG

Emilio Molinari: Osservatorio Astronomico, Merate, Italy
Jeffrey Munn: Space Telescope Science Institute, Baltimore, MD, USA
Robert W. O'Connell: Leander McCormick Observatory, Charlottesville, VA, USA
R.B. Partridge: Department of Astronomy, Haverford College, Haverford, PA, USA
Eric Persson: Mount Wilson and Las Campanas Observatories, Pasadena, CA, USA

Andrew Pickles: La Palma Observatory, Santa Cruz de la Palma, Tenerife, Spain
Christopher J. Pritchet: University of Victoria, Victoria, BC, Canada
Alvio Renzini: Dipartimento di Astronomia, Università di Bologna, Bologna, Italy
Brigitte Rocca-Volmerange: Institut D'Astrophysique, Paris, France
James A. Rose: University of North Carolina, Chapel Hill, NC, USA

Roberto P. Saglia: Scuola Normale Superiore, Pisa, Italy
John Scalo: Astronomy Department, University of Texas, Austin, TX
Smita Shanbhag: Tata Institute, Bombay, India
L. Stanghellini: Department of Astronomy, University of Illinois, Urbana, IL, USA
Monica Tosi: Osservatorio Astronomico, Bologna, Italy

Dario Trevese: Istituto Astronomico, Università di Roma, Roma, Italy
J. Anthony Tyson: Bell Laboratories, Murray Hill, NJ, USA
A. Vallenari: Dipartimento di Astronomia, Università di Padova, Padova, Italy
Paolo Vettolani: Istituto di Radioastronomia, Bologna, Italy
David Weinberg: Princeton University Observatory, Princeton, NJ, USA

Tomohiko Yamagata: Tokyo Astronomical Observatory, Mitaka, Japan
Gianni Zamorani: Osservatorio Astronomico, Trieste, Italy
H.K.C. Yee: Departement de Physique, Université de Montreal, Montreal, Canada
Esther Zirbel: Astronomy Department, Yale University, New Haven, CT, USA
V. Zitelli: Dipartimento di Astronomia, Università di Bologna, Bologna, Italy

OPTICAL AND INFRARED STUDIES OF STELLAR POPULATIONS: THE GALACTIC NUCLEAR BULGE

Jay A. Frogel
Kitt Peak National Observatory
950 North Cherry Avenue, P. O. Box 26732
Tucson, AZ 85726

ABSTRACT. Giants in the Galactic nuclear bulge are different from solar neighborhood giants in terms of their observed photometric and spectroscopic characteristics and their mean metallicity. These bulge stars are the best available sample to use in stellar synthesis models of early-type galaxies. Observations from the main sequence to the top of the giant branch show no evidence for a component of the bulge population much younger than 10 Gyr. Several independent lines of evidence indicate a strong metallicity gradient in the bulge as would be expected if it formed dissipatively.

1. INTRODUCTION

In order to synthesize a realistic population model for an unresolved stellar system three rather obvious and important criteria have to be met: the stars in the model must have the correct physical properties, *e.g.* age and chemical composition; the distribution of the stars over mass and luminosity must be correct; the possibility that the first two criteria can vary with position has to be allowed for. A key test of a stellar synthesis model should be its ability to reproduce in detail the integrated light of a system for which the individual stars can be observed (cf. Renzini and Buzzoni 1986 and Searle 1986). With the exception of models based on the Yale isochrones and used to predict optical and infrared colors and indices of galactic globular clusters (Aaronson, *et al.* 1978; Frogel, Cohen, and Persson 1983), such a test has not been carried out.

A more empirical approach to model building would start with simple resolvable systems and use them as building blocks for more complex, unresolvable systems. Such an approach avoids to a certain extent the rather nasty problems associated with the determination of physical properties and luminosity functions for stars, but does require some means of evaluating the appropriateness of the blocks chosen. For example, many galaxy models use the solar neighborhood as a basic building block and rely on empirically determined colors and luminosities of nearby stars. The star clusters in the Magellanic Clouds with their large range in age and metallicity offer an excellent set of such basic blocks; they have been used in

1

R. G. Kron and A. Renzini (eds.), Towards Understanding Galaxies at Large Redshift, 1–14.
© *1988 by Kluwer Academic Publishers.*

analyzing the integrated light of the nuclei of late-type spiral galaxies (Frogel 1985).

A critical element in the understanding of galaxies at large redshift is knowledge of how their appearance changes with lookback time due to the effects of stellar evolution. It is for this reason that the construction of "correct" stellar population models for nearby galaxies is so important.

In this review I will summarize ongoing studies of stars in the nuclear bulge of the Milky Way, both optical and infrared, and argue that these stars are the best sample we currently have for use as the basic constituents in models of spheroidal stellar systems, i.e. early-type galaxies and the bulges of spirals. Recently published studies of K and M giants in Baade's Window at b= -3.9° show that these stars have significantly different infrared colors and luminosities than corresponding stars in synthesis models (Frogel, Whitford, and Rich 1984, Frogel and Whitford 1987, hereafter FW). Optical color magnitude diagrams and band strength indicators for the bulge stars also differ markedly from those for their solar neighborhood counterparts (Rich 1986, Terndrup 1986, Cook 1987, Whitford 1985 and 1986). Finally, Blanco, Terndrup, Whitford, and I have just finished an examination of M giants in the bulge between b = -8° and -12°. We find strong gradients in the colors and band strengths of the M giants.

2. JUSTIFICATION FOR THE USE OF BULGE GIANTS

Although the range of physical parameter space occupied by the two types of stellar systems may differ somewhat, elliptical galaxies and the bulges of spirals have quite similar broad band colors and spectral line strengths. Also, with the exception of the nuclear regions of late-type spirals, stellar synthesis models of the two types of systems do not seem to have any striking differences. There may be, though, differences in the dynamical characteristics of ellipticals and spiral bulges. Furthermore, the nuclear regions of spirals of type Sbc and later show increasing amounts of activity as evidenced by ultraviolet or infrared excess emission, non-thermal continua, etc. This activity is usually attributed to recent star formation or to more exotic energy sources. Although upon careful observation a high percentage of ellipticals show such nuclear activity (e.g. Phillips, *et al.* 1986) it is at a substantially lower level than that seen in late-type spirals.

The nearest spiral bulge we can observe is that of our own galaxy. At low latitudes the bulge is heavily obscured or invisible at optical wavelengths due to extinction from dust in the disk of the Galaxy, but Baade (1963) pointed out a number of "windows" in this nearby absorbing material through which the bulge can be seen even at low latitudes. The most studied of these is at -3.9° and is usually called Baade's Window (BW). Qualitative work by Morgan (1956) and a quantitative study by Whitford (1978) demonstrated that the strengths of many of the strongest spectral features in the integrated light of BW are indistinguishable from that seen in bulges of other spirals and in the central regions of E and S0 galaxies.

In the near infrared the integrated light of E and S0 galaxies and the bulges of early type spirals has a strong contribution from M stars (Frogel, *et al.* 1978 and references therein). Surveys by Stebbins and Whitford (1948) and Johnson (1966) showed that although these M stars contribute only about 10% of the light at V, they dominate at wavelengths longward of 1.0μm. In the nuclear bulge of the Galaxy objective prism and grism surveys by Nassau and Blanco (1958) and Blanco, McCarthy, and Blanco (1984) also revealed the presence of large numbers of M giants.

These two results - similarity in spectral features and presence of large numbers of cool giants - suggest that the stars in the Galactic nuclear bulge are a population closely representative of that to be found in early-type galaxies. A caveat to bear in mind, though, is that although the bulge appears to be a well defined physical entity (*e.g.* Habing, *et al.* 1985, but see also Mould 1986) it is not clear how contaminated any sample of "bulge" stars will be by disk stars at similar galactocentric distances. A careful delineation of the observable characteristics and spatial density of the disk stars in the inner region of the Galaxy is needed.

3. DIFFERENCES BETWEEN BULGE AND LOCAL GIANTS

A key physical difference between giants in the Galactic bulge and those in the solar neighborhood is metallicity. In his thesis, Rich (1986; preliminary results reported by Whitford and Rich 1983) derived abundances for 88 K giants in BW. He found a range in [Fe/H] of from -1 to nearly +1 with a mean at 0.3, twice the solar value. Although calibration of the metal-rich end of the abundance scale is rather uncertain, a straightforward comparison of bulge stars with local standards demonstrates that nearly one-quarter of the former stars observed by Rich are stronger lined, hence more metal rich, than the most metal rich K giant near the sun. Recently, Arimoto and Yoshii (1987) have shown that such a range in metallicity is to be expected from their model for the early chemical enrichment of a galaxy with supernovae-driven galactic winds. Rich also successfully models the [Fe/H] distribution with a considerably simpler model of chemical evolution. Since they evolve from K giants, the M giants in the nuclear bulge should be at least as metal rich as the K's.

The velocity dispersion of the giants in the bulge may be a function of [Fe/H] depending on the rate of collapse and chemical enrichment of the proto-bulge. Rich (1986) may have found such a dependence for the BW K giants in the sense that the high [Fe/H] stars have a velocity dispersion 34 km/sec lower than that of the low [Fe/H] stars. This is a less than two sigma result, however, and considerably more work needs to be done.

Infrared colors and luminosities of bulge giants differ considerably from their solar neighborhood counterparts and from giants in globular clusters. FW have shown that the CO band strengths of essentially all of the M giants they observed are stronger than the mean for solar neighborhood giants of the same color. The *JHK* colors of the bulge M giants also differ significantly from ones in the solar neighborhood and even more so from globular cluster giants. K giants in the bulge differ in a similar fashion from nearby field stars (Frogel, Whitford, and Rich 1984). These CO and color differences are in the expected sense for a metal rich population.

There are three distinct characteristics to the infrared color-magnitude diagram of the bulge M giants: 1) There is a considerable number of stars with luminosities more than half a magnitude greater than the most luminous stars found in globular clusters, but at a given spectral type the

4

mean luminosities of the bulge M giants are one to two magnitudes fainter than the values derived for solar neighborhood giants and used in stellar synthesis models. 2) The giant branch is quite broad, consistent with the range in [Fe/H] deduced from optical spectra even though the scatter in color-color relations for the stars is consistent with that arising from observational uncertainties alone. 3) The giant branch's color overlaps those of 47 Tucanae and M67 rather than being much redder as would be expected from the high mean metallicity. FW conclude that some as yet unidentified blanketing agent (H_2O plus an admixture of other molecules are reasonable candidates) is responsible for the peculiar colors of the stars.

The optical color-magnitude diagrams constructed by Terndrup (1986) for stars in BW and three other fields up to b = $-10°$ are unlike any others. The main sequence turnoff, for an age of 11 ± 3 Gyr, implies a metallicity consistent with Rich's mean spectroscopic value. The location of the giants, however implies a mean metallicity several tenths of a dex lower. As in the infrared, the optical colors of the bulge giants are bluer than expected for their metallicity. The main sequence displays a width consistent with the derived range in metal abundance. Finally, Terndrup finds the fraction of stars with an age of 5 Gyr or less to be insignificant.

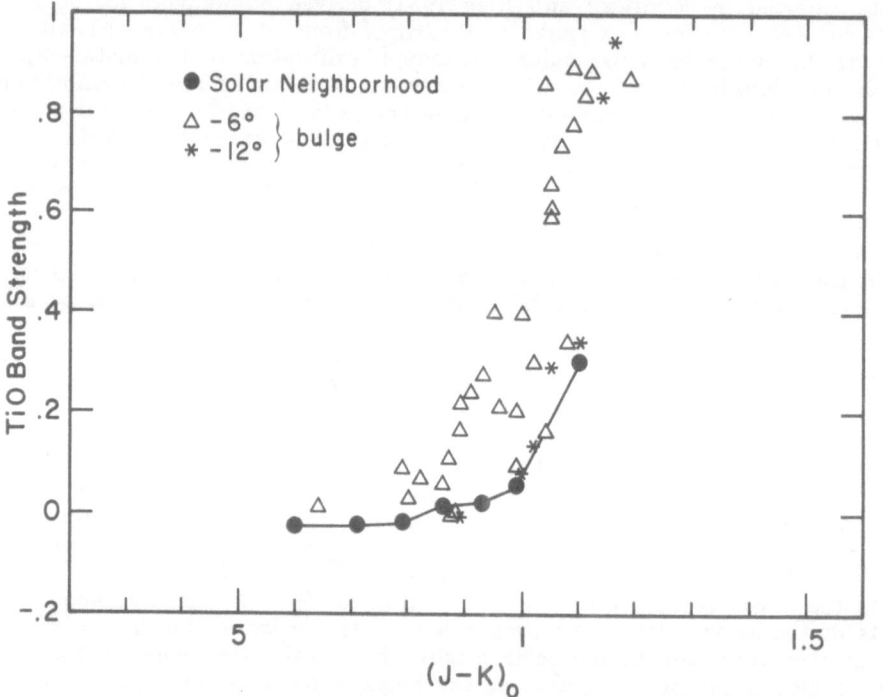

FIGURE 1. The depth of the TiO band near 7800 Å measured with respect to continua points on either side of it in local field giants and in bulge giants at b = -6 and $-12°$.

Terndrup, Whitford, and I have obtained CCD spectra for several hundred bulge M giants. One of our main goals is to determine the metallicity distribution and gradient within the bulge by measuring relative strengths of TiO bands. As an example Figure 1 shows a measure of a strong band near 7800 Å as a function of color for stars in a field at b=-6° compared to the strength of the same band in local field giants. Clearly the bulge giants have much stronger band strengths than field stars of the same color. Kem Cook (1987) in his thesis employs two narrow band filters to measure the same TiO band by imaging fields in globular clusters and the bulge onto a CCD. When plotted against $V-I$, he finds a steady progression of band strength at constant color with the BW stars having the strongest - considerably greater than that seen in local field giants.

4. RADIAL GRADIENTS IN THE NUCLEAR BULGE

Recent work by Blanco, Terndrup, Whitford, and myself strongly establish the existence of a gradient in a number of independent observable quantities for bulge K and M giants. Analysis and interpretation of these gradients should yield important clues for understanding the collapse and chemical evolution of the Galaxy. I will summarize the data as they currently stand.

4.1. Optical Photometry and Spectroscopy

Van den Bergh and Herbst (1974) called attention to the fact that their V, $B-V$ giant branch for the -8° window was somewhat bluer than that determined by Arp (1965) in BW at -3.9°, probably indicative of a lower mean metallicity at the higher latitude. Frogel, Blanco, and Whitford (1984) noted that the higher latitude window had a considerably smaller fraction of luminous red stars than did the lower latitude one. Blanco and Blanco (1986) presented a rather dramatic looking diagram which showed that the surface number density of M5 and later giants between galactic latitudes -2 and -12.6° fell off nearly *three orders of magnitude* faster than would be expected from a stellar density distribution that followed a Hubble or a de Vaucouleurs surface brightness distribution. Both of these recent results can be accounted for at least in part by the existence of a steep metallicity gradient; such a gradient will act in two ways. First the Hayashi track for giants shifts to higher temperatures, i.e. to the blue and earlier spectral types, as metallicity decreases. Second, there is the Mould and McElroy (1978) effect which, as discussed by FW, has its origin in the fact that the band strength of a simple diatomic molecule like TiO will vary as the number of heavy atoms squared. The effect is that in BW the M giants appear to be several spectral sub-classes too late for their effective temperature. If metallicity enhancement decreases with increasing latitude, this effect will decline in importance with a resulting shift to earlier spectral types. It is also possible that the Blanco and Blanco data indicate the presence of a component to the galaxy with an exceptionally steep density law and small scale height. As we will see, evidence from luminosity function data favors the metallicity gradient interpretation.

Terndrup (1986) has BVI c-m diagrams for BW and for fields at b = -6, -8, and -10°. He finds from fits to metal rich isochrones of VandenBerg

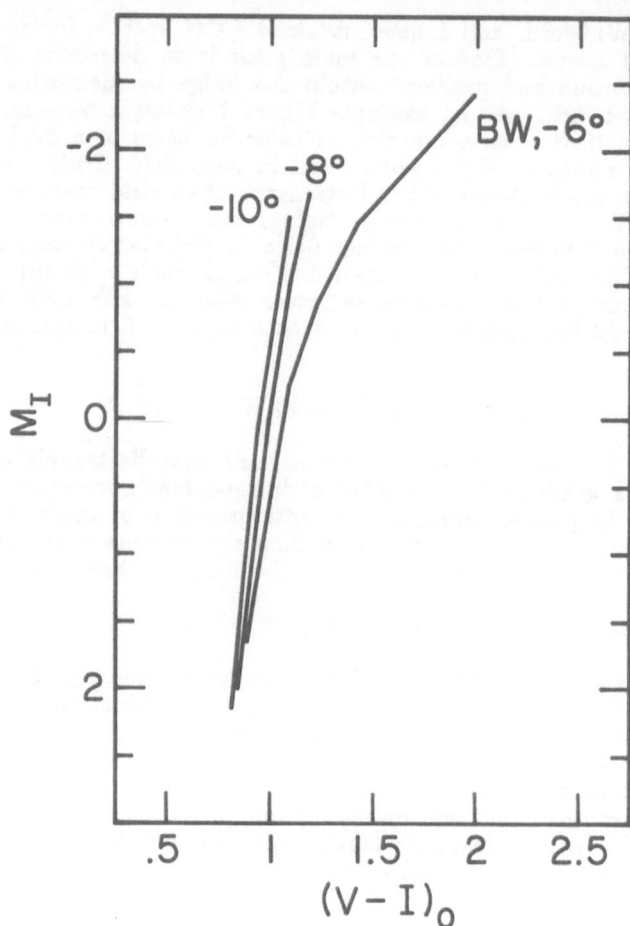

FIGURE 2. The principal giant sequences for 4 latitudes in the bulge from Terndrup's (1986) thesis. The data have been reddening corrected.

(1983) that the *mean* value of [Fe/H] decrease by 0.4 dex over this range of latitude. This shift is large enough so that it is not just the colors of the sequences in his c-m diagrams that are changing, but the shapes as well as may be seen in Figure 2. At each latitude he finds a significant width to the main sequence implying a substantial range in metallicity as pointed out above. Also Terndrup finds evidence from the distribution of $(V-I)_0$ color in each of his fields that the number of metal rich stars is declining more rapidly with radius than the number of metal poor stars.

The CCD spectra that my collaborators and I have been getting include the fields studied by Terndrup plus ones at –3 and –12°. As may be seen from Figure 1 based on only a fraction of the total data base, the TiO band strengths of the M giants in the –12° field are similar to those measured for nearby field stars of presumedly solar abundance. TiO bands

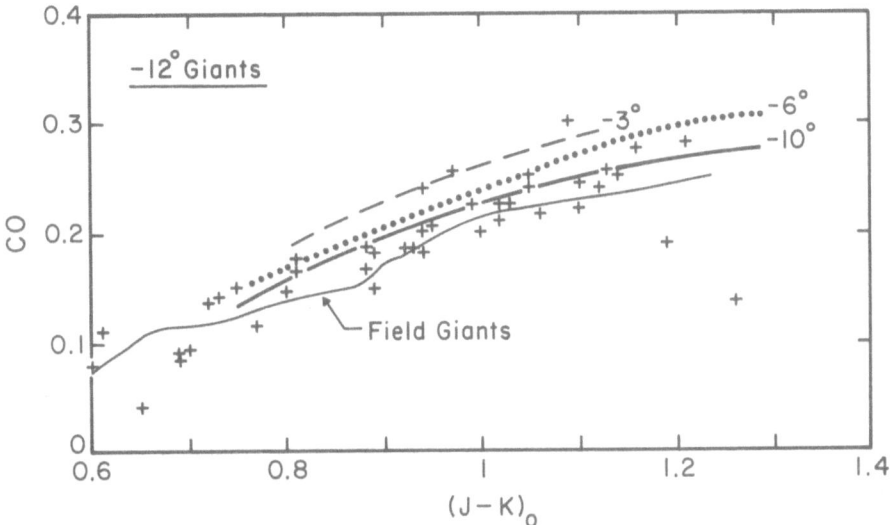

FIGURE 3. The 2.2μm CO band strength as a function of color for bulge M giants at b = -12°. The mean location of giants from three other fields is indicated. The mean relation for field giants is also shown.

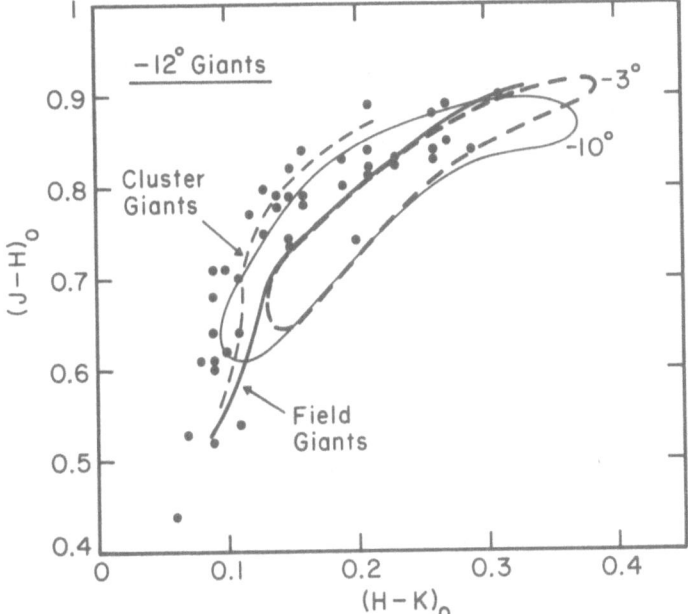

FIGURE 4. $J-H$, $H-K$ colors for bulge M giants at b = -12°. The locations of bulge giants from fields at -3 and -10° are indicated. Mean relations for field and globular cluster giants giants are also shown.

for M giants from latitudes not shown in Fig. 1 exhibit a monotonic change
in strength. The range in strength is consistent with Terndrup's estimate
for the range in [Fe/H] from c–m diagrams and with Rich's (1986)
calibration of the mean [Fe/H] for the BW giants.

4.2. Infrared Photometry

Infrared photometry for about 470 M giants in the six fields for which we
have obtained optical spectra reveals strong gradients in the observed
properties of the stars. Figure 3, for example, shows that the stars from the
$-12°$ field have CO band strengths that in the mean are close to the mean
for solar neighborhood field giants, consistent with the optical TiO band
strengths. For latitudes closer to the Galactic center the M giants show a
steady increase in the mean CO strength. In BW and the $-3°$ field nearly
all of the bulge giants have CO strengths greater than those of field giants
of similar color as shown schematically in Fig. 3.

In a $J–H$, $H–K$ plot (Figure 4 is representative of these data), M
giants from the -3 and $-3.9°$ (BW) fields lie almost exclusively on one side
of the mean line for field giants – the side opposite to that on which
globular cluster stars lie. Stars from the $-10°$ window straddle the mean
field line, whereas ones from the $-12°$ field reach over to the mean line for
globular clusters. Although the trend is consistent with what would be
expected from a simple metallicity gradient, the details are not understood;
they are another aspect of the general problem of the effects of blanketing
agents at these wavelengths (FW).

No obvious trends with galactic latitude are visible in the K, $J–K$ c–m
diagrams for the 6 fields. This is not entirely unexpected since Fig. 4 shows
that changes in $J–H$ are approximately canceled by changes in $H–K$.
Furthermore, preliminary analysis of the data show no significant dependence
of the luminosity function of the M giants on latitude. This suggests that
the rapid fall-off in the reddest and latest spectral type M giants is due
primarily to a shifting of these stars to bluer colors and earlier spectral
types.

5. STELLAR SYNTHESIS MODELS

As a test of the efficacy of the Galactic bulge giants for stellar synthesis
models, Whitford and I (in FW) examined the effects of replacing the giant
branch in the Tinsley and Gunn (1976, hereafter TG) model for an E galaxy
by the giant branch as defined by the M stars in BW. In the region of
overlap, the luminosity function of our giant branch agreed with the
unmodified one of TG, but their's included considerably more luminous stars.
Also, as we have seen, the colors of the BW M giants differ significantly
from those of field giants such as those used by TG. The net result of the
replacement was a substantially better fit to observed galaxy colors than
that achieved with TG's unmodified giant branch, but not one that could be
called "really good". FW argued that the chief source of the remaining
disagreement lay in the colors of the lower luminosity stars in the model
since they too were based on field stars of solar abundance. When their
colors were changed by small amounts to correspond to a mean metallicity of

twice solar, excellent agreement was achieved between the colors of the model and those observed in real galaxies (*e.g.* Frogel, *et al.* 1978, Impey, Wynn-Williams, and Becklin 1986) - *the broad band energy distribution of an average E galaxy from V to 10μm could be closely reproduced by a model whose giant branch was determined empirically from observations of stars in the Galactic nuclear bulge.*

Since it is now clear that there are strong gradients in the photometric and spectroscopic properties of giants in the Galactic bulge, the close agreement found by FW must be somewhat fortuitous. Remember that the colors of E galaxies generally refer to measurements made with photometric apertures that admit only a fraction of the total light of the galaxy. So the average of the light being measured in a "mean elliptical" must be strongly weighted toward the metallicity distribution which characterizes the stars in BW. Also, although the BW giants put into the model reflect the range in metallicity found at one position in the bulge, the other constituents of the model do not; rather, they must be regarded as mean values.

It has been suggested in the literature that some elliptical galaxies contain a "significant" population of stars with an age considerably less than the age of the galaxy as a whole. Arguments in support of this proposal have been based on the presence of a UV excess or the strength of a small number of spectral features. It has also been suggested that the Galactic bulge has a young stellar component on the basis of the presence of long period variables (LPVs) with luminosities considerably in excess of that expected for core helium flash and the derivation of pulsation masses for these LPVs. Whitford (1986) and FW, on the other hand, have argued against such a proposal (see also Whitelock, Feast, and Catchpole 1986) since a) VandenBerg and Laskarides (1986) have shown that in a metal rich environment stellar evolution proceeds at a slow enough rate that even in a 10 Gyr old population stars on the giant branch will have had an initial mass several tenths greater than solar allowing them to ascend to higher luminosities than can metal poor stars; b) the LPVs in the bulge are not any more luminous than LPVs in globular clusters; c) the temperatures needed to compute the pulsation mass are very uncertain as they depend on the infrared colors which are strongly affected by blanketing; and, finally, d) the optical c-m studies of Terndrup (1986) and Rich (1986) give no evidence for a young population. Hence, insofar as the stellar population of the nuclear bulge is representative of that found in early-type galaxies, the latter do not contain a significant population of relatively young stars.

The above arguments notwithstanding, it is instructive to examine graphically the contribution of various stellar groups to the colors of the integrated light of early-type galaxies and, in particular, the implications of including a young component. Although it may not be any more straightforward to interpret the broadband colors than certain spectral features, it is certainly easier to measure them. Furthermore, the requirement of matching colors from the optical to the infrared provides a simple, stringent test of any galaxy model - a test which too often is not applied.

First, Figure 5 (based on Table 5 of FW) shows how the total luminosity of Galactic bulge M giants in photometric band passes *from* V to 10μm is divided amongst three spectral groups. (It is important to emphasize that this figure and all other results in FW are based on an

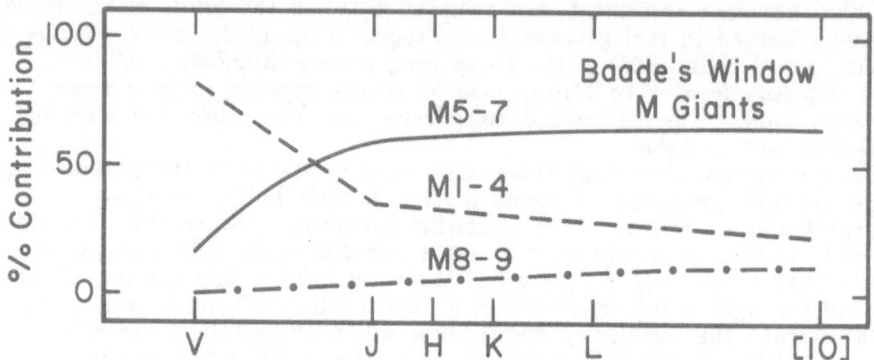

FIGURE 5.   The relative contributions of groups of bulge M giants to the total bulge M giant light in various bandpasses is shown. These results are from the complete surveys by Blanco, *et al.* (1984, 1986) and the data in FW.

unbiased selection of stars from Blanco, Mccarthy, and Blanco's (1984) and Blanco's (1986) M giants in BW. These surveys are not magnitude or distance limited.) While the early M's dominate in the visual (80% of the light), the middle M's dominate longward of 1.0μm, contributing 65% of the light at 10μm. The latest M giants make small contributions even at the longest wavelengths. Attempts to evaluate the relative contributions of solar neighborhood M's often have resulted in a significant overestimation of the contribution of the coolest most luminous giants because of difficulties involved in calculating incompleteness factors in surveys heavily biased in favor of finding such stars. FW found that all of the IRAS sources in BW contribute only half as much in the infrared as the M8-9 stars, i.e. about 6% at 10μm. The LPVs taken as a class by themselves, make a contribution to the infrared light about equal to that of the M8-9 giants.

Next we consider a stellar model similar to that of Tinsley and Gunn's (1976) but with the following changes: BW M giant luminosity function and colors used for stars with $M_{bol} \leq -1.2$; colors and luminosities of K giants in BW (Frogel, Whitford, and Rich 1984) used for fainter giants; metal rich isochrones of VandenBerg used as a guide to modify values given by Tinsley and Gunn for subgiants and evolving dwarfs, and, finally, for unevolved dwarfs. Figures 6a and b show the relative contributions of each of these four groups of stars for x = 1.35, a Salpeter initial mass function, and for a flatter one with x = 0.0. The $B-V$, $V-K$, $J-K$, and CO values are 0.97, 3.21, 0.83, and 0.14 for the Salpeter value and 0.96, 3.14, 0.83, and 0.16 for x = 1.0. For comparison, an average elliptical has 1.0, 3.30, 0.88, 0.16. Given the rather crude nature of the models, both are satisfactory.

Figure 6 shows that the visual light of both galaxy models is dominated by the fainter giants, subgiants, and evolving dwarfs. As in most other models constructed to date, the M giants contribute not more than 10-15% of the light shortward of 8000Å. At two microns and longward, though, the M giants contribute more than half of the light. Their contribution to the bolometric luminosity is comparable. The main difference

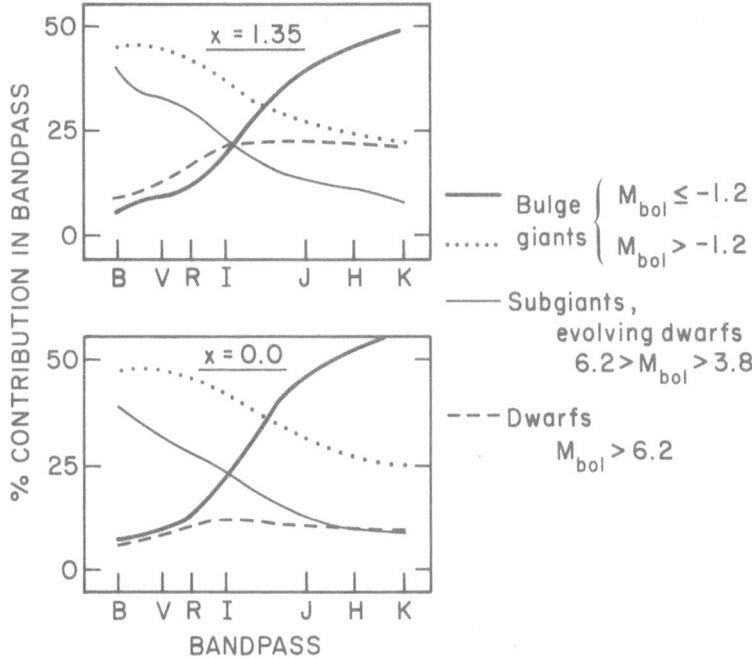

FIGURES 6a and b. The relative contributions to the light of a simple stellar
    synthesis model for a galaxy by 4 groups of stars: bright giants (Ms
    from BW), faint giants (Ks from BW), evolving dwarfs and subgiants,
    and unevolved dwarfs. Results are shown for two different initial
    mass functions.

between the models in Figure 6 is that the unevolved dwarfs ($M_{bol} \geq +6.5$)
make a substantial contribution to the infrared light of the one with the
steeper mass function but hardly any to that with the flatter mass function.
This difference illustrates why the CO and H_2O bands in the K window are
such good discriminants between giants and dwarfs in a composite
population.

Carter, Visvanathan, and Pickels (1986) assert, on the basis of the
observed strength of the infrared Na doublet, that about 40% of the light at
I in some galaxies they observed was due to late K to early M (say M3)
dwarfs. This would imply a main sequence turn off age of about 6 Gyr.
Figure 7 displays the implications of such an assertion. A 6 Gyr solar
metallicity isochrone from VandenBerg (1983) was convolved with an initial
mass function $x = -1.0$ (i.e. flat) and added to the $x = 0.0$ model of Figure
5 such that 40% of the light in the resulting model was from all of the 6
Gyr old stars. The upper line in Fig. 6 shows the relative contribution of
these stars to the other band passes; the lower one just that of the late K
to M3 dwarfs. Since these dwarfs contribute only a small fraction of the
light, if we stuck by the letter of Carter, et al.'s assertion and made them
contribute 40% of the I light, then 6 Gyr old stars would completely

12

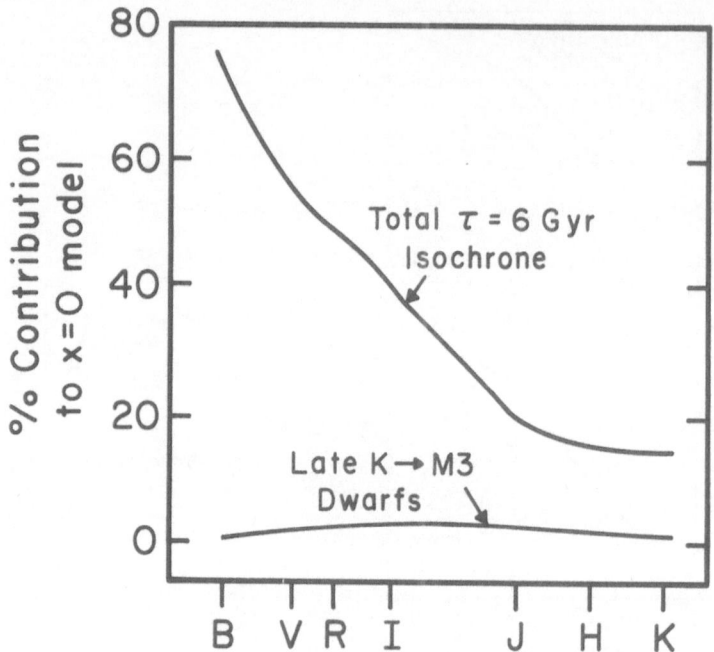

FIGURE 7. If a 6 Gyr solar metallicity isochrone from VandenBerg (1983) is given an IMF with an x of −1.0 and added to the x = 0.0 model of Figure 6 such that 40% of the light in the resulting model at I is from the young component, its relative contribution in the other bandpasses is shown by the upper line. The contribution of late K to early M dwarfs is shown by the lower line.

dominate the integrated light of the model at all wavelengths. As it is, the strength of the contribution of these 6 Gyr old stars makes the integrated $B-V$ and $V-K$ colors of the model 0.83 and 2.79, respectively − far too blue to correspond to a real elliptical. Hence, if we are to believe Carter, *et al.*'s assertion an extremely *ad hoc* IMF would be needed to not disturb the overall energy distribution of the galaxies. This is a clear example of the importance of applying the color−matching test mentioned earlier in this section.

6. SUMMARY AND CONCLUSIONS

At the beginning of this review and in FW a number of criteria were set forth which the components of a stellar synthesis model for the integrated light of a composite stellar system must satisfy. A carefully constructed model is of prime importance if the light from galaxies with large redshifts are to be understood. Whitford and I have maintained that stars from the Galactic bulge are the best known and *currently observable* candidates for these components. Photometric observations for an unbiased sample of the

M giants in BW are now available. These data provide colors and a luminosity function for the stars. They can be included in a stellar synthesis model by determining only one quantity – their total number relative to the total number of all other stars in the model. Molecular band strengths, broad band colors, and luminosities of bulge giants differ substantially from the globular cluster and solar neighborhood stars commonly used in synthesis models. These differences must be taken into account, or their effect at least evaluated, if the predictive capabilities of detailed models are to be used reliably. A first attempt at using the bulge giants in a model (FW) has resulted in integrated broad band colors from the visual to 10μm quite close to those observed in a typical elliptical galaxy.

There appears to be a substantial radial gradient in metallicity in the Galactic bulge. Such a gradient is evidence for a dissipative collapse of the gas cloud out of which the stars of the bulge formed. In addition, the stars just in BW appear to have a range in metallicity of nearly two orders of magnitude, with a mean of about twice solar. The recent models of Arimoto and Yoshii (1987) predict such a range as a natural consequence of chemical enrichment over the relatively short period of time required for the bulge to form. Observations both photometric and spectroscopic of distant galaxies encompass a sufficiently large fraction of the galaxian light that there will be a substantial range in metallicity included in the measuring slit or aperture. An evaluation of the effects of such a range should be possible via detailed spatial observations of nearby galaxies and quantitative delineation of the radial gradients in the Galactic bulge.

Optical and infrared C-M diagrams of bulge stars from the main sequence to the top of the giant branch do not contain any compelling evidence for the presence of significant numbers of stars with ages less than about 10 Gyr. In fact, given the present limitations of metal rich stellar interior models to predict observable colors, one can conclude that essentially all of the stars are coeval to within ± 2 or 3 Gyr. Refinement of this range and of the estimate for the mean age itself has to await improvements in the models. However, it is clear that inclusion into a stellar synthesis model of any substantial numbers of stars of age 6 Gyr with anything like a normal IMF is ruled out by the inability of the resulting model to match the broad band colors of an elliptical galaxy.

ACKNOWLEDGEMENTS

I appreciate the assistance received from Don Terndrup in the preparation of this review. Also I thank him, Victor Blanco, and Albert Whitford for allowing me to refer to our unpublished work.

REFERENCES

Aaronson, M., Cohen, J. G., Mould, J., and Malkan, M. 1978, *Ap. J.*, **223**, 824.
Arimoto, N., and Yoshii, Y. 1987, *Astr. Ap.*, **173**, 23.
Arp, H. 1965, *Ap. J.*, **141**, 45.

Baade, W. 1963, in *Evolution of Stars and Galaxies*, ed. C. P. Gaposhkin
    (Cambridge: Harvard University Press), p. 279.
Blanco, V. M. 1986, *A. J.*, **91**, 290.
Blanco, V. M., and Blanco, B. M. 1986, *Ap. Space Sci.*, **118**, 365.
Blanco, V. M., McCarthy, M. F., and Blanco, B. M. 1984, *A. J.*, **89**, 636
    (BMB).
Carter, D., Visvanathan, N., and Pickels, A. J. 1986, *Ap. J.*, **311**, 637.
Cook, K. H. 1987, Ph.D. thesis, Univ. of Arizona.
Frogel, J. A. 1985, *Ap. J.*, **291**, 581.
Frogel, J. A., Blanco, V. M., and Whitford, A. E. 1984, in *IAU Symposium
    105, Observational Tests of Stellar Evolution Theory*, ed. A. Maeder and
    A. Renzini (Dordrecht: Reidel), p. 571.
Frogel, J. A., Cohen, J. G., and Persson, S. E. 1983, *Ap. J.*, **275**, 773.
Frogel, J. A., Persson, S. E., Aaronson, M., and Matthews, *Ap. J.*, **89**, 636.
Frogel, J. A., and Whitford, A. E. 1987, *Ap. J.*, in press.
Frogel, J. A., Whitford, A. E., and Rich, R. M. 1984, *A. J.*, **89**, 1536.
Habing, H. J., Olnon, F. M., Chester, T., Gillett, F., Rowan-Robinson, M.,
    and Neugebauer, G. 1985, *Astr. Ap.*, **152**, L1.
Impey, C. D., Wynn-Williams, G., and Becklin, E. E. 1986, *Ap. J.*, **309**, 572.
Johnson, H. L. 1966, *Ap. J.*, **143**, 187.
Morgan, W. W. 1956, *Pub. A.S.P.*, **68**, 509.
Mould, J. R. 1986, in in *Stellar Populations*, ed. C. A. Norman, A. Renzini,
    and M. Tosi (Cambridge: Cambridge Univ. Press), p. 9.
Mould, J.R., and McElroy, D. B. 1978, *Ap. J.*, **221**, 580.
Nassau, J. J., and Blanco, V. M. 1958, *Ap. J.*, **128**, 46.
Phillips, M. M., Jenkins, C. R., Dopita, M. A., Sadler, E. M., and Binette, L.
    1986, *A. J.*, **91**, 1062.
Renzini, A., and Buzzoni, A. 1986, in *Stellar Populations*, ed. C. A. Norman,
    A. Renzini, and M. Tosi (Cambridge: Cambridge Univ. Press), p. 213.
Rich, R. M. 1986, Ph.D. thesis, Caltech.
Searle, L. 1986, in *Stellar Populations*, ed. C. A. Norman, A. Renzini, and M.
    Tosi (Cambridge: Cambridge Univ. Press), p. 3.
Stebbins, J., and Whitford, A. E. 1948, *Ap. J.*, **108**, 413.
Terndrup, D. M. 1986, Ph.D. thesis, University of Calif., Santa Cruz.
Tinsley, B. M., and Gunn, J. G. 1976, *Ap. J.*, **203**, 52.
VandenBerg, D. A. 1983, *Ap. J. Suppl.*, **51**, 29.
VandenBerg, D. A., and Laskarides, P. G. 1986, *Ap. J. Suppl.*, **64**, 103.
van den Bergh, S., and Herbst, E. 1974, *A. J.*, **79**, 603.
Whitelock, P., Feast, M. W., and Catchpole, R. 1986, *M.N.R.A.S.*, **222**, 1.
Whitford, A. E. 1978, *Ap. J.*, **226**, 777.
    _____. 1985, *Pub. A.S.P.*, **97**, 205.
    _____. 1986, in *Spectral Evolution of Galaxies*, ed. C. Chiosi and A. Renzini
    (Dordrecht: Reidel), p. 157.
Whitford, A. E., and Rich, R. M. 1983, *Ap. J.*, **274**, 723.

THE FORMATION OF GALACTIC SPHEROIDS

S. Michael Fall
Space Telescope Science Institute
3700 San Martin Drive, Baltimore, MD 21218
and
Department of Physics and Astronomy
The Johns Hopkins University
Homewood Campus, Baltimore, MD 21218

The spheroidal components of galaxies, including globular clusters, are usually assumed to form during a period of rapid collapse and high luminosity. A few years ago. I developed in collaboration with Martin Rees, a theory for the origin of globular clusters (Fall and Rees 1985. 1988). We showed that realistic density or velocity perturbations in a protogalaxy would be amplified during the collapse. The result is a two-phase medium with some gas at the virial temperature. a few $\times 10^6$K, and some gas at the temperture that hydrogen recombines, about 10^4K. The density of the hot gas can be derived from fairly general arguments, and since the two phases must be in rough pressure balance, the density and Jeans mass of the cold gas can be calculated with some confidence. We found that clouds with masses of order $10^6 \, M_{\odot}$ would collapse gravitationally, and should therefore be identified as the progenitors of globular clusters. The condition for this mass scale to be "imprinted" is that the temperatures of the clouds remain near 10^4K for at least one free-fall time. Once the protogalactic gas has been enriched in heavy elements from the first generation of stars in globular clusters, cooling becomes more efficient and much smaller clouds can collapse. thereby promoting the formation of later generations of field stars. We therefore expect that on average globular clusters should have lower abundances of heavy elements and more extended space distributions than the field stars in the spheroidal components of galaxies. As the result of various selection biases, these predictions are not easy to test for the Milky Way but they are consistent with a growing body of data for other galaxies (Forte, Strom and Strom 1981, Harris 1986, 1987, Mould 1987, Mould, Oke and Nemec 1987, Elson and Walterbos 1987).

Whether or not our theory is correct in detail, it strongly suggests that any chemically homogeneous model is likely to be a bad approximation to a real proto-galaxy. Thus most of the currently popular models for the chemical and spectral evolution of protogalaxies may give seriously misleading results. The following argument demonstrates the problem explicitly. For a protogalaxy to collapse at

R. G. Kron and A. Renzini (eds.), Towards Understanding Galaxies at Large Redshift, 15–16.
© 1988 by Kluwer Academic Publishers.

anything like the free-fall rate, the sound speed in the hottest component of the gas must be less than the collapse speed, that is $v_{collapse} \gtrsim v_{sound}$. If this inequality were reversed the protogalaxy would contract quasistatically. For the protogalaxy to be chemically homogeneous, the heavy elements produced at one location must be transported to other locations in a time shorter than, and perhaps much shorter than the collapse time; hence $v_{metals} \gtrsim v_{collapse}$. This is the condition that the system be well-mixed. Combining the previous inequalities gives $v_{metals} \gtrsim v_{sound}$. In other words, the heavy elements must move with a typical speed greater than, and perhaps much greater than the sound speed in a collapsing protogalaxy for it to be chemically homogeneous. All diffusion processes violate this condition. Moreover even the shock fronts from supernova explosions spend most of their time moving at a speed comparable to the sound speed in the ambient medium. As the result of these and other considerations, I conclude that chemically homogeneous models for collapsing protogalaxies are almost certainly not realistic.

REFERENCES

Elson, R. A. W., and Walterbos, R. 1987, *Ap. J.*, submitted.

Fall, S. M., and Rees, M. J. 1985, *Ap. J.*, **298**, 18.

Fall, S. M., and Rees, M. J. 1988, in *IAU Symposium 126, Globular Cluster Systems in Galaxies*, eds. J. E. Grindlay and A. G. D. Philip (Dordrecht: Reidel), in press.

Forte, J. C., Strom, S. E., and Strom, K. M. 1981, *Ap. J. (Letters)*, **245**, L9.

Harris, W. E. 1986, *A. J.*, **91**, 822.

Harris, W. E. 1987, *Ap. J. (Letters)*, in press.

Mould, J. 1987, in *Stellar Populations*, eds. C. A. Norman, A. Renzini and M. Tosi (Cambridge: Cambridge University Press), p. 9.

Mould, J. R., Oke, J. B., and Nemec, J. M. 1987, *A. J.*, **93**, 53.

GLOBAL STELLAR POPULATIONS OF ELLIPTICAL GALAXIES:
A. OPTICAL PROPERTIES

David Burstein, Dept. of Physics, Arizona State University
Roger L. Davies, NOAO, Kitt Peak National Observatory
Alan Dressler, Mt. Wilson and Las Campanas Observatories
S.M. Faber, Lick Observatory, U.C. Santa Cruz
Donald Lynden-Bell, Institute of Astronomy, Cambridge, England
Roberto Terlevich, Royal Greenwich Observatory
Gary Wegner, Dept. of Physics and Astronomy, Dartmouth College

ABSTRACT. The stellar population properties of elliptical galaxies are examined using a homogeneous, distance-independent set of parameters for over 400 ellipticals. The relationship between (B-V) color and the Mg_2 index shows little evidence of cosmic scatter, despite the fact that the color is measured in a signficantly larger aperture. However, the relationship between Mg_2 and central velocity dispersion, σ, does show evidence for a significant subset of ellipticals having weaker values of Mg_2 for their values of σ.

1. INTRODUCTION

Nature or Nurture? Is the formation of elliptical galaxies an intrinsically different process from that of spirals (i.e., Nature), or do the mergers of spirals or proto-spirals form ellipticals (Nurture)? The answer to this question is still central to our understanding of the evolutionary histories of elliptical galaxies. In the context of this workshop, two questions regarding the stellar populations of ellipticals are relevent: a. How closely tied are the overall stellar populations and global mass properties of ellipticals? b. What are the present-day systematics of the spectral energy distributions of ellipticals from 1200 Å to 2.5 μ, and how have these systematics evolved with time?

This paper uses optical data obtained in our survey of 455 elliptical galaxies (Davies et al. 1987; Burstein et al. 1987) to address the first question: central velocity dispersion, σ, central absorption line strength, Mg_2, and the average $(B-V)_0$ color measured within a 67" aperture. A second paper (Part B) uses a separate set of ultraviolet data to address the second question.

2. THE Mg_2 - Log σ RELATIONSHIP FOR 455 ELLIPTICALS

The recent analyses by Faber et al. (1987) and Djorgovski and Davis (1987) have shown that the core and global measures of mass-to-light

17

R. G. Kron and A. Renzini (eds.), Towards Understanding Galaxies at Large Redshift, 17-21.
© *1988 by Kluwer Academic Publishers.*

18

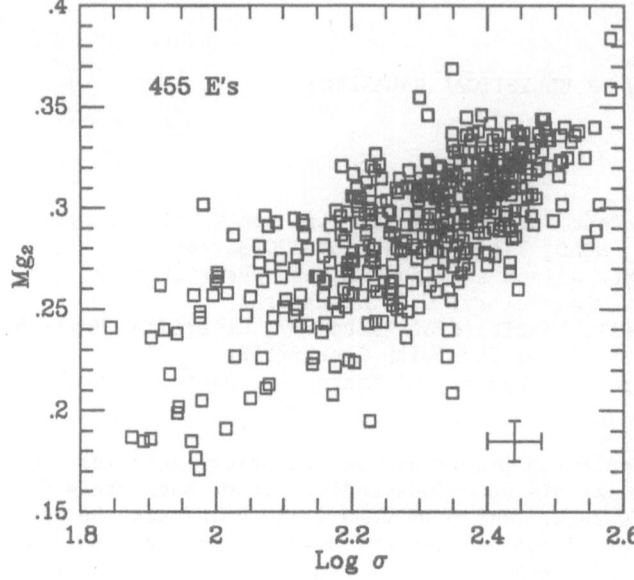

Fig. 1. (top): The relationship between the Mg₂ index and the central velocity dispersion, σ for 455 ellipticals.
Fig. 1. (bottom): The histogram of Mg₂ residuals from the Mg₂-Log σ relationship, compared to a normal distribution of points with a conservatively-large estimated standard deviation of 0.016 mag.
A significant number of ellipticals exist which have weaker values of Mg₂ at a given value of Log σ.

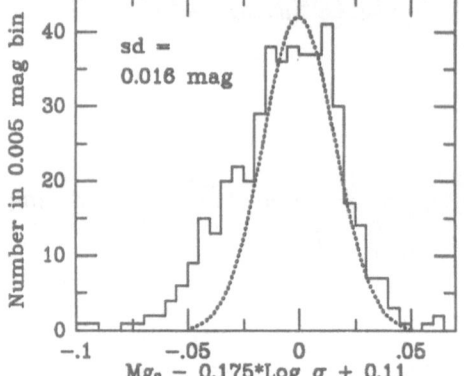

ratios (M/L) in elliptical galaxies agree well, implying that ellipticals are mainly baryon-dominated within their effective radii, and that their M/Ls are stellar in origin. Moreover, while the average value of M/L is correlated with L, intrinsic scatter in M/L at a given absolute luminosity also apparently exists.

The Mg_2 parameter and the $(B-V)_0$ color are two separate measures of the average stellar population of old stellar populations. Mg_2 is primarily sensitive to the mean metallicity of the population, and secondarily to the average age (cf. Burstein 1979; Burstein et al. 1984). $(B-V)_0$ is well-known to be approximately equally sensitive to both the mean age and mean metallicity of a stellar population.

The degree of corrlation between the mass parameters and the stellar populations among ellipticals could be tested with either Mg_2 or $(B-V)_0$; Mg_2 is the preferred quantity due to its higher measurement accuracy. Fig. 1 (top) presents the correlation between Mg_2 and Log σ for the 455 galaxies in our survey with measures of both quantities. Fig. 1 (bottom) presents the histogram of residuals, $Mg_2 - 0.175 Log \sigma + 0.11$, as determined from a least-squares fit to these data.

The expected errors, both random and systematic, are 0.01 mag in Mg_2 and 0.04 dex in Log σ (Davies et al. 1987). A Gaussian distribution in Mg_2 residuals with a standard deviation (sd) of 0.016 mag, corresponding to these errors, is also shown in Fig. 1 (bottom).

Inspection of the histogram of Fig. 1 shows that a signficant number of elliptical galaxies have weak values of Mg_2 at a given value of central velocity dispersion. The reverse is not true; the number of ellipticals having abnormally strong values of Mg_2 does not appear to be significant.

Many of the elliptical galaxies with the more extreme weak values of Mg_2 are known to be actively forming stars (e.g. N4742). It is therfore possible that the many of the more marginally-weak Mg_2 values are due to the presence of low amounts of star formation in these elliptical galaxies. It is also possible that many of the weaker values of Mg_2 are due to real differences among the chemical abundance properties of old stellar populations (cf. Burstein et al. 1984). Unfortunately, the accuracy of these data is not sufficient to reliably test whether the variations of Mg_2 with Log σ are correlated with the variations of M/L with L.

3. THE Mg_2 -$(B-V)_0$ CORRELATION: A 'POOR MAN'S' MEASURE OF STELLAR POPULATION GRADIENTS IN ELLIPTICALS

Mg_2 and Log σ are measured within the same aperture (typically 2" square) and are independent of reddening. In contrast, the $(B-V)$ color used here is the average of all available aperture measurements with an aperture size of 67" or less (Burstein et al. 1987), and must be reddening-corrected. Hence, the degree of correlation of Mg_2 with $(B-V)_0$ can place an upper limit on variations of stellar population gradients among elliptical galaxies.

Much of the scatter in the Mg_2-$(B-V)_0$ correlation could come from errors in reddenings. Tests of various methods of estimating reddenings using these data will be detailed elsewhere. For the purposes of this paper we use the method of Burstein and Heiles (1978). Since the most accurate reddening esimates in the Burstein-Heiles method are obtained for declinations north of -23°, we restrict our examination of the Mg_2 -$(B-V)_0$ relationship to the 276 galaxies with reliable measures of $(B-V)$ that lie in this declination range.

Fig. 2 (top) presents the Mg_2 - $(B-V)_0$ relation for these galaxies, Fig. 2 (bottom) presents the histogram of $(B-V)_0$ residuals, $(B-V)_0$ - $1.12Mg_2$ - 0.615, as determined from a least-squares fit with Mg_2 as the independent variable. In contrast to Fig. 1, the _a priori_ expected errors expected for $(B-V)_0$ are difficult to estimate. Instead, various values of $(B-V)_0$ errors were combined with the errors in Mg_2, and tested against the histogram of Fig. 2 (bottom). Although a reasonable Gaussian fit to the histogram is obtained with an average error of 0.02 mag in $(B-V)_0$, a small number of galaxies have both more accurate and less accurate values of $(B-V)_0$. We note that an error of 0.02 mag in $(B-V)_0$ is consistent with previous estimates of the expected error in the most accurate measures of $(B-V)$ alone (cf. Burstein et al. 1987).

The apparent lack of measurable intrinsic variation of $(B-V)_0$ with respect to Mg_2, given the order of magnitude difference in the size of measuring apertures, is somewhat surprising. This lack of observed

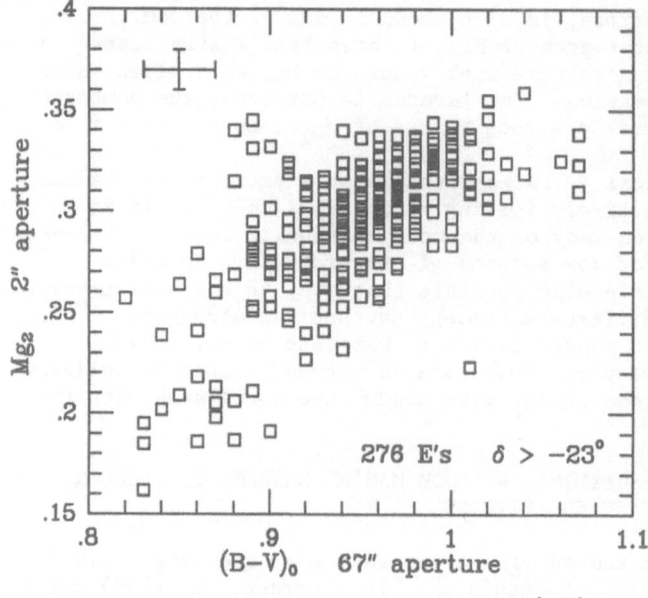

Fig. 2. (top): The relationship between $(B-V)_0$ color and the Mg_2 index for 276 ellipticals north of $-23°$ declination.
Fig. 2. (bottom): The histogram of $(B-V)_0$ residuals from the $(B-V)_0$–Mg_2 relationship, compared to a normal distribution of points with a standard deviation (sd) of 0.023 mag (=0.02 mag for $(B-V)_0$). Observational errors alone appear to account for the scatter in this relationship.

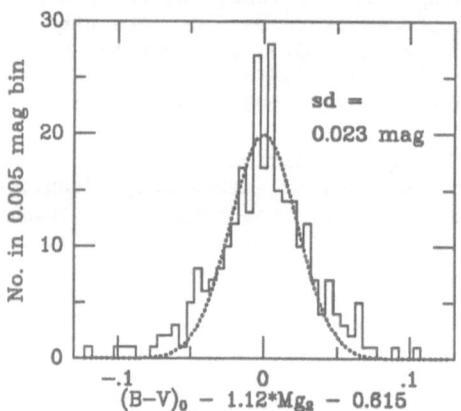

variation could be due to one or more of three possibilities: a) The true variation of Mg_2 with $(B-V)_0$ is smaller than the measurement errors. b) The scale of the gradient in stellar population within galaxies is the same for both Mg_2 and color, and proportional to the central values of Mg_2. c) Most elliptical galaxies possess little or no Mg_2 or color gradient in their central regions. Possiblity (a) must be reconciled with the fact that a variation of Mg_2 with Log σ is observed for the same sample of galaxies. Possibility (c) is inconsistent with numerous reports of both color and Mg_2 gradients in ellipticals. Possibility (b) is intriguing, and would imply a strong relationship among the stellar populations throughout an elliptical galaxy. Distinguishing among these possibilities will require direct measurements of metallicity gradients in elliptical galaxies.

4. DISCUSSION

Both in this short contribution and in other papers, our survey of elliptical galaxies has found that most ellipticals exhibit a remarkable regularity in overall structure and stellar population. This regularity can be seen in the similarity in their growth curves (Burstein et al. 1987), and the strong correlations between Mg_2 and Log σ (Fig. 1) and Mg_2 and $(B-V)_0$ (Fig. 2).

On the other hand, the data also indicate that at least 15% of ellipticals show an intrinsic range in stellar population properties (Mg_2, M/L, $(B-V)_0$) at a given absolute luminosity. The physical source (or sources) of this variation is not yet clear: If the formation process of ellipticals is intrinsic to these galaxies (i.e., Nature), then this range in stellar population at a given luminosity could plausibly be due to the accretion of more gas since formation (cf. Knapp et al. 1985). If ellipticals are the products of mergers among spiral galaxies (Nurture), the strong coupling of physical properties and stellar populations would imply that most mergers occurred when the merger components were primarily gaseous. Indeed, if both Nature and Nurture can produce elliptical galaxies, is it possible that both have operated in the real universe to produce present-day ellipticals? If true, the average physical properties of elliptical galaxies will not be simple functions of look-back time.

5. REFERENCES

Burstein, D. 1979, Ap. J. 232, 74.
Burstein, D., Davies, R.L., Dressler, A., Faber, S.M., Stone, R.P.S., Lynden-Bell, D., Terlevich, R. and Wegner, G. 1987, Ap. J. Suppl. 64, 601.
Burstein, D., Faber, S.M., Gaskell, C.M. and Krumm, N. 1984, Ap. J. 287, 586.
Burstein, D. and Heiles, C. 1978, Ap. J. 225, 40.
Davies, R.L., Burstein, D., Dressler, A., Faber, S.M., Lynden-Bell, D., Terlevich, R. and Wegner, G. 1987, Ap. J. Suppl. 64, 581.
Djorgovski, S. and Davis, M. 1987, Ap. J. 313, 505.
Faber, S.M., Dressler, A., Davies, R.L., Burstein, D., Lynden-Bell, D., Terlevich, R. and Wegner, G. 1987, in Nearly Normal Galaxies: From the Planck Time to the Present, ed. S.M. Faber (Springer-Verlag). pg. 175.
Knapp, G.R., Turner, E.L. and Cunniffe, P.E. 1985, Astron. J. 90, 454.

GLOBAL STELLAR POPULATIONS OF ELLIPTICAL GALAXIES:
B. ULTRAVIOLET ENERGY DISTRIBUTIONS

David Burstein, Dept. of Physics, Arizona State University
F. Bertola, Dept. of Astronomy, University of Padova
L. M. Buson, Astronomical Observatory, Padova
S.M. Faber, Lick Observatory, U.C. Santa Cruz
Tod R. Lauer, Princeton University Observatory

ABSTRACT. Ultraviolet energy distributions have been derived from IUE
spectra for 32 early-type galaxies, including 30 elliptical and S0
galaxies. These are combined with V magnitudes, Mg_2 indices and central
velocity dispersions (σ) to investigate a) the degree of correspondence
between optical and UV energy distributions and b) the intrinsic range
of stellar population at a given luminosity among early-type galaxies.

1. INTRODUCTION

What are the present-day systematics of the UV energy distributions
of early-type galaxies relative to their optical stellar populations and
overall size? Many previous investigations have documented a high degree of similarity in the optical and near-infrared energy distributions
of old stellar populations with similar absorption line-strengths (e.g.
Faber 1973; Whitford 1978). On the other hand, the first investigations
into the UV component of old stellar populations uncovered an unexpectedly wide range of luminosity due to one or more 'hot' components in
these galaxies (e.g. Code and Welch 1979; Bertola et al. 1980).

IUE-determined UV energy distributions and corresponding optical
data are now available for 32 early-type galaxies. A full analysis of
these data is presented elsewhere (Burstein et al. 1988); here we
summarize those data relevent to understanding the relationship of
optical and UV stellar populations in elliptical and S0 galaxies.

2. RELATIONSHIPS AMONG (1550–V) COLOR, Mg_2 and Log σ

The galaxies in this survey are a priori divided into three
catagories, based on their known optical properties: a) quiescent
(showing little or no signs of nuclear activity; 24 galaxies); b) active
(with significant nuclear activity; 4 galaxies) and c) demonstrably
star-forming (4 galaxies). A far–UV/optical color (expressed in
magnitudes) is formed from the flux observed between 1250 and 1850 Å (a
'1550' mag) with the equivalent V magnitude measured within the IUE
aperture. The relationships between (1550–V) and Mg_2, and between

23

R. G. Kron and A. Renzini (eds.), Towards Understanding Galaxies at Large Redshift, 23–28.
© *1988 by Kluwer Academic Publishers.*

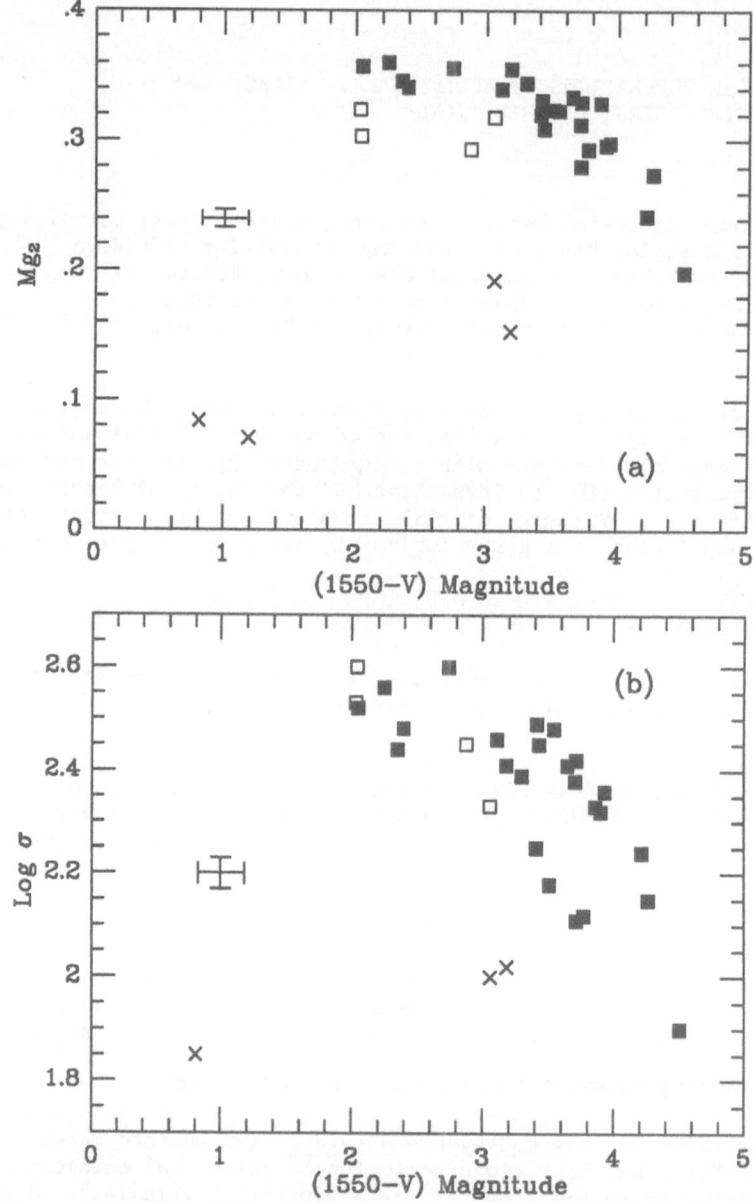

Fig. 1. (a) The Mg$_2$ index plotted vs. the UV/optical (1550−V) color (in mag) for 32 early−type galaxies. Closed squares represent 'quiescent' galaxies, open squares represent 'active' galaxies, and crosses represent galalxies with obvious on−going star formation. (b) Log of central velocity dispersion (Log σ) for 31 of these galaxies versus (1550−V) color. In contrast with Fig. 1(a), cosmic scatter exists in the Log σ−(1550−V) relationship among the quiescent galaxies.

(1550-V) and Log σ are presented in Fig. 1. The stellar populations of the 24 quiescent galaxies define a relatively tight, non-power-law relationship between Mg_2 and UV/optical color in Fig. 1(a). The existence of this relationship was first suggested by Faber (1983), and is the only known component of an old stellar population that becomes bluer with increasing metallicity. The stellar populations of active galaxies are somewhat bluer in (1550-V) color than quiescent galaxies with the same Mg_2 index, while the stellar populations in star-forming galaxies are much bluer, and have weaker values of Mg_2.

In contrast to Fig. 1(a), the relationship between (1550-V) color and Log σ (Fig. 1(b)) shows a general trend for both quiescent and active galaxies, but with signficant intrinsic scatter. As with the optical data of Paper A, these UV data also imply real differences in stellar populations at a given absolute size among early-type galaxies.

3. UV ENERGY DISTRIBUTIONS

Galaxies with simlilar values of (1550-V), Mg_2 show a remarkable degree of similarity in their overall ultraviolet energy distributions from 1250 - 3350 Å. Four such sets of energy distributions are presented here in Fig. 2 (following two pages). Descriptions of the full range of UV energy distributions that are found in the whole sample are given in Burstein et al. (1988).

The UV energy distributions are presented as ratios of the average flux (in wavelength units) in a 50 Å interval, relative to the V magnitude flux, corrected for both reddening and redshift. The four sets of spectra in Fig. 2 are segregated according to their positions in the (1550-V), Mg_2 diagram: (a) The four quiescent galaxies with the bluest (1550-V) colors (identified by NGC number); (b) five quiescent galaxies with intermediate values of (1550-V) and Mg_2; (c) the three quiescent galaxies with the reddest (1550-V) colors and (d) the two active galaxies with the bluest (1550-V) colors. The solid line drawn in each figure is a free-hand fit to the average spectrum in Fig. 2a, and is repeated in the other figures for visual reference.

Futher analysis of these data (Burstein et al. 1988) gives quantitative support to inferences one draws by eye in looking at these energy distributions: (a) Galaxies with similar values of (1550-V) and Mg_2 have essentially the same energy distributions from 1250 Å to 5500 Å. (b) The form of energy distribution of the 'hot' component of early-type galaxies is similar for all quiescent galaxies at wavelengths shorter than 2000 Å; it is the level of this component relative to the V luminosity that increases with increasing metallicity. (c) The UV light of early-type galaxies from 2000-3300 Å is a composite of flux from this hot component together with flux from stars near the main sequence turn-off. The increase of the level of this hot component with increasing metallicity greatly complicates the determination of the mean spectral energy distribution of the main sequence. For example, we estimate that the hot component and the main sequence component in a metal-rich galaxy produce equal amounts of flux at 2800 Å. d) The UV energy distributions of the two bluest active galaxies appear to exhibit excess flux between 2000 - 2500 Å, relative to the bluest quiescent galaxies. These data

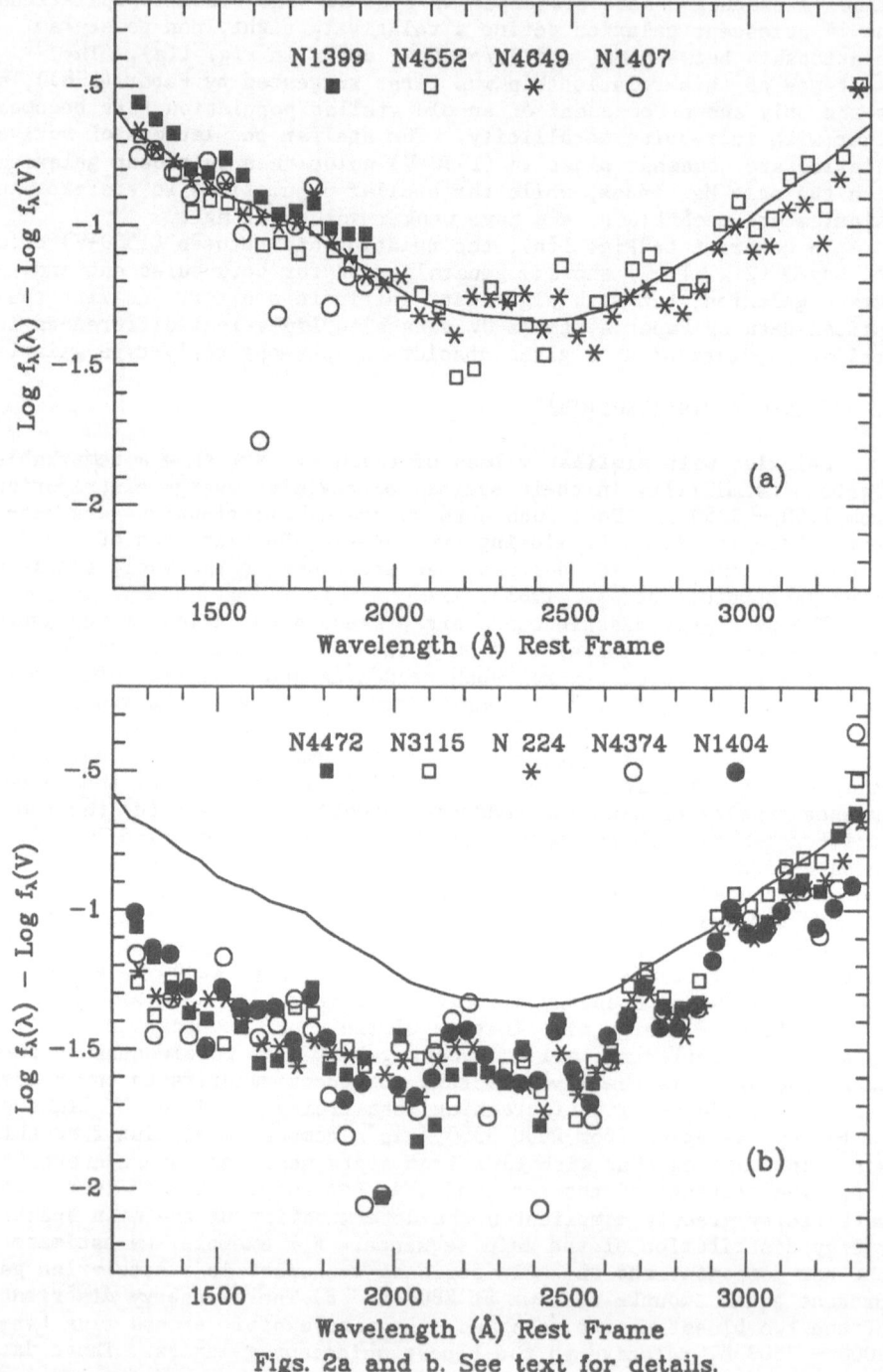

Figs. 2a and b. See text for details.

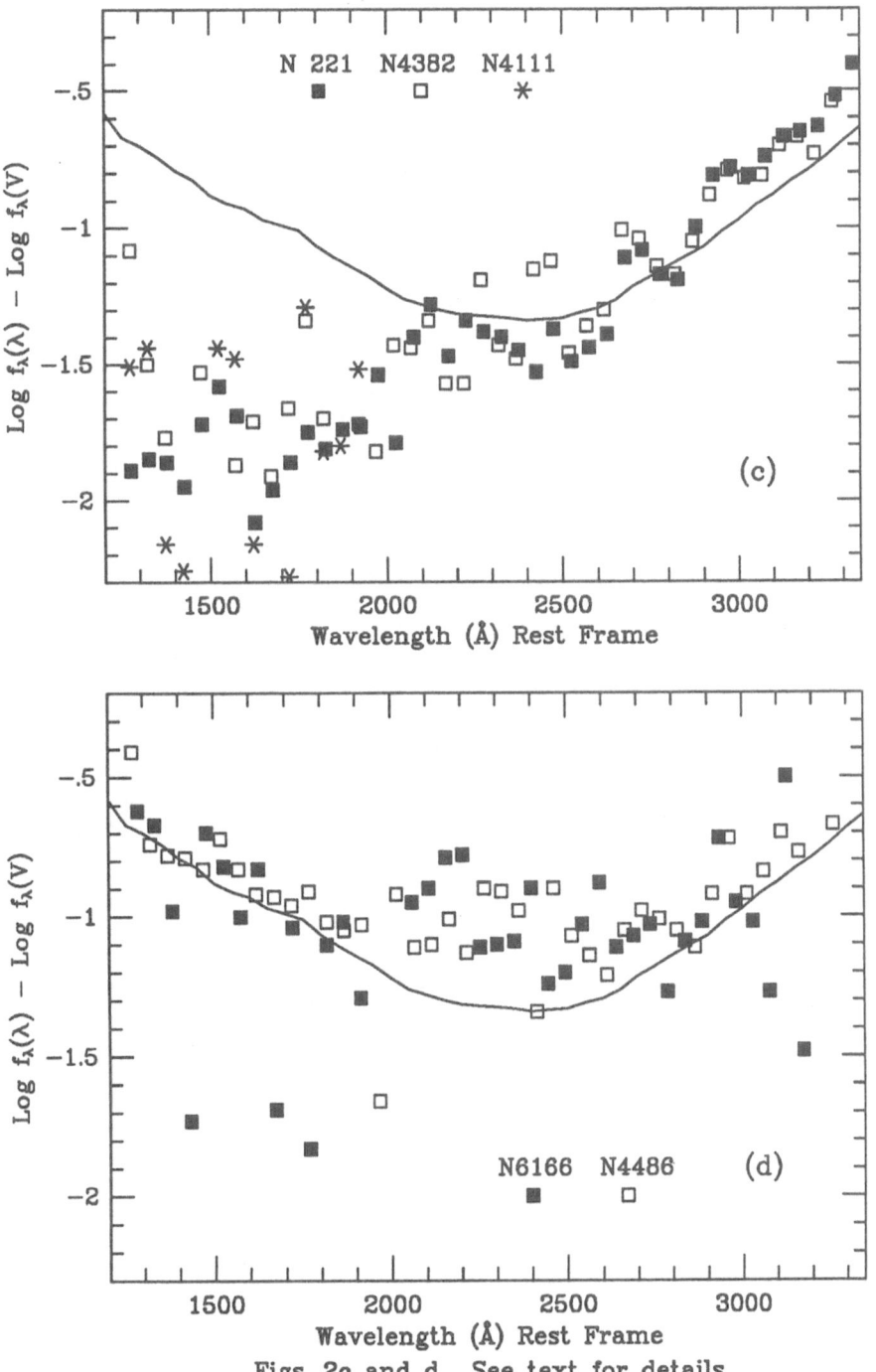

Figs. 2c and d. See text for details.

are noisy, however, and should be supplemented by additional IUE spectra of high signal-to-noise.

4. DISCUSSION

The UV spectra of the old stellar populations of early-type galaxies show a high degree of regularity between their UV and optical properties. The data presented here imply continuity from 1250 to 5500Å; data in the literature would extend this continuity into the near infrared (e.g, Whitford 1978). Disappointingly, we have also found that additional knowledge of these systematic properties of the hot component yields little additional direct information about its physical source (or sources; see discussions by Nesci and Perola 1985; Renzini and Buzzoni 1986; Burstein et al. 1988).

Given the data presented in this paper and in Paper A (preceding), it would be wise to proceed very cautiously in associating the spectral energy distributions of elliptical galaxies today with those of galaxies inferred to be ellipticals at large look-back times. Too many questions have been raised that, as yet, have no definite answers: What is the source (or sources) of the intrinsic variation in old stellar populations at given luminosity? How did these differences in stellar population evolve with time: were they bigger at earlier times, or smaller? Does the relative level of the hot component of stellar populations increase or decrease with look-back time? Since almost all sources of hot stars yield very similar UV spectra (e.g., Nesci and Perola 1985), is it possible that the source of hot stars was different at large look-back times than it is today? Finally the most basic question is still unresolved, namely, are ellipticals formed ab initio, or are they the products of mergers?

Thus, in several contexts, the properties of early-type galaxies as a function of look-back time may be needed to constrain the interpretation of present day stellar populations, rather than vice versa.

5. REFERENCES

Bertola, F., Capaccioli, M., Holm, A.V. and Oke, J.B. 1980, Ap. J. (Letters) 237, L65.
Burstein, D., Bertola, F., Buson, L.M., Faber, S.M. and Lauer, T.R. 1988, Ap. J., in press.
Code, A.D. and Welch, G.A. 1979, Ap. J. 228, 95.
Faber, S.M. 1973, Ap. J. 179, 731.
Faber, S.M. 1983, Highlights of Astronomy 8, 165, ed. R.M. West (Dordrecht: Reidel).
Nesci, R. and Perola, G.C. 1985, Astron. Ap. 145, 296.
Renzini, A. and Buzzoni, A. 1986, Spectral Evolution of Galaxies, ed. C. Chiosi and A. Renzini, (Dordrecht: Reidel).
Whitford, A.E. 1978, Ap. J. 226, 777.

INTERPRETING INTEGRATED SPECTRA

A.J. Pickles & P.C. van der Kruit
Kapteyn Laboratorium
Postbus 800
9700 AV Groningen
The Netherlands

ABSTRACT. Bright galaxies in clusters covering the redshift range 0 < z < 0.4 have been synthesised with a set of composite spectra based on VandenBerg's isochrones. All fits require some distribution of stellar colours around the main sequence turnoff; if this intrinsic width is interpreted as being due purely to metallicity distribution, then the mean metallicity around the main sequence turnoff is lower than that on the giant branch. Optimal fits are interpreted as requiring a distribution of ages as well as of metallicity. First results indicate no firm trend of age distribution with redshift however, indicating that most stars formed in these cluster dominating galaxies before the lookback time appropriate to z=0.4.

1. INTRODUCTION

The interpretation of integrated spectra of normal galaxies has had a long and successful history. Early applications included objects such as M31 and M32 in the local group (Spinrad and Taylor 1971), which are now accessible to direct observations of the individual stars dominating their light output; thus permitting an extended comparison of synthesised populations with directly observed components.

The next few years should produce much more information about the characteristic populations within local group galaxies. But in view of the complications already revealed, we take the view that this magnified perception will probably not fully reveal their temporal and chemical evolution, or the timescales of galaxy formation. Most probably this will require a much larger sample, of different galactic morphologies and of a range of lookback times. We conclude that direct results must be accompanied by a clear understanding of how the revealed components combine to produce the integrated light, and how this knowledge may be inverted to interpret the populations of galaxies outside the local group, which can only be seen by their integrated light.

The nature of stellar populations in composite systems has been extensively reviewed recently (eg. Burstein 1985; Whitford 1985;

29

R. G. Kron and A. Renzini (eds.), Towards Understanding Galaxies at Large Redshift, 29–41.
© *1988 by Kluwer Academic Publishers.*

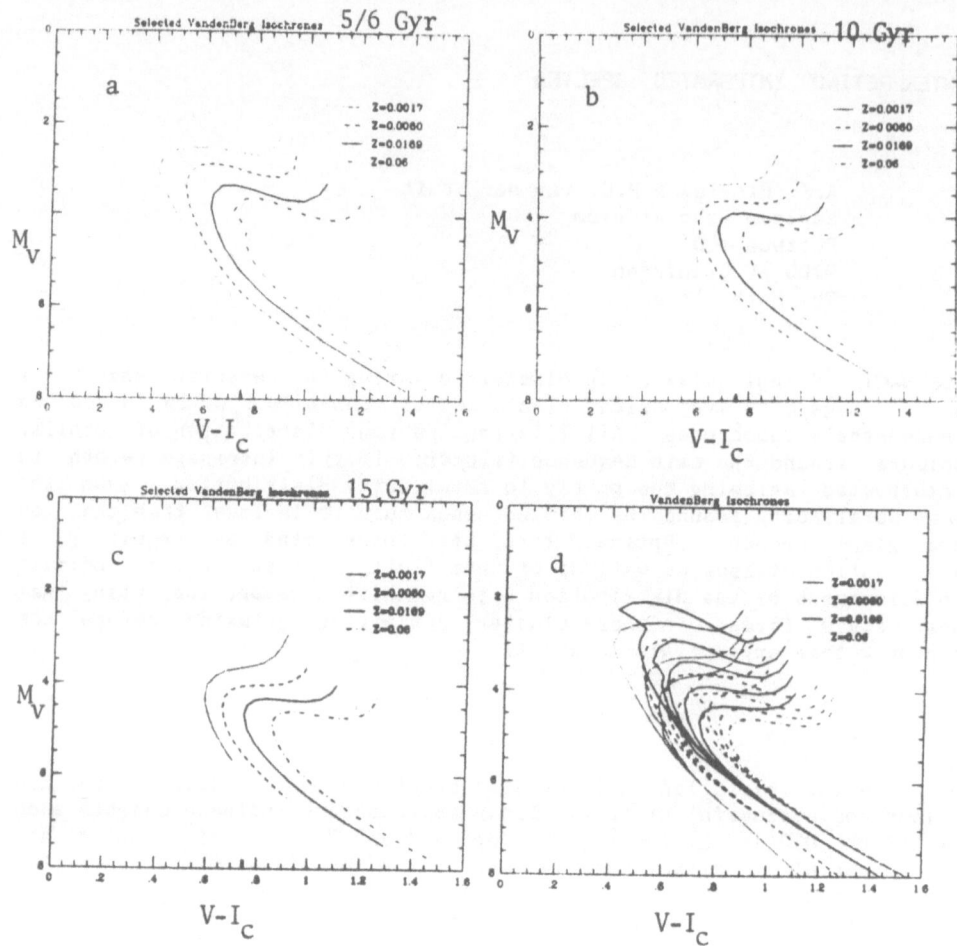

Figure 1(a,b,c): show the selected isochrones for four metalicities at three ages of 5/6, 10 and 15 Gyr. The helium abundance is constant at Y=0.25, and the ratio of mixing length to pressure scale height is either 1.5 or 1.6 (see table).
Figure 1d: is a composite of these three diagrams to illustrate how easily an observed broad CMD may encompass distributions of both age and metallicity.

O'Connell 1986, 1987; Pickles 1987). Apart from the well known trend of increasing mean stellar metallicity with increasing galactic mass, it is now clear that there is a large spread of metallicity within each galaxy (Whitford and Rich 1983, Arimoto and Yoshii 1987). It is therefore not surprising that colour magnitude diagrams (CMDs) for the Galactic bulge (Terndrup 1987), the halo of M31 (Mould and Kristian 1986) and M32 (Seitzer and O'Connell, preprint) show broad colour distributions around the main sequence turnoff, subgiant and giant branch regions.

Figure 1 shows the isochrones used later to define the synthetic model spectra; figures 1(a,b,c) illustrate the intrinsic width expected for a given single age, and a large metallicity range. It is important to note however that these broad distributions could encompass a range of ages as well as of metallicity. Figure 1(d) is a combination of the other three to make the point clearly that isochrones of many ages as well as of many metallicities can be fit to an observed broad CMD. The problem of age and metallicity fitting is much more severe than for globular clusters, as there is no compelling a priori reason here to expect a single age. It is further apparent that some form of population decomposition will be necessary even when direct observations are available.

For the Galactic centre it is clearly possible to measure line strengths directly, and for main sequence stars as well as for giants. If the observed broadening of the CMD is due to metallicity alone, then the stars defining its blue edge should be exclusively metal-weak and conversely for the red edge. If a substantial age spread exists, as has been claimed for ellipticals from population synthesis analysis, then there should be a range of line strengths at each point in the CMD. In view of the doubts still surrounding this issue, this clarification is sorely missed. For distant composite populations which must be studied by their integrated light, the corresponding test depends on population synthesis alone - which is the topic of this contribution.

In what follows, we investigate the extent to which the integrated light of ellipticals can be produced by a single, old generation of stars with a large spread of metallicity (cf. Renzini 1986). In line with the conclusion from this that ellipticals do indeed contain a significant fraction of intermediate age stars, and quite possibly a range of stellar ages, we also present first results from our observational program to differentially synthesise bright galaxies in clusters up to redshift z = 0.4.

2. ISOCHRONE SELECTION AND SYNTHETIC ISOCHRONE SPECTRA

A set of isochrones was selected from those recently published by VandenBerg (1985 - V85) and VandenBerg and Laskarides (1987 - VL87). The parameters of the selected isochrones are (Y = 0.25 throughout):

Age (Gyr)	Z	[Fe/H]	mixing length ratio	Ref.
5,10,15	0.0017	-1.0	1.6	V85
5,10,15	0.0060	-0.5	1.6	V85
2,4,6,10,15	0.0169	0.0	1.6	V85
6,9,12,15	0.060	0.6	1.5	VL87

For each selected isochrone, the stellar numbers and light fractions appropriate to each mass point were calculated. The relevant spectra from the stellar library were then co-added in these proportions to form a synthetic spectrum appropriate to the chosen isochrone. The methodology involved in this procedure is detailed below.

i) Take the listed M_{bol} and Log Te values for each mass point for a selected isochrone. Compute

$$Log\ Te' = Log\ Te - \delta$$

where δ = 0.01 for metallicity (Z) values of solar or less, and δ = 0.00 for metal rich isochrones; this is the correction recommended in VL87.

ii) For convenience compute $(V-I)_C$ from Log Te' using :

$$(V-I)_C = 1512.55 - 1165.41.(LogTe') + 300.052.(LogTe')^2 - 25.809.(LogTe')^3$$

which is a chi-squared fit to the data from Bessell (1979) and Ridgeway et. al. (1980). The formal fit has a maximum deviation of only 0.04 mag. in the range $0 < (V-I)_C < 3.5$, and actually gives $(V-I)_C$ colours very similar to those tabulated in V85. In fact the computed main sequence colours are the same as those tabulated by V85 to within 0.01 mag. except for the youngest, most metal-weak isochrone, where the computed colour is 0.03 mag. redder at the turnoff than that listed in V85. Computed subgiant and giant colours are redder than the listed values by between 0.00 and 0.05 mag.

iii) Extrapolate the lower main sequence by adding points with the $(V-I)_C$ colours of the library groups. Compute appropriate values of Mass and M_{bol} by linear extrapolation on $(V-I)_C$.

iv) Compute stellar numbers appropriate to each mass point by the formula given by Miller and Scalo (1979).

$$Log\ \in(log\ m) = 1.59 - 0.87.log\ m - 0.5.(Log\ m)^2$$

v) Compute the Bolometric Correction for each point from:

$$BC = 0.3 - 0.62.(V-I)_C + 0.14.(V-I)_C^2$$

which is a chi-squared fit to the data from Bessell and Wood (1984).

vi) Calculate V light fractions appropriate to each mass point from M_{bol}, BC and stellar numbers, and add library spectra appropriately.

This procedure and an example of a resulting composite isochrone spectrum (CIS) are illustrated in figure 2. Note that the resulting CIS lack their full complement of giant stars, since these are not in the new isochrones. The missing giant groups are therefore included as extra free parameters in the syntheses.

3. STELLAR LIBRARY

The stellar library used here is a combination of three existing stellar libraries, together with additional EFOSC spectra of stars in NGC6522 (Baade's Window). The stellar library from Pickles (1985) has reasonably good coverage of the spectral types expected to be present in large composite systems, and wavelength coverage extending from 3600 A to 10000 A. This library has spectral resolution of 12 A, and is defined on a grid of 3 A per pixel. The photometric quality is poor bluewards of 3800 A however; this has now been improved and extended in three ways.

In the first stage, spectra were selected from the library of Jacoby, Hunter and Christian (1984) on the basis of spectral type, colour and metal-line strength, so as to match the spectral groups of the P85 library. This library has resolution of 7 A and is defined on a grid of 1.4 A per pixel. A wavelength region from 3520 A to 6800 A has been used, avoiding the earth atmospheric features present in these spectra redwards of 6800 A.

In the second stage, the spectra were extended to 3000 A in the blue using spectra from the Vilnius library (Straizys and Sviderskiene 1972). This library is defined on a grid of 50 A per pixel, and hence is useful for defining the flux level, but not the lines, in this region.

We have obtained EFOSC spectra of several stars in NGC6522 (Arp 1965), including some of the metal-rich stars studied by Whitford and Rich (1983). The line equivalent widths measured from our spectra correlate well with those tabulated by Whitford and Rich, but are systematically lower by about 10%. This is probably a resolution effect since their band definitions are very narrow. The spectra have been allocated to the appropriate library group on the basis of their measured colours and line-strength, where we have preferred to use the band definitions of Faber et. al. (1985).

The spectral combination was done using a program COMBINE, which takes an arbitrary number of flux-calibrated spectra, defined on arbitrary wavelength grids, interpolates them all to a chosen common

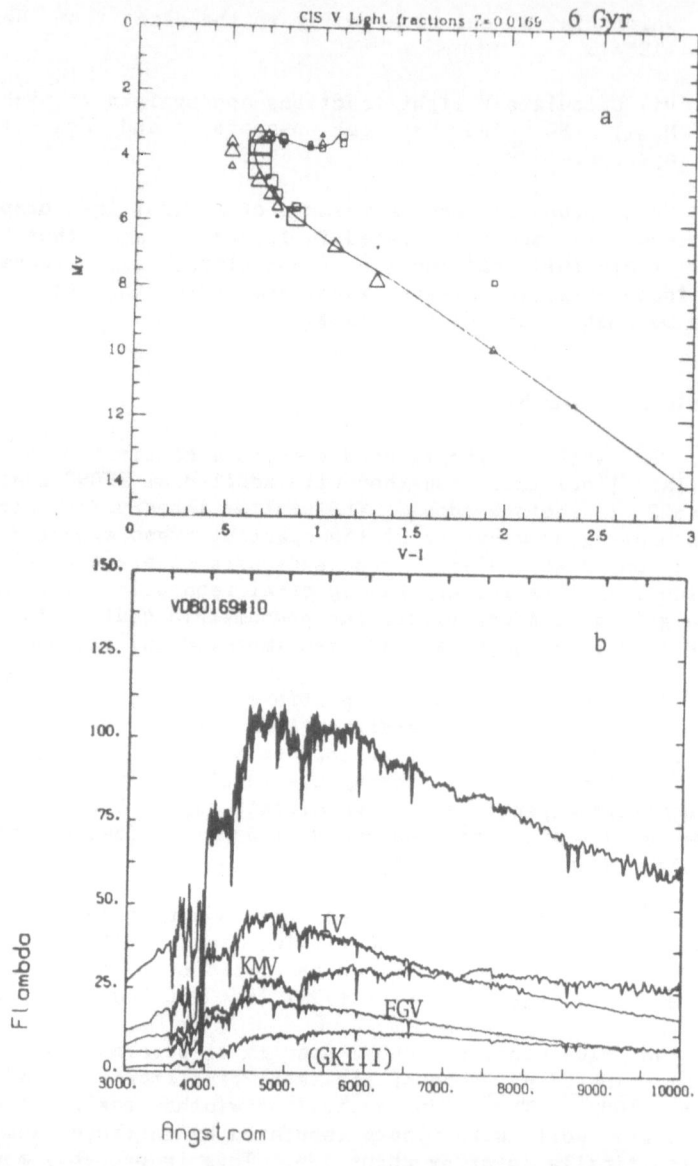

Figure 2a: illustrates the procedure for constructing a CIS from standard spectra to represent a selected isochrone (here solar metallicity, age 6 Gyr). Triangles and rectangles represent the location in the (M_V, $V-I_C$) plane of the blue and red contributing groups respectively; their area represents their contribution in V light to the final composite spectrum. Figure 2b: illustrates one of the derived CIS (solar metallicity, 10 Gyr), and the relative contributions from dwarf, subgiant and (partial) giant stars.

output grid, and performs a weighted average. A 3σ clip rejection loop after the first average was used to reject edge effects and bad pixels in single input spectra. The final library is defined on a grid of 6 A per pixel between 3000 and 10000 A.

4. OBSERVATIONS

Abell clusters covering a range of redshift were selected from those previously studied by Butcher and Oemler (1984) and Couch and Newell (1984), and CCD photometry then obtained with the European Faint Object Spectrograph and Camera (EFOSC) attached to the 3.6m telescope at La Silla. The CCD photometry through Cousins BVR and Gunn I filters was calibrated by reference to E region fields, and galaxy colours then measured inside an 8 arcsec diameter circular aperture. The Fornax cluster (Pickles and Visvanathan 1985) has also been included as a low redshift reference point.

Cluster	mean galaxy redshift	reddening removed E_{B-V}	No. of coadded galaxies
Fornax	0.005	0.05	2
AC122	0.210	0.06	2
A2397	0.220	0.07	4
A1525	0.260	0.05	3
A370	0.373	0.06	1
C0024+16	0.393	0.07	3

EFOSC was also used to obtain medium resolution spectra of the brighter galaxies. Two grisms were used, giving a combined useful observing range from 4000 A to 9000 A, at 7 A per pixel. Most galaxies were observed through a 1.5 arcsec wide long slit, giving effective resolution of 15 A. The spectra were flat-fielded, sky subtracted, extracted according to the optimal extraction algorithm of Horne (1986), and flux calibrated. Repeated observations were then coadded with relative weighting as advocated by Robertson (1986). Synthetic broad band BVRI colours measured from the final spectra agree well with direct CCD colours, indicating no significant losses due to atmospheric refraction or serious problems with colour gradients. Galaxy spectra were dereddened, de-redshifted by amounts determined from cross-correlation with a K giant spectrum, and then coadded for each cluster separately.

5. AGE RANGE vs. METALLICITY RANGE

All cluster spectra have been synthesised in two ways to try to establish whether intermediate age populations are indeed present in the galaxies studied, or whether their blue light could come from a purely old population with substantial metallicity range (Renzini 1986).

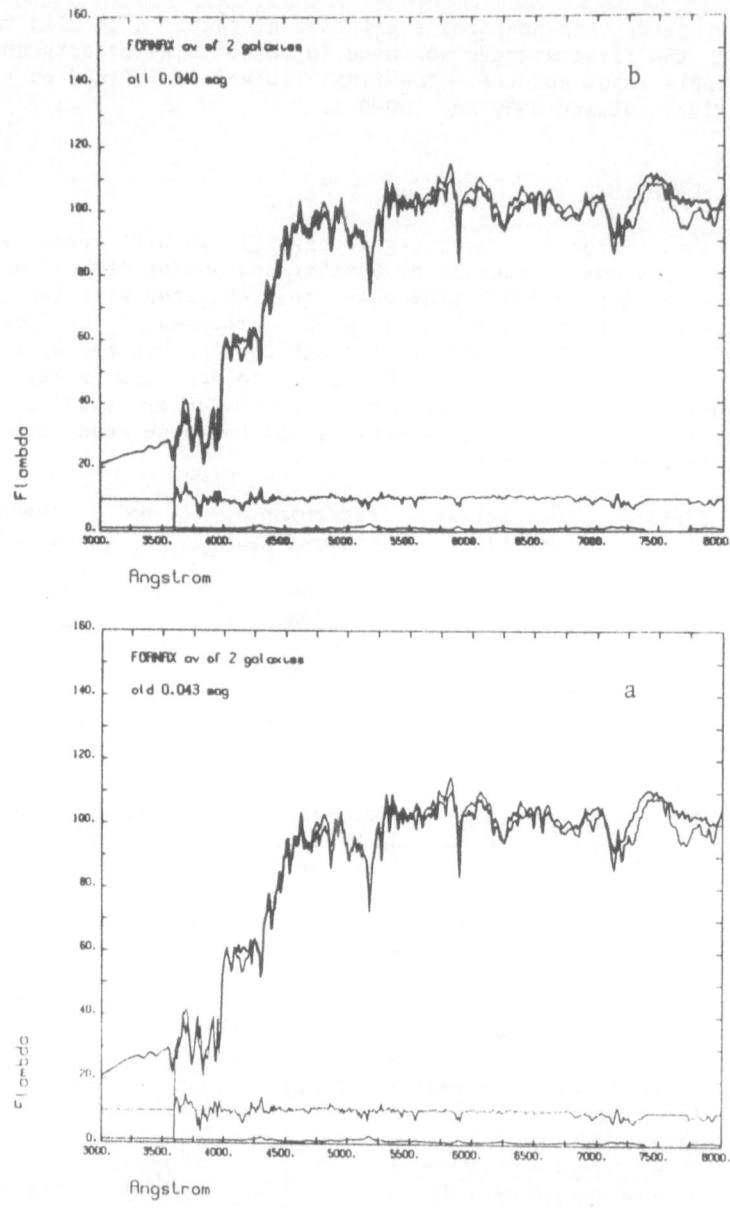

Figure 3a: The observed, coadded spectrum of the two brightest ellipticals in the Fornax cluster (thin line) and the synthetic fit (thick line) obtained using a library containing only old CIS and HB and red giant spectra. The lower trace is an amplified residual. Figure 3b: shows the corresponding fit obtained using the complete CIS library.

In the first, restricted method, we have fitted cluster spectra with a library devoid of young or intermediate age CIS, and containing only old (15 Gyr) CIS together with horizontal branch giant spectra and red giant spectra of differing metallicity. No constraints were applied other than that all contributions should be positive. The subgiant and main sequence constraints are of course built into the CIS, but no connectivity constraints were applied to the horizontal branch or red giant group contributions. This has the unfortunate consequence of removing age resolution from the giant branch components, but is unavoidable without reliable theoretical data for the giant branch where mass loss can be expected to play an important part. Reasonably good fits were obtained as illustrated for the Fornax cluster in figure 3a.

In the second method, we have fitted cluster spectra with a library containing CIS of all ages together with giant groups and, as expected because of the extra degrees of freedom, obtain somewhat better fits. This is illustrated in figure 3b, again for the Fornax cluster. The parameters of these two fits can be summarised as follows:

Method	CIS Turnoff Group	other groups	[Fe/H]	Age (Gyr)	% V Light contribution
restricted	F7-8 V		-0.5	15	10
	G5-8 V		0.0	15	60
		GK giants	+0.6	-	22
		M giants	-	-	7
		HB giants	-	-	1
all ages	G0-4 V		0.0	10	30
	G5-8 V		+0.5	12	42
		GK giants	+0.6	-	20
		M giants	-	-	6
		HB giants	-	-	1

There are several points to note about these fits, which hold generally for all the cluster spectra fitted here:

i) The bluest main sequence group is the same regardless of the method used. In the restricted (old) fit, the necessary blue light comes from metal-weak CIS, whereas it comes from young or intermediate age populations if they are available. This illustrates a point which deserves to be made strongly; that population synthesis is very reliable in fitting colours and components from the HR diagram. Different astrophysical interpretations can always be made with these components, but this implies a degree of non-uniqueness which is hardly more severe than that associated with fitting isochrones to globular cluster CMDs.

ii) The mean metallicity found in the restricted fits is *lower* for the CIS groups (which give the main sequence light) than for the red giant groups. It is clear that the cluster spectra *can* be well

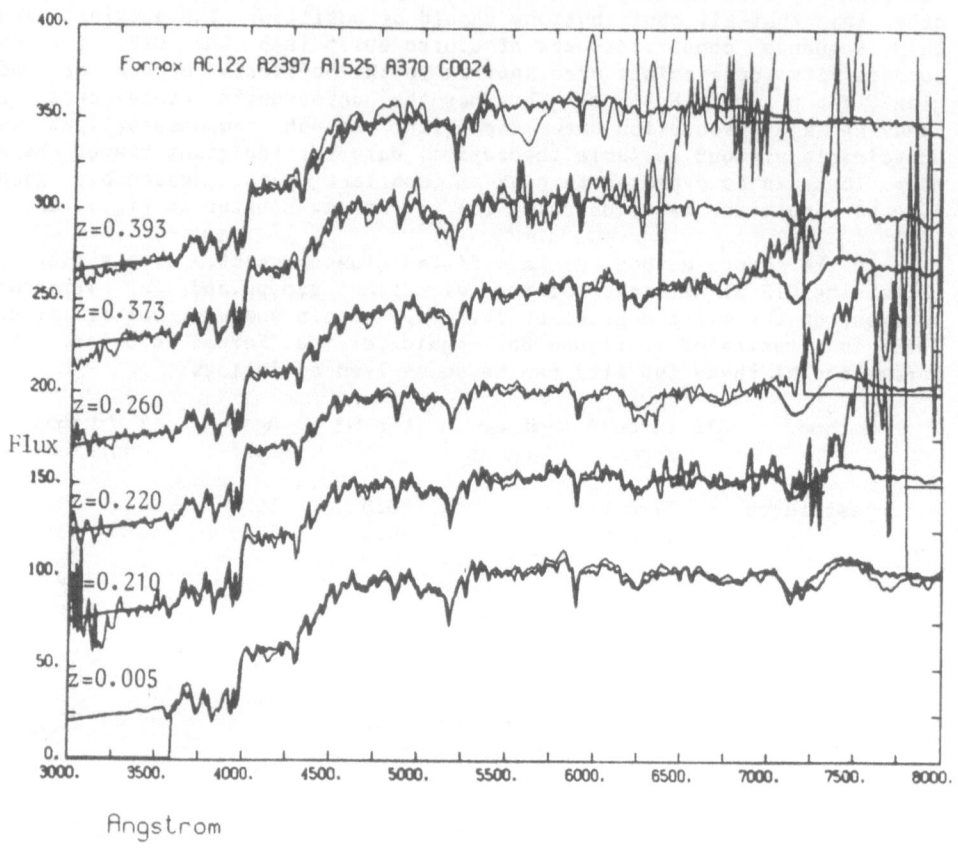

Figure 4: Observed coadded (and deredshifted) spectra (thin lines) of the brightest galaxies in 6 clusters of varying redshift, and their optimal synthetic fits (thick lines). The red wavelength limit over which each spectrum is fitted decreases slightly with redshift, but the red spectrum of C0024+16 was not used in the synthesis.

fitted with purely old (and metal-weak) CIS, together with a substantial metal-rich component on the giant branch. But there is a metallicity discontinuity around the subgiant region in this case which is astrophysically untenable.

iii) The old, restricted fit is typically 10-15% worse than the free fit using all ages, and is mainly worse in terms of fits to the stronger line features. This difference is insufficient to reject the old fit (which has less degrees of freedom) at the observed resolution of 15 A. But observing at higher resolution generally implies less spectral coverage, which leads in turn to poor colour resolution. Purely old fits therefore **are** possible (cf. Buzzoni, this conference), but the preference here for young or intermediate age remains as one of astrophysical interpretation.

6. DIFFERENTIAL SYNTHESIS OF CLUSTER DOMINATING GALAXIES

Optimal syntheses of all the cluster spectra are presented in figure 4, and results of these syntheses displayed schematically in figure 5, where the filled symbols represent optimal fits (all ages) and the open symbols the restricted, old fits. The first point to note concerns the metallicity discontinuity of the old fits discussed above for the Fornax cluster. It is clear by comparing the metallicity distribution of the old fits (CIS - open circles) with that on the giant branch (open squares on right hand side) that the giant branch is typically more metal-rich than the corresponding old CIS. Exceptions to this are cluster A2397, which is noticeably weak lined (see figure 4), and C0024+16 whose poor spectral quality and restricted wavelength coverage mean that this cluster should be given low weight here. A2397 may offer a good example of bright elliptical galaxies containing only old stars (and such clusters will presumably exist if the history of star formation is widely variable), but the other 4 well observed cluster spectra show clear evidence for giant branch metallicity distributions incompatible with purely old stellar populations. There is no such discontinuity present in the optimal fits when comparing the metallicity distributions of the CIS (filled circles) and giant branch groups (filled squares). In what follows we will assume that young or intermediate age populations are present in all cluster ellipticals, and look for trends in the optimally determined age distribution with cluster redshift.

Looking at the optimal fits only in figure 5 (filled symbols), a slight trend of age distribution towards younger mean ages with increasing redshift can be seen for the 4 clusters in the redshift range $0 < z < 0.3$, but this then vanishes for the 2 highest redshift clusters. At this stage it is clearly not possible to say whether there is a real trend which is obscured by the inferior quality of the highest redshift observations, or whether the apparent trend at low redshift is purely fortuitous.

40

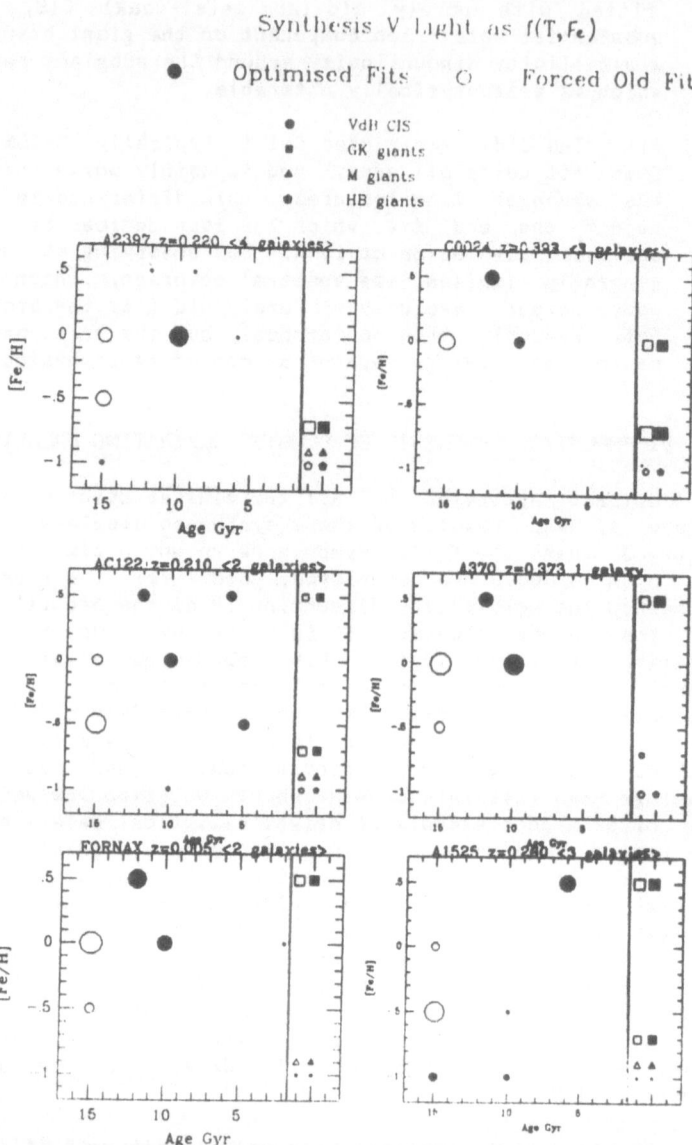

Figure 5: Schematic representations of the synthesis results for each cluster. Filled symbols represent optimised fits using all CIS, open symbols represent restricted fits using only old CIS; their area represents the relative V light contributions of the various components. The abscissa shows the age in Gyr, and the ordinate shows the metallicity. HB, GK and M giants (which have ascribed metallicity but no age in this treatment) are put to the right of each panel.

It is appropriate here to stress that the galaxies synthesised so far are the brightest ones, dominating each cluster. These galaxies are *not* part of the sample displaying the Butcher-Oemler effect, for example, and are not expected to show the same effects observed for the fainter, bluer cluster members by Couch and Sharples (preprint - discussed by Ellis, this conference) and Gunn (this conference). Nor are they similar to the case of M32, for which a very strong observational case for intermediate age populations now exists (O'Connell 1986, Burstein - this conference). Intermediate age (8-10 Gyr) populations, if present in giant ellipticals, should be clearly visible at redshifts z > 0.5 and marginally visible at lower redshifts (depending on cosmology) however. The most reasonable statement to make from this work, therefore, is that there is no firm trend of stellar age distribution with redshift, indicating that most stars formed in these cluster dominating galaxies before the lookback time appropriate to z = 0.4.

REFERENCES

Arimoto N, & Y. Yoshii, 1987, *Astr. Ap.* 173, 23
Arp H, 1965, *Ap. J.* 141, 45
Bessell M.S, 1979, *P.A.S.P,* 91, 589
Bessell M.S. & P.R. Wood, 1984, *P.A.S.P.* 96, 247
Burstein D, 1985, *P.A.S.P.* 97, 89
Butcher H. & A. Oemler, 1984, *Ap. J.* 285, 426
Couch W.J. & E.B. Newell, 1984, *Ap. J. Suppl.* 56, 307
Faber S.M, E.D. Friel, D. Burstein & C.M. Gaskell, 1985, *Ap. J. Suppl.* 57, 711
Horne K, 1986, *P.A.S.P,* 98, 609
Jacoby G.H, D.A. Hunter & C.A. Christian, 1984, *Ap. J. Suppl,* 56, 257
Miller G. & J. Scalo, 1979, *Ap. J. Suppl.* 41, 513
Mould J & J. Kristian, 1986, *Ap. J.* 305, 591
O'Connell R.W. 1986, in *Stellar Populations,* eds. C. Norman, A. Renzini & M. Tosi (Cambridge: Cambridge Univ. Press), p 167.
O'Connell R.W. 1987, in *Starbursts and Galaxy Evolution,* eds. T. Montmerle and T. Thuan (Paris: Editions Frontieres), in press
Pickles A.J. 1985, *Ap. J. Suppl.* 59, 33
Pickles A.J. 1987, in *Structure and Dynamics of Elliptical Galaxies,* (IAU Symposium No. 127) ed. T. de Zeeuw (Dordrecht: Reidel)
Pickles A.J. & N. Visvanathan, 1985, *Ap. J.* 294, 134
Renzini A. 1986, in *Stellar Populations,* eds. C. Norman, A. Renzini & M. Tosi (Cambridge: Cambridge Univ. Press), p 213.
Ridgeway S.T, R.R. Joyce, N.M. White & R.F. Wing, 1980, *Ap. J.* 235, 126
Robertson J.G, 1986, *P.A.S.P,* 98, 1220
Spinrad H. & B.J. Taylor, 1971, *Ap. J. Suppl.* 22, 445
Straizys V. & S. Sviderskiene, 1972, *Vilnius Obs. Bull.* 35, 23
Terndrup D.M. 1987, Thesis, Lick Obs.
VandenBerg D.A, 1985, *Ap. J. Suppl.* 58, 711
VandenBerg D.A. and P.G. Laskarides, 1987, *Ap. J. Suppl,* 63
Whitford A.E. 1985, *P.A.S.P.* 97, 205
Whitford A.E. & R.M. Rich, 1983, *Ap. J.* 274, 723

INTEGRATED COLORS OF GALAXIES OF COMPOSITE METALLICITY

Nobuo Arimoto
Tokyo Astronomical Observatory
Mitaka, Tokyo 181
Japan

ABSTRACT. Chemical and photometric properties of galaxies of composite metallicity are studied by using an evolutionary method of population synthesis. We find reasonable consistency between the model and the observation, except for irregular galaxies. Ellipticals, S0's and spirals are systems large enough to allow a statistical model treatment ignoring spacial and temporary fluctuation. If statistically continuous one-zone picture is applicable, these galaxies are well characterized by the universal initial mass function (IMF) and the simple star formation rate (SFR) per unit gas mass which decreases from ellipticals to late spirals. A wind separates galaxies into two distinct sequences of the gas-poor galaxies and the gas-rich ones. High SFR's induce a supernova (SN) driven galactic wind at $t < 1$ Gyr in ellipticals and at 1 Gyr $< t < (T_G-1)$ Gyr in S0's, whereas in early and late spirals the explosion rates of SN are too low to expel the gas from the systems within a galactic age of $T_G=15$ Gyr. It is suggested that the gas-poor galaxies such as ellipticals, dwarf ellipticals, and globular clusters are essentially formed as a one-parameter family of mass and that the Hubble sequence of galaxies should reflect a certain initial correlation between mass and angular momentum of protoclouds.

1. INTRODUCTION

Galaxies are divided into the gas-poor galaxies and the gas-rich ones, although they exhibit a rather diverse variety of chemical and photometric features. The sequence of the gas-poor galaxies such as ellipticals and dwarf ellipticals is clearly separated from that defined by the gas-rich spirals and irregulars in the $(U-V, V-K)$ diagram (Aaronson, 1978; Persson et al. 1979; Caldwell, 1983; Bothun et al., 1985). Tully et al. (1982) also pointed out that the gas-poor and the gas-rich galaxies in the Virgo Cluster occupy quite separate domains in the $(B-H, H^{abs})$ diagram. Whereas, two sequences are nearly identical in the $(J-K, H^{abs})$ diagram (Persson et al., 1979; Bothun et al., 1984).

The gas-poor color-magnitude (CM) relation has been proposed to be a metallicity effect (Aaronson et al., 1978; Tinsley, 1978). Since the J-K color is an excellent indicator of the average metallicity of stellar system, the similarity of both CM relations in the $(J-K, H^{abs})$ diagram implies that the

43

R. G. Kron and A. Renzini (eds.), Towards Understanding Galaxies at Large Redshift, 43–59.
© *1988 by Kluwer Academic Publishers.*

metallicity effect also works in the gas-rich CM relation. On the other hand, the distinct separation of the gas-poor and the gas-rich sequences in the (U-V,V-K) diagram surely reflects the absence (presence) of young stars in the gas-poor (gas-rich) galaxies. Therefore, the gas-rich CM relation should be caused by the relative mix of old and young stellar populations as well as the metallicity effect.

Apparently, mass decreases along the gas-poor sequence from giant ellipticals to dwarf ellipticals, and globular clusters. While, on an average, early spirals are more luminous than late ones, and late spirals than irregulars. Therefore, there also seems to be a monotonical trend of mass along the gas-rich sequence, although Sandage (1986) pointed out that the mass range of all Hubble type galaxies earlier than Sd's overlapps each other.

If both sequences of the gas-poor and the gas-rich galaxies defined in the CM diagram are characterized by galactic mass, it is of great interest to know what is the key factor which divides galaxies into two distinct groups and introduces significant changes in their chemical and photometric properties along the sequences.

The integrated colors of galaxies are very sensitive to the population structure of stars with different mass, age, and chemical compositions. Since almost all galaxies are composed of stars with different metallicities, it is essential to incorporate the effects of a metallicity dispersion to predict the integrated colors. Arimoto and Yoshii (1986, hereafter AY86) developed the evolutionary method of population synthesis taking explicitly into account the change in the stellar metallicity following the chemical evolution process. In this paper, we discuss the integrated properties of the gas-poor and the gas-rich galaxies by using the method of AY86. In Sect.2, we construct the galactic wind model for the gas-poor galaxies and show that all the spheroidal systems are a one-parameter family of their mass. In Sect.3, we try to understand why galaxies are divided into the gas-poor and the gas-rich sequences. Our conclusions are given in Sect.4.

2. THE GAS-POOR SEQUENCE: SPHEROIDAL SYSTEMS AS A ONE-PARAMETER FAMILY OF THEIR INITIAL MASS

Spheroidal systems undoubtedly experienced intensive formation of stars and multiple supernovae (SN) explosions in the early era, and this results in a low gas and dust content. As soon as the released thermal energy from SN explosions exceeds the binding energy of the residual gas, a large amount of the gas is expected to leave the system in a galactic wind. The chemical evolution is stopped and the system expands thereafter (Larson,1974; Ikeuchi,1977; Saito, 1979b).

Arimoto and Yoshii (1987, hereafter AY87) constructed a SN-driven wind model using an evolutionary method of population synthesis, and successfully derived the CM relation from normal to giant ellipticals. Applying the wind model to much smaller systems of dwarf ellipticals and globular clusters, we demonstrate in this section that all the properties of spheroidal systems can be explained by a common relation between binding energy and mass at birth of the system. A part of this section has discussed by Yoshii (1987) and will appear in Yoshii and Arimoto (1987).

2.1 The SN-driven wind model

It is known that a scaling law between the binding energy Ω_G and the mass M_G exists common to giant ellipticals and compact globular clusters (Saito,1979a)

$$\Omega_G = 1.66 \ 10^{60} \ (M_G/10^{12}M_\odot)^\eta \ \text{ergs} \ (\eta=1.45). \quad (1)$$

As long as galaxies are in virial equilibrium, the above relation gives a set of structural quantities such as the average radius, density, and velocity dispersion in terms of M_G. We assume that these quantities are not greatly different from those at the onset of global star formation.

The metal enrichment in a system is mainly determined by the initial stellar mass function (IMF) and the star formation rate (SFR). In AY87, the IMF is assumed to have a power-law mass spectrum, $\Phi(m) \propto m^{-\mu}$, with the mass range of $0.05 \leqslant m/M_\odot \leqslant 60$, independent of the initial mass M_G of a galaxy. The slope index $\mu=0.95$ is adopted such that the model of giant ellipticals reproduce the observed UBVRIJK colors at a galactic age of $T_G=15$ Gyr [for the Salpeter IMF, $\mu=1.35$].

The SFR per unit mass is assumed to be proportional to the fractional mass f_g of gas, $C(t)=\nu/\nu_o f_g(t)$, where ν is the constant rate coefficient in star formation and is normalized by an assumed value $\nu_o=6.08 \ 10^{-18} \ \text{s}^{-1}$ for solar neighbourhood.

The time scale t_{SF} of star formation is assumed to be proportional to the shorter of the free-fall time t_{ff} and the collision time t_{col} of fragmentary clumps. Substituting the structural quantities derived from Eq.(1) into the definition of t_{ff} and t_{col} (AY87), we have

$$t_{ff} = 6.97 \ 10^7 (M_G/10^{12}M_\odot)^{(5-3\eta)/2} \ \text{yr}, \quad (2)$$

and

$$t_{col} = 7.18 \ 10^6 (M_G/10^{12}M_\odot)^{3-2\eta} \ \text{yr}. \quad (3)$$

Applying an instantaneous recycling approximation, we define the time scale t_{SF} as $t_{SF}=-dt/d\ln f_g \propto \nu^{-1}$. Since t_{ff} and t_{col} have the same order of magnitude at $M_G \sim 10^9 M_\odot$, we finally obtain

$$\nu/\nu_o \propto M_G^{2\eta-3}, \qquad M_G \gg 10^9 M_\odot,$$
$$\propto M_G^{(3\eta-5)/2}, \quad M_G \ll 10^9 M_\odot. \quad (4)$$

The proportionality constants in both limiting cases are adjusted such that the models with $M_G=10^{12}$ and $10^6 M_\odot$ reproduce the observed magnitudes of giant ellipticals ($M_V=-23$ mag) and globular clusters ($M_V=-7$ mag), respectively, at an age of $T_G=15$Gyr; i.e., $\nu/\nu_o=45$ for $M_G=10^{12}M_\odot$ and $\nu/\nu_o=800$ for $M_G=10^6 M_\odot$.

When the chemical evolution proceeds in a closed zone with such a high SFR, the metallicity in the gas grows rapidly. While active star formation accelerates metal enrichment, it also increases the frequency of SN explosions which generate a galactic wind at the time t_{GW}, and the formation of new stars is stopped thereafter. Under the assumed IMF and SFR, the time evolutions of metal abundance $Z_g(t)$ and fractional mass $f_g(t)$ of gas are computed up to t_{GW} with the initial values $Z_g(0)=0$ and $f_g(0)=1$, while the evolution of photometric quantities is computed up to $T_G=15$ Gyr.

For details of basic equations, conditions for a galactic wind, and stellar ingredients for population synthesis, the reader should consult with AY86 and AY87.

2.2 Chemical and dynamical properties of the model

The SFR coefficient ν in Eq.(4) is proportional to a negative power of M_G. With such a SFR, a galactic wind occurs later and chemical evolution proceeds more effectively in larger galaxies.

The frequency distribution of stellar metallicity for the models of $M_G=10^7$ to $10^{12} M_\odot$ are shown in Fig.1. Evidently this figure shows that both giant and dwarf ellipticals are true composite stellar systems, and that larger galaxies are composed of stars with higher metallicities.

The binding energy of systems with $M_G<10^6 M_\odot$ is so small that a wind occurs as soon as massive stars of the earliest generation explode. Consequently, such systems experience no metal enrichment. This means that there exists a critical M_G which divides small systems into dwarf ellipticals and globular clusters. Actually, globular clusters are composed of stars of the same metallicity (e.g.,M15, Sandage and Katem,1977), whereas dwarf galaxies have a spread in stellar metallicities (e.g.,NGC147; Mould et al.,1983, NGC205; Mould et al., 1984, Sculptor; Da Costa,1984).

The metallicity of composite stellar system is characterized by the luminosity-weighted average $[Fe/H]_\ell$ and by the mass-weighted average $[Fe/H]_m$. The former is defined as

$$[Fe/H]_\ell = \Sigma_{ij} n_{ij} [Fe/H]_i L_j / \Sigma_{ij} n_{ij} L_j, \qquad (5)$$

where n_{ij} is the number of stars binned into the intervals centered on $[Fe/H]_i$ and $\log L_j$ (AY87). The latter is defined as, according to Pagel and Patchett (1975),

$$\begin{aligned}[Fe/H]_m = {}& \log Z_g(t_{GW})/Z_\odot \\ & + \log\{[1-[f_{GW}/(1-f_{GW})]\ln f_{GW}^{-1}]/\ln f_{GW}^{-1}\}, \qquad (6)\end{aligned}$$

Figure 2 shows evolutions of $[Fe/H]_\ell$, $[Fe/H]_m$, and $\log Z_g/Z_\odot$ for the model of $M_G=10^{12} M_\odot$. Both $[Fe/H]_\ell$ and $[Fe/H]_m$ increase rapidly and attain maxima $[Fe/H]_\ell \simeq +0.7$ and $[Fe/H]_m \simeq +0.4$, respectively, just before an onset of a galactic wind. Then $[Fe/H]_\ell$ drops suddenly to the level of $[Fe/H]_\ell \simeq +0.2$, whereas $[Fe/H]_m$ remains constant. This occurs because, for a slope $\mu=0.95$ of the IMF, the metallicity $[Fe/H]_\ell$ corresponds to the average metallicity of luminous giants, while $[Fe/H]_m$ corresponds to late dwarfs. Thus, the very metal-rich young supergiants and AGB stars are responsible for an extremely high value of $[Fe/H]_\ell$ at the final stage of chemical enrichment. On the other hand, metal-poor red giants which dominate the light of the system give a low value of $[Fe/H]_\ell$ at the present time of $T_G=15$ Gyr.

The metallicity of a system we observe is essentially weighted by the stellar luminosity. Therefore, the observed metallicities should be compared with $[Fe/H]_\ell$, although the average metallicity of composite stellar system is more properly characterized by $[Fe/H]_m$.

Figure 3 shows the average metallicity $[Fe/H]$ versus absolute V-magnitude at $T_G=15$ Gyr. The theoretical $[Fe/H]_\ell - M_V$ relation is consistent

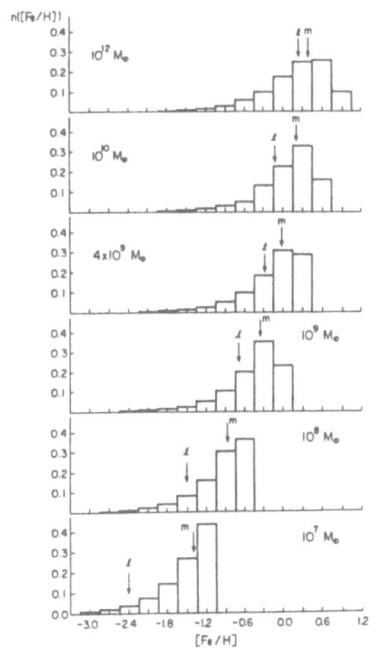

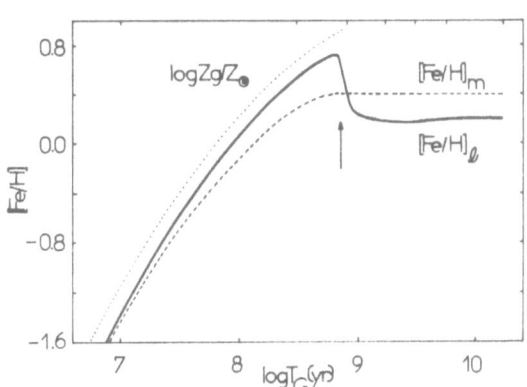

Fig.2. Evolutions of [Fe/H]$_\ell$, [Fe/H]$_m$, and log$Z_g/Z_\odot$ for the model of $M_G=10^{12}M_\odot$. The arrow indicates an epoch of an occurence of a galactic wind.

Fig.1. Metallicity distribution of stars in the models with $M_G=10^7$ to $10^{12}M_\odot$. The arrows with m and ℓ indicate the average stellar metallicities [Fe/H]$_m$ and [Fe/H]$_\ell$ weighted by stellar mass and luminosity, respectively.

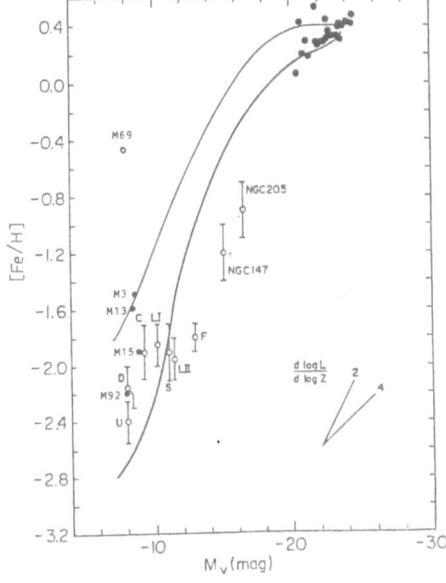

Fig.3. Metallicity versus absolute V-magnitude. The thin and thick lines represent the [Fe/H]$_m$-M$_V$ relation and the [Fe/H]$_\ell$-M$_V$ relation, respectively. The data for ellipticals, dwarfs, and globular clusters are distinguished by the size of circles. Ellipticals are represented by filled circlres of the largest size and dwarfs by open circles of the intermediate size. Galactic globular clusters in three metallicity ranges of [Fe/H]$\geqslant$-0.7, $-0.7\geqslant$[Fe/H]$\geqslant$-1.5, and $-1.5\geqslant$[Fe/H] are represented by open, crossed, and filled circles of the smallest size, respectively.

with the data for giant and dwarf elliptical galaxies. However, the predicted metallicities for globular clusters are smaller than the observed lower limit. If we adopt an initial metallicity of $[Fe/H] \sim -2.5$ in the chemical evolution, the faint end of the theoretical $[Fe/H]_\ell - M_V$ relation shifts up to the region of extremely metal-poor globular clusters. Thus, we may conclude that ellipticals, dwarfs, and globular clusters obey a monotonical sequence of their mass, and that the globular clusters now observed with $[Fe/H] > -2$ were formed from the gas already chemically processed during the early collapse of the Galaxy (Gunn, 1980).

It must be noted that our conclusion is different from those of previous authors who stated that the simple model cannot reproduce the low metallicities of dwarf ellipticals in the Local Group. To remove this serious difficulty, Vader (1986a) proposed the model of metal-enhanced galactic winds which carry away a large fraction of metal produced by SN explosions. Dekel and Silk (1986) proposed the model of mass loss in a system with dominant halo, assuming the relation $M_G \propto R_G^{2.5}$ presently observed only for dwarf galaxies [Compare the relation $M_G \propto R_G^{1.8}$ adopted in the present section]. Their relation predicts lower mass density and hence lower efficiency of chemical evolution toward lower-mass galaxies. However, it should be noted that dwarf galaxies are composite stellar systems similar to ellipticals. The observed metallicity of such a composite system is a luminosity-weighted average metallicity $[Fe/H]_\ell$ dominated by luminous, metal-poor red giants. The currently used mass weighted metallicity $[Fe/H]_m$ is larger than the luminosity-weighted metallicity $[Fe/H]_\ell$ and the difference becomes increasingly large for small and metal-poor composite systems, amounting in particular up to a factor of 1.0 dex for dwarf galaxies in Local Group. If this effect is taken into account, the previous criticism is not fatal to the simple wind model of mass loss.

The $(U-V)-M_V$ and $(J-K)-M_K$ diagrams are particularly important because the integrated U-V and J-K colors correspond mainly to the blue and the red stellar populations, respectively. The U-V color is sensitive to stellar metallicity as well as to the presence of young main-sequence stars and/or blue HB stars, while the J-K color is only sensitive to metallicity.

In the $(U-V)-M_V$ diagram (Fig.4a), the synthesized colors are systematically redder than the observed ones for a given absolute magnitude M_V. As shown in Fig.4a, the difference between the synthesized and the observed colors amounts to $\Delta(U-V) \sim 0.2-0.3$ mag for dwarf galaxies. This difference may be attributed to the omission of blue HB stars in our program of population synthesis (AY86). In composite stellar systems, there are blue HB stars whose progenitors are metal-poor red giants of different generations. Since such progenitors increase in number with decreasing average stellar metallicity, the resulting U-V color is expected to shift blueward as M_G decreases if blue HB stars are properly included in the wind model.

The $(J-K)-M_K$ diagram (Fig.4b) is useful in evaluating the average metallicity of composite stellar system, free from the uncertainties due to young stars and blue HB stars. Figure 4b shows that the theoretical CM relation is in perfect agreement with the observations from giant ellipticals to metal-poor globular clusters.

The metallicities of ellipticals and dwarfs can reasonably be predicted from their J-K color, i.e.,

$$[Fe/H]_{\ell} = 7.94(J-K)-6.66, \quad -1.60 \leqslant [Fe/H]_{\ell} \leqslant +0.40. \qquad (7)$$

Equation (7) gives $[Fe/H]_{\ell}=-0.63\pm0.47$ for the mean color $J-K=0.76\pm0.06$ of 8 dwarf ellipticals in Virgo Cluster; thus, the J–K color implies that the metallicity of dwarf ellipticals is similar to that of the metal–rich galactic globular clusters such as M69 and M71 (cf. Zinnecker et al.,1985).

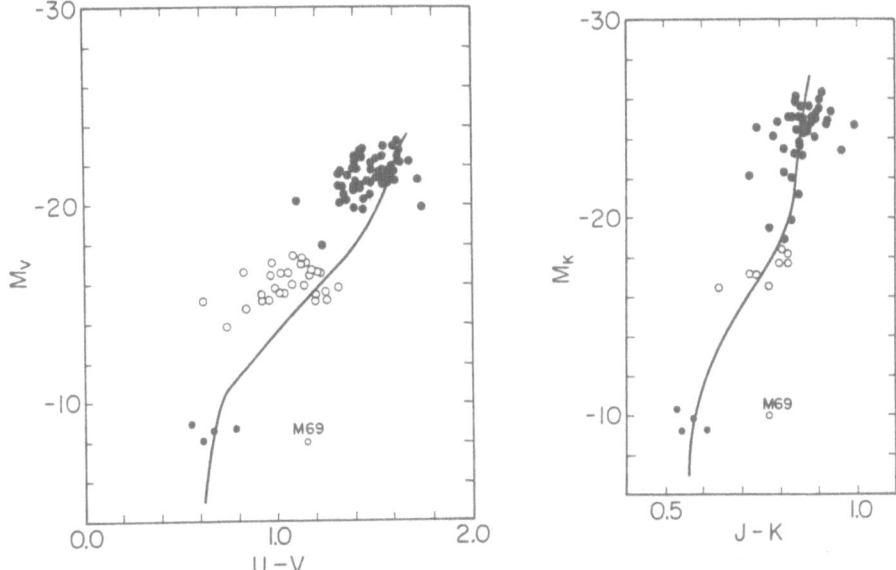

Fig.4. Color–magnitude diagram. The thick line represents the theoretical CM relation. The symbols of the data are the same as in Fig.3. (a) $(U-V)-M_V$ diagram. (b) $(J-K)-M_K$ diagram.

Observationally, the mass–to–luminosity ratio is usually evaluated by using the dynamical mass. Therefore, the observed ratio should be compared with the computed ratio M_*/L_B, where M_* includes the remnant mass M_{rem} in addition to the stellar mass M_s. In Fig.5, we have plotted M_*/L_B against M_*. As M_* increases, the M_*/L_B ratio increases steadily and then remains nearly constant for $M_G \geqslant 10^{10} M_\odot$. This trend of M_*/L_B ratio with M_* can be understood if we remember that the M_*/L_B ratio can be divided into the M_s/L_B ratio and the M_{rem}/L_B ratio. The former is usually called the "photometric mass–to–luminosity ratio" and is an increasing function of M_* because red giants with cooler effective temperature dominate in larger galaxies due to the higher metallicity (Tinsley,1978,1980). On the other hand, the ratio M_{rem}/L_B decreases as M_* increases, because the contribution of non–luminous, dead remnants decreases monotonically as M_* increases owing to the larger evolutionary time scale of metal–rich stars. Therefore, the ratio M_*/L_B shows less clear trend with M_* than the photometric mass–to–luminosity ratio.

While the model M_*/L_B ratio is nearly equal to 23 for -20 mag $> M_B > -23$ mag, the observed average for the same magnitude range is about 8 (Faber and Jackson,1976; Michard,1980; Schechter,1980). Thus, the model M_*/L_B ratio is systematically larger than the observed average. This overestimate could come from adopting a small value $m_\ell=0.05M_\odot$ of lower mass limit in our analysis. It is clear that a larger m_ℓ lowers the M_*/L_B ratio without introducing serious change into other quantities. Therefore, the model M_*/L_B can be adjusted to coincide with the observed average $M_*/L_B=8$ at $M_B=-22$ mag.

The present model gives $M_*/L_B\sim5-6$ for dwarf galaxies after this adjustment. Recently, by measuring directly the velocity dispersion of stars, the mass-to-luminosity ratio has been obtained for the dwarf galaxies in the Local Group (Aaronson,1983; Faber and Lin,1983; Seitzer and Frogel,1985). Sculptor and Carina have values consistent with the present model within the observational uncertainties, whereas Ursa Minor and Draco have the values much larger than predicted. If these observed values are taken for granted, it seems that Ursa Minor and Draco possess large amount of non-luminous matter other than dead remnants.

The present model gives $M_*/L_B\sim4$ for globular clusters, the value of which is larger than the observed value $M_*/L_B=0.9-3.0$ (Illingworth,1976). Ostriker et al. (1972) suggested that dynamical evolution of globular clusters will lead to preferential loss of low-mass stars. In such a case, the mass-to-luminosity ratio will decrease from the initial value.

2.3 Dynamical response to large-scale mass loss

The fractional mass f_{GW} of gas which remains at the epoch t_{GW} is not monotonically related to the initial mass M_G of galaxies (Fig.6). In case of larger galaxies ($M_G\sim10^{11-12}M_\odot$), the hot gas heated by SN explosions cannot escape from the system owing to their larger binding energy. The star formation proceeds until almost all the gas is turned into stars. In the other extreme case of compact globular clusters ($M_G\sim10^{5-6}M_\odot$) which have considerably high SFR's, most gas is converted into stars before massive stars of the earliest generation explode as SNs. For this reason, only the dwarf galaxies of intermediate mass between ellipticals and globular clusters can lose significant fraction of their initial mass.

We assume that the star-gas system was initially in a state of virial equilibrium and that, after the gas is lost, the stellar system recovers the virial equilibrium as a more loosely bound system. It is known, however, that the system expands without recovering the equilibrium if more than half of the initial mass ($f_{GW}>0.5$) is lost on a time scale t_{loss} smaller than a crossing time t_{cross} of a star in the system. On the other hand, if t_{loss} is larger than t_{cross}, the system remains gravitationally bound regardless of the amount of mass lost in a wind (e.g.,Hills,1980; Mathieu,1983).

As a result of dynamical response of mass loss, the system expands as long as it is in virial equilibrium. The increase in diameter D depends on the amount of mass loss f_{GW} and the ratio t_{loss}/t_{cross}. In the extreme case of $t_{loss}/t_{cross}\gg1$, it follows that

$$\Delta\log D = -\log(1-f_{GW}), \qquad (8)$$

while, when $t_{loss}/t_{cross}\ll1$,

$$\Delta\log D = \log[(1-f_{GW})/(1-2f_{GW})]. \qquad (9)$$

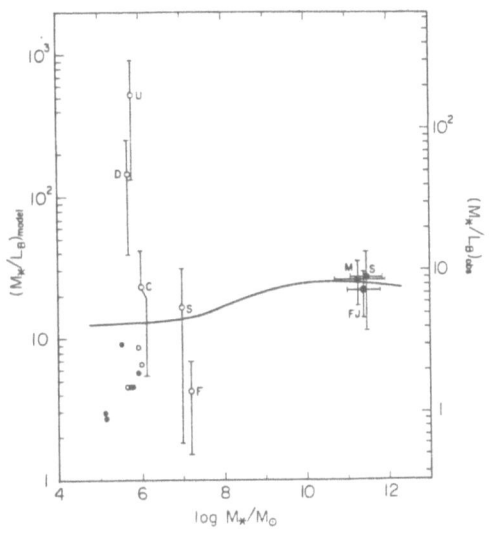

Fig.5. Mass to luminosity ratio versus total stellar mass. The scales on the left ordinate are for the model M_*/L_B ratios, and those on the right ordinate for the observed M_*/L_B ratios. The model M_*/L_B ratio reduced by a constant factor is used for comparison with the observations. The symbols FJ, M, and S for ellipticals indicate the average M_*/L_B ratios taken from Faber and Jackson (1976), Michard (1980), and Schechter (1980), respectively. The other symbols are the same as in Fig.3.

Fig.6. Fractional mass of the residual gas f_{GW} at the epoch of occurence of a galactic wind t_{GW} plotted against the intial mass M_G of the system.

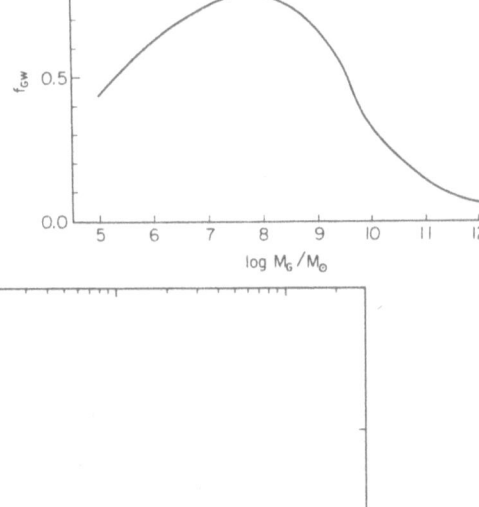

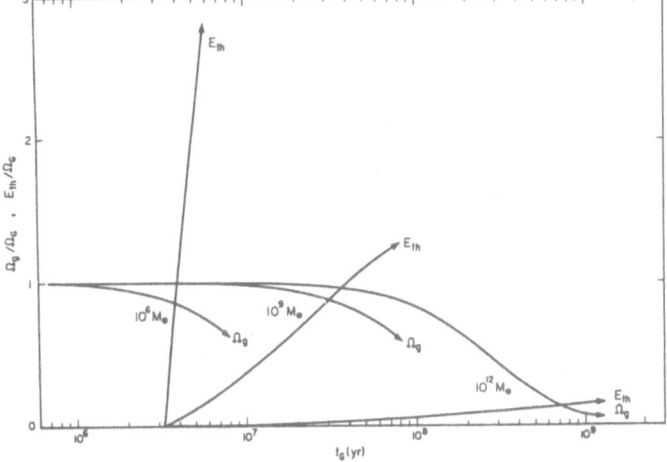

Fig.7. Evolutions of thermal energy E_{th} and binding energy Ω_g of the residual gas for the systems with $M_G = 10^6$, 10^9, and $10^{12} M_\odot$. The energies are relative to the intial binding energy Ω_G of the system.

As described in AY87, mass loss occurs as a wind when the sound velocity c_s of gas heated by SN explosions exceeds the escape velocity. Thus, the ratio of time scales t_{loss}/t_{cross} is approximately expressed as $t_{loss}/t_{cross} \sim v/f_{GW}c_s$, where v is the velocity dispersion of the system.

Figure 7 shows evolutions of thermal energy E_{th} and binding energy Ω_g of the residual gas. This figure shows that E_{th} for $M_G=10^6 M_\odot$ grows rapidly to exceed Ω_g by more than an order of magnitude soon after $t=t_{GW}$ and hence the gas begins to escape from the system at the sound velocity enormously larger than the virial velocity of stars, i.e., $t_{loss}/t_{cross} \ll 1$. Therefore, such a small system favors the condition for rapid gas removal. In the other cases of $M_G=10^9$ and $10^{12} M_\odot$, E_{th} remains slightly larger than Ω_g even after $t=t_{GW}$, so that the sound velocity is of the same order of magnitude as the virial velocity. Taking into account that f_{GW} decreases progressively as M_G increases from $M_G=10^8 M_\odot$ (Fig.6), we have $t_{loss}/t_{cross} \sim 1$ for $M_G=10^9 M_\odot$ and $t_{loss}/t_{cross} \gg 1$ for $M_G=10^{12} M_\odot$. Therefore, such large systems favor the condition for slow gas removal.

We show the results of the wind model in a diamter D versus surface brightness SB diagram. Since the time scale of expansion of a system is much shorter than the time scale of luminosity evolution, the increase in diameter results in the decrease in surface brightness; i.e., $\Delta SB/\Delta \log D = 5$. Using Eqs. (8) and (9) with the values of f_{GW} given by the wind model (Fig.6), we calculate the differential quantities of $\Delta \log D$ and ΔSB. The theoretical D–SB relations realized before and after the gas removal are drawn in Fig.8. The thin lines with arrows represent evolutionary tracks which connect the initial state of a system to the present–day state. It is clear that, in either slow or rapid gas removal, the region to the right of the terminal D–SB line is a forbidden area.

The terminal D–SB line for slow gas removal is continuous from ellipticals to globular clusters. Particularly, it has a negative slope in the domain of dwarf ellipticals. On the other hand, the terminal D–SB line for rapid gas removal is discontinuous at $M_G=10^{6-9}M_\odot$ owing to $f_{GW}>0.5$. The galaxies in this region expand without recovering a new equilibrium state.

The observational data for ellipticals and dwarfs in Virgo Cluster and Galactic globular clusters are taken from Ichikawa et al. (1986). It is evident from Fig.8 that each of the three categories (ellipticals, dwarfs, and globular clusters) has its own empirical sequence, and that the theoretical D–SB relation without correcting for the dynamical response of the system deviates from all these empirical sequences.

The theoretical D–SB relation deviates from the observed sequence for the ellipticals. Since there is little gas left to be lost in a wind, the ellipticals are not expected to expand appreciably. In order to fit the observed and theoretical sequences, a slight increase of the M_*/L_B ratio toward the bright end of the sequence (Faber and Jackson,1976; Michard,1983; Vader,1986b) is required. This indicates that two parameters seem to be necessary at least for giant ellipticals, although the mass is evidently a dominant one (Terlevich et al.1981).

Most dwarf elliptical galaxies reside in the forbidden area for rapid gas removal, but within the permitted area for slow gas removal. Since these galaxies have the mass of $M_G \sim 10^{7-9} M_\odot$, the remaining gas heated by SN explosions is expected to leave the system in the following condition of

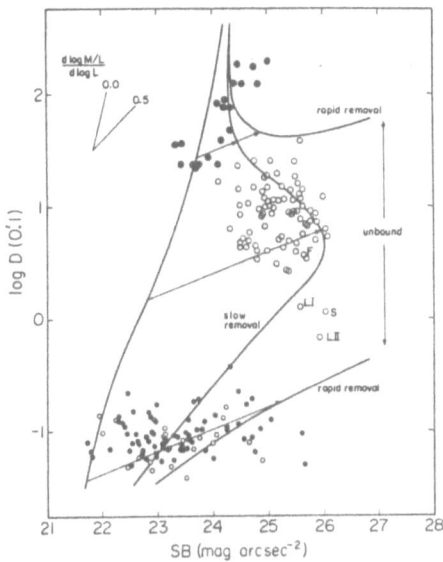

Fig.8. Diameter versus surface brightness. The thick lines are the theoretical D-SB relations with and without correcting for the effect of expansion of the system. The thin lines with arrows are evolutionary tracks which connect the initial state of a system to the present-day state predicted by the wind model. The symbols are the same as Fig.3.

$t_{loss}/t_{cross} \sim 1$. This condition favors, more or less, slow gas removal; thus consistent with the distribution of dwarf ellipticals within the permitted area for slow gas removal.

Most globular clusters reside within the permitted area for rapid gas removal and are distributed along a sequence with a characteristic slope $\Delta SB/\Delta \log D \sim 5$. Thus, it is likely that globular clusters expand after rapid removal of the remaining gas.

Through Sects.2.2 and 2.3 we have shown that the wind model successfully accounts for the observed chemical, photometric, and dynamical properties of ellipticals, dwarf ellipticals, and globular clusters under a universal IMF and a common relation between binding energy and mass. This strongly suggests that all the spheroidal systems are essentially formed as a one-parameter family of their initial mass.

3. TOWARD UNDERSTANDING THE GAS-RICH SEQUENCE

3.1 Intrinsic dispersion in the CM relation of elliptical galaxies

As is shown in Figs.4a-4b of this paper and in Figs.7a-7e of AY87, elliptical galaxies exhibit considerable scatter around the empirical CM relation. The width of scatter is greater than observational error and seems to be intrinsic (Faber,1977; Visvanathan and Sandage,1978). Pickles (1985) suggested that variations in the number of recently formed stars could account for the intrinsic scatter around the CM relation in the $(U-V)-M_v$ diagram. Theoretically, there is a possibility that intermittent star formation can occur up to the present time (Ikeuchi,1977). Figure 9 shows evolutions of E_{th} and Ω_g for the case of $M_G=2~10^{12}M_\odot$ computed by taking into account the possible star formation after a wind [The computation is economized by using a rough time step of $\Delta t=10^8$yr, so that a wind occurs much earlier than the time t_{GW} given in table 5 of AY87]. After the gas is completely expelled out in a wind, the gas shed from low mass giants begins to accumulate in a galaxy and is used again in the formation of new generation stars. Then in a short period, a subsequent burst of SN explosion occurs and ejects the gas out of the system. Such intermittent bursts of star formation and SN explosions would play a role in producing the considerable scatter around the CM relation in the $(U-V)-M_v$ diagram. Some static method of population synthesis also suggest that elliptical galaxies contain the intermediate age populations (e.g., O'Connell,1976; Pickles,1985; Rose,1985).

However, from data by Persson et al. (1979), we have found that there is a tight correlation between the deviations of color from the average CM relation in the $(U-V)-M_v$ diagram and those in the $(V-K)-M_v$ diagram. Figure 10 demonstrates that, at a fixed absolute magnitude, ellipticals redder in the U-V color are always also redder in the V-K color. Thus, it is unlikely that young stars are the cause of the scatters in the CM relations, because the V-K color is essentially free from contamination by young stars (AY86).

Instead of recent star formations, we suggest the second possibility that an intrinsic dispersion in the SFR at a fixed galactic initial mass M_G could explain the observed scatter of colors. At a fixed value of M_G, the lower SFR reduces the rate of SN explosions, which induces a wind much later. As a result, the average stellar metallicity increases and the integrated colors become redder, whereas the absolute magnitude remains nearly constant. The predicted width of scatter around the CM relations in the $(U-V)-M_v$ and the $(V-K)-M_v$ diagrams are shown in Figs.9a-9b of AY87, which give satisfactory agreement with the observed ones.

3.2 S0 galaxies

The integrated colors of S0 galaxies are quite similar to those of elliptical galaxies (Persson et al.,1979). This indicates that galaxies of both Hubble types should have experienced nearly the same history of star formation, especially the same IMF. On the other hand, S0's have disks and HI gas is preferentially detected in S0's, which implies at least that the star formation continued longer in S0's than in ellipticals. In Fig.11 we show evolutions of E_{th} and Ω_g for the models with various values of ν/ν_0. This figure explicitly indicates that at a fixed M_G a wind occurs later if the smaller SFR is adopted. This suggests that if we adopt slightly smaller values of ν/ν_0 for S0's than those for ellipticals, star formation can continue long enough to form a disk but eventually can induce a wind which expels the gas out of the system. Arimoto (1987) showed that, under the same IMF as that of ellipticals ($\mu=0.95$), the integrated colors of the models with arbitrary values of t_{GW} agree with the observed colors of S0's so far as $t_{GW}\lesssim(T_G-1)$Gyr. In other words, if a wind occured more than 1 Gyr ago, the resulting system is recognized as a S0 galaxy. Thus, the wind model reasonablly reproduces the

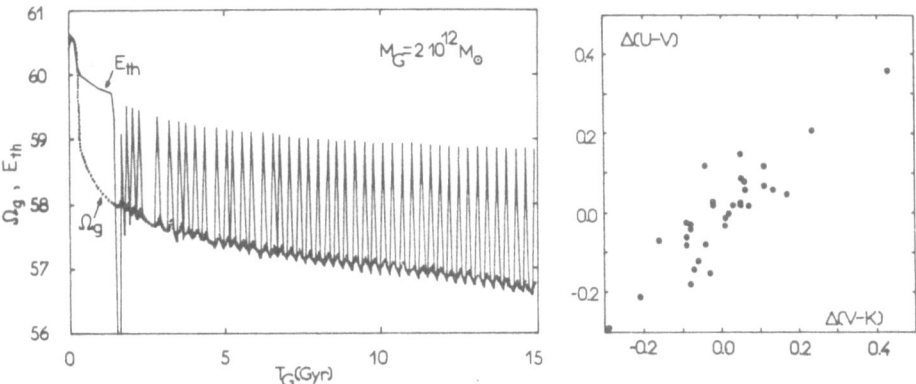

Fig.9. Evolutions of thermal energy E_{th} and binding energy Ω_g of the residual gas for the system with $M_G=2\ 10^{12}M_\odot$ before and after the galactic mass loss.

Fig.10. Correlation between the color deviation $\Delta(U-V)$ from the average CM relation in the $(U-V, M_V)$ diagram and $\Delta(V-K)$ in the $(V-K, M_V)$ diagram.

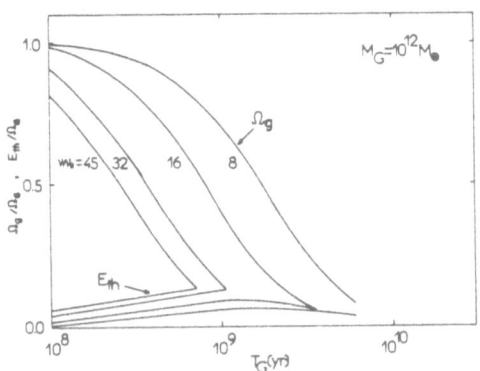

Fig.11. Evolutions of thermal energy E_{th} and binding energy of the residual gas Ω_g for the systems with $\nu/\nu_o=8$, 16, 32, 45; $\mu=0.95$; $M_G=10^{12}M_\odot$. The energies are relative to the initial binding energy Ω_G of the system.

observed properties of S0 galaxies, and an intrinsic dispersion in the SFR can divide protoclouds into ellipticals and S0's.

3.3 Gas-rich galaxies

Discussions in Sects.3.1 and 3.2 suggest that the SFR's for galaxies in the gas-rich sequence should be smaller than those for ellipticals and S0's, otherwise the explosion rate of SN is high enough to transmute the gas-rich galaxies into the gas-poor ones. Figure 11 indicates that, in the case of $M_G=10^{12}M_\odot$ and $\mu=0.95$, a wind does not occur within a galactic age of $T_G=15$ Gyr so far as $\nu/\nu_0 \lesssim 8$. Therefore, in such a case the systems become the gas-rich galaxies such as early and late type spirals.

As demonstrated by Aaronson (1978), a separation of the gas-poor and the gas-rich sequences is well defined in the (U–V)–(V–K) diagram. The very young stars mainly contribute to the U–V color, whereas the average stellar metallicity of old stars is essential to the V–K color. Since the relative number of young stars and the average metallicity of the system are strong functions of ν/ν_0 and μ, loci of models with varying ν/ν_0 and constant μ show clear separation in the (U–V)–(V–K) diagram (AY86). Thus, the (U–V)–(V–K) diagram is most suitable to investigate what kinds of the SFR and the IMF are appropriate for the gas-rich galaxies.

Arimoto (1987) computed an extensive grid of composite model with various ν/ν_0 and μ, and showed that only models with $\mu \simeq 1$ and ν/ν_0 decreasing 16 to 1 [$\nu/\nu_0 = 45$ for ellipticals, see Sect.2.1] can reproduce satisfactorily the continuous change of the observed UVK colors and a monotonical increase of the hydrogen gas mass-to-luminosity ratio $\log M_H/L_B$ from early to late spirals [see Figs.6 and 7 in Arimoto (1987)]. It should be noted that the values of ν/ν_0 are nearly consistent with the condition that a wind is not induced within a Hubble time. Thus, we suggest that spiral galaxies have essentially the same IMF as ellipticals and S0's, whereas the SFR's per unit mass are smaller than those for the gas-poor galaxies and decrease from early to late spirals.

Figure 6 of Arimoto (1987) shows that the observed colors of irregulars are consistent with the models of $\nu/\nu_0 = 0.01 - 0.1$; $\mu = 0.95$. The predicted abundance of heavy elements of the gas is $\log Z_g/Z_\odot = -0.2 \sim -1.2$, which also agrees with the observed ones. However, Fig.7 of Arimoto (1987) indicates a significant discrepancy that the models give much higher values of $\log M_H/L_B$ than the observed ones. Irregulars show too low fractionary gas mass for their estimated metal abundances. This suggests that the star formation history of small galaxies is directly controlled by the local IMF of individual star forming regions, in contrast to sufficiently large galaxies where the chemical and photometric evolution can be described by the overall and time-constant IMF defined as an ensemble of the local IMF's. In such a case, photometric properties of irregulars are mainly determined by several regions presently forming stars, and their chemical properties are given as the integrated results of the past star formations. Thus, statistically continuous one-zone picture is not applicable for irregular galaxies. Detailed discussions for the star formation history of irregulars will appear elsewhere (Arimoto and Tarrab,1987).

3.4 Intrinsic dispersion in the SFR

Under the universal IMF, the properties of sufficiently large galaxies are well explained if the SFR decreases along the Hubble sequence from ellipticals to late spirals. The high SFR induces a wind in ellipticals within the collapse time (~1 Gyr) of a galaxy and in S0's after the formation of disks. Whereas in the gas-rich galaxies, the SFR is too low to induce a wind within the Hubble time and gives monotonical trends of decreasing bulge-to-disk ratio and of increasing the relative fraction of gas and young stars from early to late type spirals.

Let's consider why the SFR decreases from ellipticals to late spirals.

It is noticed that forms of ellipticals are essentially supported by random motion of stars, whereas gas and stars in irregulars rotate almost rigidly. These facts suggest that the random component decreases monotonically with respect to the rotational motions from early to late Hubble type galaxies. In ellipticals, the random motion is so dominant that the frequency of cloud collisions and hence the SFR during the stage of bulge formation is extremely high. From ellipticals to late spirals the rotational motion becomes prominent more and more, which surpresses the cloud collisions due to random motion and gives the smaller SFR during bulge formation. As a result, the SFR per unit mass in the bulge decreases along the Hubble sequence from ellipticals to spirals.

Since a density wave well developes in the disk of a sufficiently large spiral which rorates differentially, the disk component stars are mainly formed in the regions compressed by the shocks. Then, the rate coefficient of SFR in the disk could be given as $\nu/\nu_o \propto k(\omega-\omega_p)$, where k is the number of arms, ω the angular velocity of material in circular orbit, and ω_p the pattern speed (Talbot and Arnett,1975). Since the maximum velocity of a spiral galaxy along its rotation curve decreases gradually from early to late spirals (Giraud,1984), the SFR during disk formation would be smaller in late spirals than in early ones.

Above discussions indicate that the SFR's during both stages of bulge and disk formation decrease from ellipticals to spirals. If mass of a galaxy is constant regardless of its morphology, it could be said that the Hubble sequence is a sequence of galaxies with decreasing angular momentum per unit mass. However, a galactic mass actually tends to decrease from early to late Hubble types. Therefore, a certain correlation between mass and angular momentum of a galaxy could have been an initial condition for galaxy formation.

4. CONCLUSIONS

Integrated colors of galaxies of composite stellar metallicity are investigated by using an evolutionary method of population synthesis.

A supernova–driven wind model is constructed for the gas–poor galaxies. An assumed relation $\Omega_G \propto M_G^\eta$ ($\eta=1.45$) for protoclouds gives the SFR per unit mass which is proportional to a negative power of mass. With such a SFR and a universal IMF, we successfully reproduced the structural and chemical properties of elliptical galaxies, dwarf elliptical galaxies, and globular clusters. The chemical and photometric properties of the systems are monotonically related to their mass. Whereas, the expansion of a system due to galactic mass loss changes the structural properties most significantly for dwarf elliptical galaxies and brings about the distinct sequences of ellipticals, dwarfs, and globular clusters in the diameter–surface brightness (D–SB) diagram. It is suggested that the spheroidal systems are one–parameter family of mass at their birth, although an extra parameter seems to be necessary for bright ellipticals to fit the sequence in the D–SB diagram.

Under the universal IMF, the chemical and photometric properties of sufficiently large galaxies are well explained if the SFR per unit mass decreases along the Hubble sequence from ellipticals to late spirals. The high SFR's induce a wind at t<1 Gyr in ellipticals and at 1 Gyr<t<(T_G-1) Gyr in

58

SO's, whereas in early and late spirals the SFR's are too low to induce a wind within a galactic age of T_G=15 Gyr. This intrinsic dispersion in the SFR explains naturally the reason why there are two distinct sequences of the gas-poor and the gas-rich galaxies. Such dispersion in the SFR is possible if a certain correlation between mass and angular momentum per unit mass of a protocloud was an initial condition for galaxy formation.

Acknowledgement: I am very grateful to Prof.A.Renzini for providing financial support to attend this quite interesting workshop.

References

Aaronson,M.: 1978, Astrophys.J.Lett. 221, L103.
Aaronson,M.: 1983, Astrophys.J.Lett. 266, L11.
Aaronson,M., Cohen,J.G., Mould,J., Malkan,M.: 1978, Astrophys.J. 223, 824.
Arimoto,N.: 1987, in Proc. of Japan-France Seminar, Chemical Evolution of Galaxies with Active Star Formation, ed. K.Takakubo (Faculty of Science, Tohoku University, Sendai, Japan), p.381.
Arimoto,N., Tarrab,I.: 1987, in preparation.
Arimoto,N., Yoshii,Y.: 1986, Astron.Astrophys. 164, 260 (AY86).
Arimoto,N., Yoshii,Y.: 1987, Astron.Astrophys. 173, 23 (AY87).
Bothun,G.D., Mould.J.R., Wirth,A., Caldwell,N.: 1985, Astron.J. 90, 697.
Bothun,G.D., Romanishin,W., Strom,S.E., Strom,K.M.: 1984, Astron.J. 89, 1300.
Caldwell,N.: 1983, Astron.J. 88, 804.
Da Costa,G.S.: 1984, Astrophys.J. 285, 483.
Dekel,A., Silk.J.: 1986, Astrophys.J. 303, 39.
Faber,S.M.: 1977, in The Evolution of Galaxies and Stellar Populations, eds. B.M.Tinsley and R.B.Larson , (Yale University Observatory, New Haven), p.157.
Faber,S.M., Jackson,R.E.: 1976, Astrophys.J. 204, 668.
Faber,S.M., Lin,D.N.C.: 1983, Astrophys.J.Lett. 266, L17.
Giraud,E.: 1984, Ph.D.thesis, University of Montpellier, France.
Gunn,J.E.: 1980, in Proc. of NATO Advanced Study Institute, Globular Clusters, eds. D.Hanes and B.Maeder (Cambridge University Press, Cambridge), p.301.
Hills,J.G.: 1980, Astrophys.J. 225, 986.
Ichikawa,S., Wakamatsu,K., Okamura,S.: 1986, Astrophys.J.Suppl. 60,475.
Ikeuchi,S.: 1977, Prog.Theore.Phys. 58, 1742.
Illingworth,G.: 1976, Astrophys.J. 204, 73.
Larson,R.B.: 1974, Mon.Not.Roy.Astron.Soc. 169, 229.
Mathieu,R.D.: 1983, Astrophys.J.Lett. 267, L97.
Michard,R.: 1980, Astron.Astrophys. 91, 122.
Michard,R.: 1983, Astron.Astrophys. 121, 313.
Mould,J.R., Kristian,J., Da Costa,G.S.: 1983, Astrophys.J. 270, 471.
Mould,J.R., Kristian,J., Da Costa,G.S.: 1984, Astrophys.J. 278, 575.
O'Connell,R.W.: 1976, Astrophys.J. 206, 370.
Ostriker,J.P., Spitzer,L., Chevalier,R.A.: 1972, Astrophys.J.Lett. 176, L51.
Pagel,B.E.J., Patchett,B.E.: 1975, Mon.Not.Roy.Astron.Soc. 172, 13.
Persson,S.E., Frogel,J.A., Aaronson,M.: 1979, Astrophys.J.Suppl. 39, 61.

Pickles,A.J.: 1985, Astrophys.J. 296, 340.

Rose,J.A.: 1985, Astron.J. 90, 1927.

Saito,M.: 1979a, Publ.Astron.Soc.Japan 31, 181.

Saito,M.: 1979b, Publ.Astron.Soc.Japan 31, 193.

Sandage,A., Katem,B.: 1977, Astrophys.J. 215, 62.

Schechter,P.: 1980, Astron.J. 85, 801.

Seitzer,P., Frogel,J.A.: 1985, Astron.J., 90, 1796.

Talbot,R.J., Arnett,W.D.: 1975, Astrophys.J. 197, 551.

Terlevich,R., Davies,R.L., Faber,S.M., and Burstein,D.: 1981, Mon.Not. Roy. Astron. Soc. 196, 381.

Tinsley,B.M.: 1978, Astrophys.J. 222, 14.

Tinsley,B.M.: 1980, Fundamental of Cosmic Physis 5, 287.

Tully,R.B., Mould,J.R., Aaronson,M.: 1982, Astrophys.J. 257, 527.

Vader,J.P.: 1986a, Astrophys.J. 305, 669.

Vader,J.P.: 1986b, Astrophys.J. 306, 390.

Visvanathan,N., Sandage,A.: 1977, Astrophys.J. 216, 214.

Yoshii.Y.: 1987, in Proc. of Japan-France Seminar, Chemical Evolution of Galaxies with Active Star Formation, ed. K.Takakubo (Faculty of Science, Tohoku University, Sendai, Japan), p.371.

Yoshii,Y., Arimoto,N.: 1987, Astron.Astrophys. in press.

Zinnecker,H., Cannon,R.D., Hawarden,J.G., MacGillivary,H.T.: 1985, in Proc. of ESO Workshop, Virgo Cluster of Galaxies, eds. O.-G.Richter and B.Binggeli (ESO, Garching), p.135.

EVOLUTIONARY POPULATION SYNTHESIS: A NEW GRID OF MODELS

Alberto Buzzoni
Osservatorio Astronomico di Brera
Via Brera 28
20121 Milano
Italy

ABSTRACT. A new set of synthetic models of stellar populations is presented. Evolutionary synthesis has been performed for a variety of chemical compositions and initial mass functions in an age range from 4 to 18 Gyrs. Important refinements have been introduced in respect to previous works such as the full treatment of post Main Sequence stellar evolution, through Horizontal Branch, Asymphotic Giant Branch and Planetary Nebula phases. As a check of the reliability of the computational approach an application to the globular cluster M3 is made and further extensions to the extragalactic domain are suggested concerning the study of the evolution of the galaxies in the cosmological framework.

1. INTRODUCTION

The basic purpose of Evolutionary Population Synthesis (EPS) is to decode the status of stellar populations by properly accounting for their emitted Spectral Energy Distribution (SED). In its natural applications synthesis closely deals with the study of stellar systems in a wide dimensional range, from small stellar associations to galaxies as a whole.

EPS rests on the fundamental requirement that a reliable representation of the evolutionary phenomena occurring in stars is made: in this sense a careful theoretical approach has to be pursued in order to put into the code the right (or at least the main) ingredients. There are in our opinion at least three new relevant contributions that require now an important revision and a substantial update in computing EPS models in respect to previous works like for instance those of Bruzual (1983).

The first one is a fully new set of isochrones by VandenBerg (1983, 1985) VandenBerg and Bell (1985) and

R. G. Kron and A. Renzini (eds.), Towards Understanding Galaxies at Large Redshift, 61–71.
© *1988 by Kluwer Academic Publishers.*

VandenBerg and Laskarides (1987). These isochrones are now available for a variety of ages and chemical mixtures and refinements have been introduced over the previous set by Ciardullo and Demarque (1977), the most important being an observational calibration of the mixing length parameter, a crucial requisite for the application of isochrones to EPS. For instance we are now able to perform a more reliable picture of the C-M diagrams of galactic globular clusters, that are well known to be a canonical example of a "pure" coeval stellar population.

The second improvement concerns the study of the late evolutionary phases in the life of the stars. A better knowledge has been reached in understanding mechanisms which determine evolution along Horizontal Branch (HB), Asymphotic Giant Branch (AGB), Post-AGB and more consistent hints allow us now to guess the effectiveness of these phases in contributing to the SED and the energetic balance of old populations. In particular, for what concerns HB, a wide grid of evolutionary stellar tracks are now available (Sweigart and Gross 1976, Sweigart 1987, Seidel et al. 1987) despite of our still imperfect knowledge of the mechanism(s) leading to the observed spreads in the morphology of the HBs in our globular clusters (the well known "second parameter" dilemma).

AGB and Planetary Nebula event (Post-AGB) still rise up as formidable problems so many and intriguing phenomena occur in determining evolution. No detailed grids of stellar tracks are known to be available up to now, with the only remarkable exceptions of the models of Gingold (1974, 1976), Paczynski (1971), Schönberner (1979, 1981). Nevertheless important clarifying contributions already enable us to remove trouble in approaching crucial problems like nucleosynthesis and lifetimes in these phases (Renzini and Voli 1981, Iben and Renzini 1983).

As a third relevant occurrence (last but not least for our specific aims) it is worth noting the development of a comprehensive approach to the energetics of the stellar populations. SED of stellar populations must be directly related to their bolometric emission or, in other words, to their total energy available (Tinsley and Gunn 1976, Renzini 1981, Renzini and Buzzoni 1983, 1986).
A fundamental check of inner consistency of the EPS models must then hold: it is the so called "Fuel Consumption Theorem".

This work attempts to explore some implications stemming from this new scenario.

2. THE MODELS

A complete set of more than three hundred synthetic models

of stellar populations has been calculated. It fully
explores a "phase space" described by a set of four primary
parameters: they are age (t), metallicity (Z), initial mass
function (s) and stellar mass loss efficiency (η).

In this work we consider only old populations, with an
age ranging from 4-8 Gyrs to 15-18 Gyrs depending on the
assumed metallicity; they are dominated by low mass stars
(tipically less than 1.3 M⊚). In order to roughly mimic the
chemical enrichment of the primeval mixture two values for
the helium abundance in mass (Y) have been taken: Y= 0.23
for metal poor models, the value commonly accepted for
galactic globular clusters (Buzzoni et al. 1983), and
Y=0.25 for solar-like and super metal rich populations.

For what concerns the initial mass function (IMF) a
canonical power law N(M) α M^{-s} is assumed with the exponent
's' varying from 1.35 to 3.35 (being the classical
Salpeter's value of 2.35); a lower cutoff at 0.1 M⊚ in the
Main Sequence (MS) has been always adopted in the models.

Also mass loss has been considered via the treatment
given by Reimers (1975) by means of his dimensionless
parameter η , directly related to the mass loss rate. Its
commonly accepted value turning to be around 0.4 (as an
empirical calibration on galactic globular clusters, Iben
and Renzini 1983), a range from 0.3 to 0.5 has been
explored. Table 1 lists the age and composition
combinations explored by the present grid of models.

TABLE 1
Adopted ages for the models at
different chemical composition

(Z , Y)			Age (Gyrs)					
0.0001 0.23			6	8	10	12.5	15	18
0.001 0.23			6	8	10	12.5	15	18
0.01 0.25	4	5	6	8	10	12.5	15	
0.017 0.25	4	5	6	8	10	12.5	15	
0.03 0.25	4	5	6	8	10	12.5	15	

Three different morphologies for HB have been
tentatively introduced. They refer to present day
populations (assumed to be 15 Gyrs old) and resemble tipical
observed distributions for our globular cluster: a red
clump like in 47 Tucanae, a spread distribution like in M3
and a blue HB like in NGC 6752. Evolution from these
different distributions is followed back in time via a
rather complicated procedure, fully described in its
computational details in Buzzoni (1987).

However, as clearly shown in Fig. 1, whatever is the

assumed present day HB morphology, evolution always takes place quickly in "bunching" stellar distribution redward so that only red HBs are present in young populations.

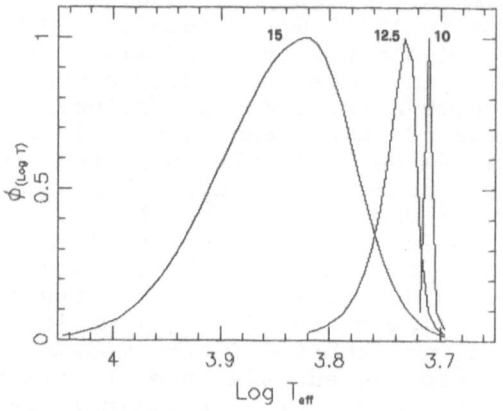

Figure 1. Evolution in HB morphology for models with Z=0.001. Each distribution is normalized to its maximum and labelled with its age in Gyrs.

In constructing isochrones we have connected MS and Subgiant Branch (SGB) from VandenBerg (1983, 1985), VandenBerg and Bell (1985) and VandenBerg and Laskarides (1987) with Red Giant Branch (RGB) from Sweigart and Gross (1978), HB as above stated, AGB from Gingold (1974, 1976), Renzini and Voli (1981) and Iben and Renzini (1983) and Post-AGB according to Paczynski (1970, 1971).

All the "canonical" stellar evolutionary phases have been taken into account in deriving integral SED of populations but no binary stars have been considered in the computations. This could be a limit of the models; however we have to remember in this respect that only "merged" or strongly interactive stars may perturb the convenctional scenario, while the majority of the binary systems can be regarded as normal stars from the photometric point of view. Moreover, one can guess that the "perturbing" effect of such exotic contributors may be negligible, like in the case of M3 (Buonanno et al. 1986, 1987) in which a consistent number of Blue Stragglers (commonly assumed to be [merged] binaries, Nemec and Harris 1987) contribute with only a few percent (even less than 1% in bolometric) to the total light of the cluster.

Isochrones have been all constructed in the theoretical plane LogL-LogT and a subsequent matching with libraries of stellar model atmospheres has been made in order to compute the integral SED of the populations. The two most popular sources of theoretical model atmospheres have been adopted: the wide grid published by Kurucz (1979) and the set of models of Bell and Gustafsson (1980). Due to the lack of models for cold stars (T<4000 K) an extrapolation has been

necessary to extend the grid to the zone of the brightest
AGB stars and to that of red dwarfs. In doing this of
course there has been no attempt to take into account
spectral peculiarities that arise at very cold temperatures.
For instance Carbon stars, possibly present at the tip of
the AGB, always contribute to the integral SED of our models
as normal AGB members.

3. AN OBSERVATIONAL CHECK: THE CASE OF M3

First of all, to fully appreciate the reliability of the
models we have to meet some specific requirements coming
from the observational frame. Essentially four questions
deserve our attention:

- how does stellar populations emit;

- in what a manner light is shared among stars in the
  various evolutionary phases;

- how do stochastic fluctuations in the population affect
  its integral SED;

- what does C-M diagram of a specific stellar population
  look like.

     Study of local stellar clusters like globulars in the
Galaxy and in the Magellanic Clouds already enable us to
closely explore all these questions and perform a confident
comparison between theory and observations. This must be
regarded as a first unavoidable step to better try the
extrapolation to farther objects like possibly high redshift
galaxies.
     Just as an explanatory example we will consider here
the case of the well known globular cluster M3, referring to
the extended work carried out by Buonanno et al. (1986,
1987). Fig. 2 shows their observed C-M diagram of a sample
of 9879 stars, complete in magnitude down to V=21.5.
Starting from these observations a synthetic model of the
cluster has been constructed, assuming the following set of
distinctive parameters: (t, Y, Z, s, η) = (16., 0.23,
0.0004, 0., 1/3).
     In Fig. 3 we have plotted SED as it results from the
observations (Faber 1973, Frogel et al. 1980, De Boer 1985).
As a reference also SEDs obtained by our population model
and by simply summing up contributions from each individual
star in the observed sample displayed in Fig. 2 are shown.
Just a glance to the figure supports the effectiveness of
the suggested computational approach.

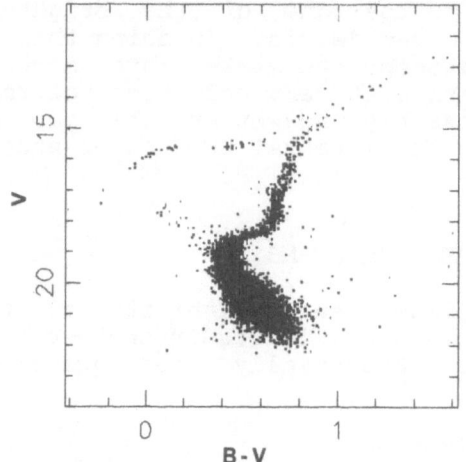

Figure 2. Color-Magnitude diagram of the globular cluster M3 (from Buonanno et al., 1987).

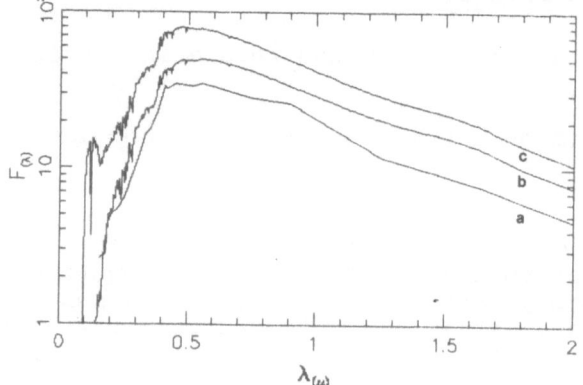

Figure 3. Spectral energy distribution of M3 as it results from three different procedures:
(a) observations
(b) synthetic model
(c) star-by-star synthesis using the sample of Fig.2.

Actually, some discrepancies are evident only in the UV tail of the cumulative SED of Fig. 3 in respect to the theoretical one: this is a typical stochastic effect, which arises in the star-by-star synthesis procedure, due to the discreteness of the observed stellar sample, with only a few hot blue HB stars dominating the far UV emission. Let assume for instance a sample of stars having individual luminosities (at a given wavelength) $l(i)$, $i=1,...N$, being N the total number of stars. The poissonian uncertainty to be related to the luminosity of each single star obviously equals luminosity itself so that summing up all individual contributions we can state the following relation for the relative variance of the total luminosity L_{tot} at that wavelength:

$$\Delta L_{tot} / L_{tot} = (\sum_{i}^{N} l(i)^2)^{1/2} / \sum_{i}^{N} l(i) \qquad (1)$$

As a particular case, if all stars have the same luminosity (i.e. $l(i) = L_{tot}/N$), one has:

$$\Delta L_{tot}/L_{tot} = 1/\sqrt{N} \qquad (2)$$

In a more general way we define

$$\Delta L_{tot}/L_{tot} = (\sum_{i}^{N} l(i)^2)^{1/2} / \sum_{i}^{N} l(i) = 1/\sqrt{N_{eff}} \qquad (3)$$

where N_{eff} can be regarded as the "effective" number of stars emitting at a given wavelength. By definition N_{eff} can never exceed N.

Once the function $N_{eff}(\lambda)$ is known, statistical uncertainties in the SED of real stellar samples can be easly derived from eq. (3). This function for our model of M3 is given in Fig. 4: from that figure one estimates an intrinsic variance of the sampled light in the UV flux around 1500 Å of the order of 0.2 magnitudes. By the same way we derive an intrinsic variance of $\pm$ 0.03 and $\pm$ 0.02, respectively, for the B-V and U-B of the whole cluster.

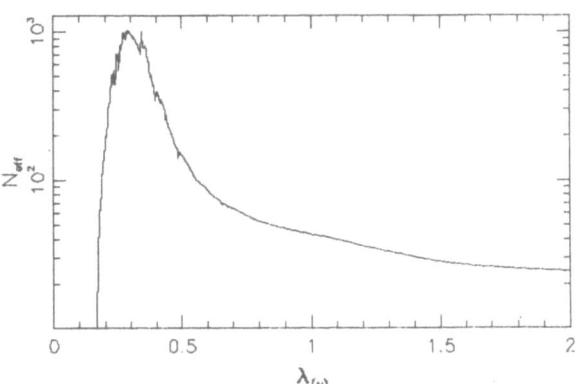

Figure 4. The function $N_{eff}(\lambda)$ for the synthetic model of M3. Values are scaled for the total luminosity of the sample of Fig.2 (i.e. $3*10^4$ L⊙).

4. GALAXY SYNTHESIS

An important information to be retained in computing integral SED concerns the relative partition of the contributions from stars in the different evolutionary phases. The better knowledge of this relation becomes of paramount importance for instance to consistently link observed SEDs of galaxies with theoretical predictions about their stellar composition.

The main problem which we have to deal with in this regard is the discrimination of the effects due to the age from those due to the chemical composition. As a glance to Fig. 5 and 6 confirms, MS stars are always expected to be the main contributors to the emitted light of galaxies around the 4000 Å break. Then the direct connection with the contribution of the stars around the Turnoff point enable us to derive information about age of such systems. Unfortunately this simple idea seems to crash into at least two serious difficulties.

The first one is related to the well known dependence of the Turnoff temperature on metallicity of the stellar population since colors of high-metallicity isochrones resemble those of older, low-metallicity ones.

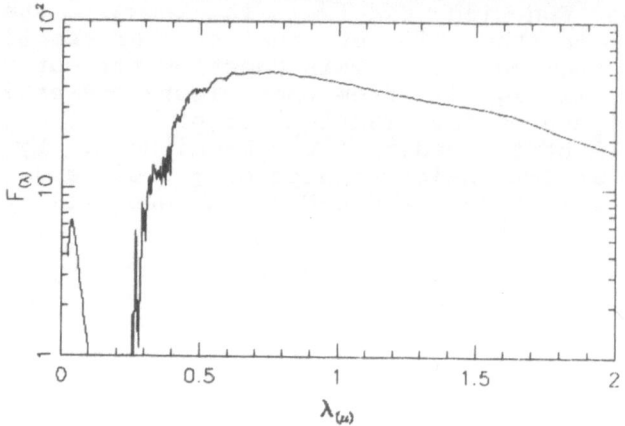

Figure 5. Computed spectral energy distribution for a synthetic model with (t, Z, Y, s, η) = (15, 0.017, 0.25, 2.35, 0.5) and a red HB. It closely resembles normal elliptical galaxies.

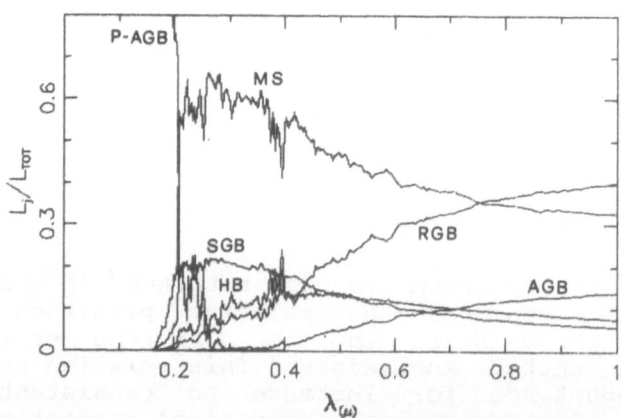

Figure 6. Partition of the contributions from the different stellar evolutionary phases to the total light of the model of Fig. 5.

In principle, the difficulty could be overcome once spectral features or convenient narrow-band photometric indices are calibrated versus [Fe/H].

A relevant contribution in this sense has come from the works of Spinrad and Taylor (1969, 1971), Faber (1973), Faber et al. (1977), Burstein (1979), who have been able to rank a variety of objects (single stars, globular clusters, galaxies) versus their metallicity. However untill a reliable quantitative calibration for some primary indices, like for instance the Mg index at 5200 Å (Mould 1978), will not be effective there is no hope to firmly solve the problem.

Metallicity might indeed complicate the game in an even more perverse way. Just looking at Fig. 6 one notes that the contribution of HB to the spectral break is not negligible: even in the assumed case of a red clump morphology it turns to be of the order of 15%. Now, metallicity might influence the HB morphology also via the mass loss mechanism. Actually there is no special reason to expect high metallicity populations in the galaxies to have red HBs: if mass loss efficiency increases with metallicity we might guess blue distributions to be possible. It is clear that if we miss this contribution in the synthetic models we need some bluer younger isochrones to fit observed SEDs.

Beside a full investigation of the present-day galaxies, one of the goals of the EPS approach is the study of the evolution of galaxies back in time. A variety of cosmological problems are involved in approaching this topic and an important theoretical effort is now required to support the more and more refined observational tests.

Fig. 7 shows the expected apparent color evolution in B-V and V-K of a model resembling an elliptical galaxy, up to redshift z=1.0.

One sees that no important changes seem to occur in the observed V-K color; this does not mean of course that no changes took place in the emitted flux but only that fluxes changed at the same rate. Actually if we look for instance at the intrinsic V flux of the same population, as shown in Fig. 8, we clearly appreciate an important increase of the brightness back in time.

Once more, from Fig. 7, we remark that only colors involving the break of the SED at short wavelengths are effective to possibly date galaxies while (infra-)red magnitudes better work as distance tracers in attempting to mark cosmological space-time dimension.

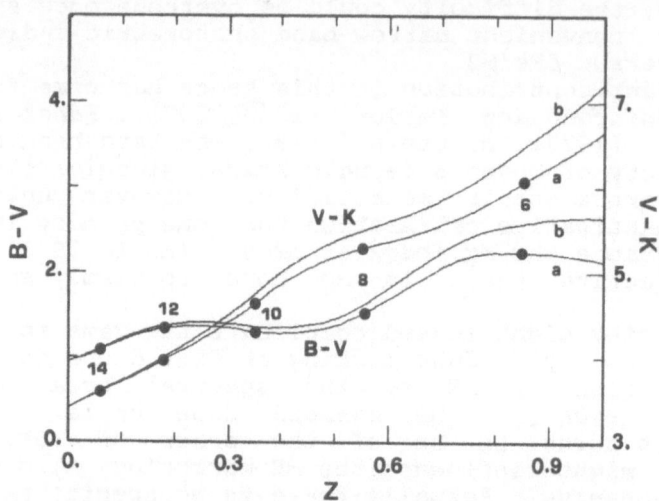

Figure 7. Expected apparent B-V and V-K colors vs. redshift
(z) for the synthetic model of Fig. 5. Curves (a) take into
account evolution of the stellar population while curves (b)
simply represent the present-day model passively shifted at
different z. Labels on curves (a) refer to the age of the
model: an (Ho, qo) = (50., 0.) cosmology is assumed.

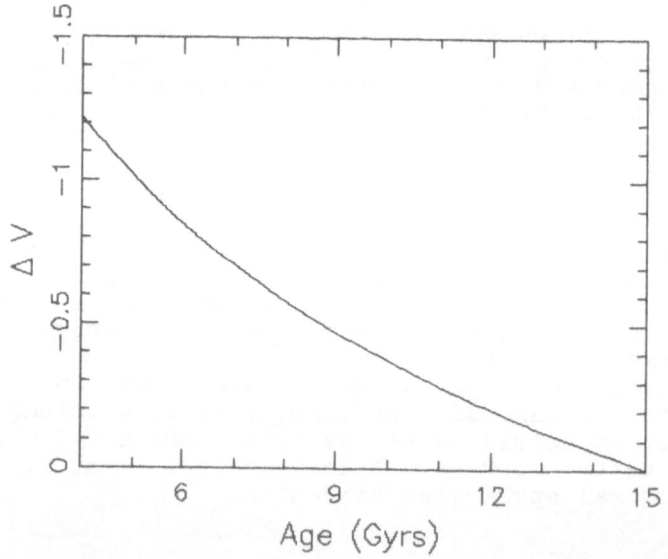

Figure 8. Evolution of the intrinsic V magnitude vs. age
for the model of Fig. 5.

REFERENCES

Bell, R.A., Gustafsson, B. 1978, Astron. Astrophys. Suppl., **34**, 229.

Bruzual, G. 1983, Ap.J., **273**, 105.

Burstein, D. 1979, Ap.J. **232**, 74.

Buonanno, R., Buzzoni, A., Corsi, C.E., Fusi Pecci, F., Sandage, A.R. 1986, Mem. Soc. Astron. It., **56**, 391.

Buonanno, R., Buzzoni, A., Corsi, C.E., Fusi Pecci, F., Sandage, A.R. 1987, in preparation.

Buzzoni, A. 1987, in preparation.

Buzzoni, A., Fusi Pecci, F., Buonanno, R., Corsi, C.E. 1983, Astron. Astrophys., **128**, 94.

Ciardullo, R., Demarque, P. 1977, Trans. Astron. Obs. Yale Univ., **35**.

De Boer, K.S. 1985, Astron. Astrophys., **142**, 321.

Faber, S.M. 1973, Ap.J., **179**, 731.

Faber, S.M., Burstein, D., Dressler, A. 1977, A.J. **82**, 941.

Frogel, J.A., Persson, S.E., Aaronson, M., Matthews, K. 1978, Ap.J., **220**, 75.

Gingold, R.A. 1974, Ap.J., **193**, 177.

Gingold, R.A. 1976, Ap.J., **204**, 116.

Kurucz, R. 1979, Ap.J. Suppl., **40**, 1.

Iben, I.Jr., Renzini, A. 1983, Ann. Rev. Astron. Astrophys., **21**, 271.

Mould, J. 1978, Ap.J., **220**, 434.

Nemec, J.M., Harris, H.C. 1987, Ap.J., **316**, 172.

Paczynski, B. 1970, Acta Astron., **20**, 47.

Paczynski, B. 1971, Acta Astron., **21**, 417.

Reimers, D. 1975, Mem. Soc. R. Sci. Liege, 6-th Ser. **8**, 369.

Renzini, A. 1981, Ann. Phys. Fr., **6**, 87.

Renzini, A., Buzzoni, A. 1983, Mem. Soc. Astron. It., **54**, 739.

Renzini, A., Buzzoni, A. 1986, in Spectral Evolution of Galaxies, ed. C.Chiosi, A.Renzini, Dordrecht: Reidel, p.195.

Renzini, A., Voli, M. 1981, Astron. Astrophys., **94**, 175.

Schönberner, D. 1979, Astron. Astrophys., **79**, 108.

Schönberner, D. 1981, Astron. Astrophys., **103**, 119.

Seidel, E., Demarque, P., Weinberg, D. 1987, Ap.J. Suppl., **63**, 917.

Spinrad, H., Taylor, B.J. 1969, Ap.J., **157**, 1279.

Spinrad, H., Taylor, B.J. 1969, Ap.J. Suppl., **22**, 445.

Sweigart, A.V. 1987, Ap.J. Suppl., **65**, 95.

Sweigart, A.V., Gross, P.G. 1976, Ap.J. Suppl., **32**, 367.

Sweigart, A.V., Gross, P.G. 1978, Ap.J. Suppl., **36**, 405.

Tinsley, B.M., Gunn, J.E. 1976, Ap.J., **203**, 52.

VandenBerg, D.A. 1983, Ap.J. Suppl., **51**, 29.

VandenBerg, D.A. 1985, Ap.J. Suppl., **58**, 711.

VandenBerg, D.A., Bell, R.A. 1985, Ap.J. Suppl., **58**, 561.

THE INTEGRATED SPECTRA OF GLOBULAR CLUSTERS

James A. Rose
Department of Physics and Astronomy
University of North Carolina
Phillips Hall 039A
Chapel Hill, NC 27514

ABSTRACT. The problems in interpreting the integrated spectra of globular clusters are briefly summarized.

Since a great deal of attention is devoted in this workshop to the interpretation of integrated spectra of galaxies, it is worthwhile to consider what may be learned from studying the integrated spectra of globular clusters. It is now generally acknowledged that the study of the integrated spectra of galaxies is an exceedingly complicated task, given that even elliptical galaxies undoubtedly contain a range in stellar abundances and perhaps a substantial range in stellar ages as well. In contrast, it is often pointed out that globular clusters are simple stellar systems, and hence provide excellent test cases for the interpretation o fmore complex stellar systems in integrated light. The belief in globular clusters as simple systems comes from color-magnitude diagrams and spectroscopy of individual giants in the outer regions of clusters as well as broad-band integrated light measurements obtained in the central regions. Those studies all tend to indicate that, with the problems of horizontal branch morphology and of CNO abundance anomalies aside, globular clusters are well-behaved systems whose range of properties are fully explained by the differences in heavy element abundance between clusters. On the other hand, a close examination of the integrated spectra of both Galactic and M31 globular clusters reveals them to be much more complex than anticipated, and it is the purpose here to briefly review the difficulties encountered so far.

The earliest indication that globular cluster integrated spectra cannot be simply described as a single parameter family differing only in heavy element abundance is the seminal work of van den Bergh (1969). In the latter study he pointed out systematic differences at a given integrated spectral type between M31 globular clusters and Galactic globulars. He also mentioned that for the most strong-lined Galactic globulars, there appeared to be a significant spread in Balmer line strengths. This early indication of the differences between M31 and Galactic globular cluster spectra was expanded on in several subsequent studies (Spinrad and Schweizer 1972, Rabin 1980, O'Connell

R. G. Kron and A. Renzini (eds.), Towards Understanding Galaxies at Large Redshift, 73–75.
© 1988 by Kluwer Academic Publishers.

1983), culminating in the fundamental study by Burstein et al. (1984). The latter authors found for metal-rich clusters that at a given value of their Mg$_2$ index, both Hβ and CN strengths are stronger in M31 globulars than for Galactic globulars; the situation is less clear in the case of metal-poor clusters. The Burstein et al. (1984) study seriously brings into question whether globular clusters in all galaxies are indeed similar in origin to those in our Galaxy (Burstein 1987).

Recently, Rose and Tripicco (1986) found that the integrated spectra of the metal-rich Galactic globular clusters do not form a simple one-parameter family. Specifically, at a given value of their Hδ/Fe I index (roughly indicating spectral type) they find a large range in Sr II/Fe I and 3888/3859 indices. The Sr II/Fe I index has been shown to discriminate surface gravity, independent of [Fe/H], for normal composition stars. A straightforward interpretation of the spread in cluster Sr II/Fe I indices at a given Hδ/Fe I is that the luminosity-weighted mean surface gravity (i.e., dwarf-to-giant ratio) varies from one cluster to another. Alternatively, a non-standard chemical composition (i.e., compared with the solar neighborhood) perhaps could explain the observed spread. Rose and Tripicco rule out the possibility that the spread could be caused by differences in the hot star content between clusters (on the basis of the observed Ca II H + Hε/Ca II K indices). The spread in the 3888/3859 cluster indices indicates a corresponding dispersion in the mean CNλ3883 band strengths.

In his Ph.D. dissertation, Tripicco (1987) has studied a number of metal-rich clusters in M31 using similar methods as in Rose and Tripicco (1986). He finds that the M31 clusters are at least as extreme in Sr II/Fe I and (in agreement with Burstein et al. 1984) more extreme in CN indices than the Galactic globular clusters that are themselves anomalous in those indices. Moreover, on the basis of the Ca II H + Hε/Ca II K indices, he finds that, in general, the hot star content of M31 globulars is much too small to account for the large Hβ indices observed by Burstein et al. (1984) and others.

To summarize, the integrated spectra of metal-rich M31 and Galactic globular clusters show a spread in behavior that is not understood. Whether there are fundamental differences between the stellar populations (i.e., distribution in the HR diagram) in clusters or whether non-standard chemical compositions can explain the observations is not clear at present. It is hoped that a comprehensive study of the cluster 47 Tuc, which exhibits both the Sr II/Fe I and CN anomalies in its central region, will lead to a better understanding of the integrated spectra of metal-rich globulars.

Finally, Rose, Stetson, and Tripicco (1987) have studied 12 metal-poor Galactic globulars in integrated light. On the basis of fitting Hδ profiles with Kurucz (1979) model profiles and of measurements of the gravity-sensitive Balmer discontinuity, they infer an unusual stellar content for the cores of M30 and NGC4147, as originally proposed by Zinn and West (1984). In particular, it appears that while the total number of evolved stars (relative to main sequence stars) in the latter two clusters is roughly normal, the ratio of horizontal branch stars to giant branch stars is anomalously high.

To conclude, it appears that the integerated spectra of globular clusters are not suitable building blocks for understanding integrated light observations of ostensibly more complicated systems, such as early-type galaxies. On the other hand, the globular clusters spectra have proven sufficiently challenging to interpret, that, once they are indeed understood, we should be able to approach elliptical galaxies in integrated light with a greater degree of confidence.

REFERENCES

Burstein, D. (1987). In Nearly Normal Galaxies, edited by S. M. Faber, Springer-Verlag), p. 47.
Burstein, D., Faber, S. M., Gaskell, C. M., and Krumm, N. (1984). Astrophys. J. 287, 586.
Kurucz, R. L. (1979). Astrophys. J. 40, 1.
O'Connell, R. W. (1983). In Highlights of Astronomy, 6, 147.
Rabin, D. M. (1981). Ph.D. Thesis, California Institute of Technology.
Rose, J. A., and Tripicco, M. J. (1986). Astron. J. 92, 610.
Rose, J. A., Stetson, P. B., and Tripicco, M. J. Astron J. (in press).
Spinrad, H., and Schweizer, F. (1972). Astrophys. J. 173, 619.
Tripicco, M. J. (1987). (In preparation).
van den Bergh, S. (1969). Astrophys. J. Suppl. 19, 145.

A NEW APPROACH FOR STELLAR POPULATION SYNTHESIS IN GALAXY NUCLEI : A LIBRARY
OF STAR CLUSTERS

E. Bica, D. Alloin
Observatoire de Paris
92195 Meudon Principal Cedex
France

ABSTRACT. We develop a new approach for population synthesis in galaxy
nuclei, which makes use exclusively of a library of integrated spectra of star
clusters. This method has given interesting results for nearby galaxy nuclei
in terms of dating successive stellar generations and detecting bursts of
star formation, as well as of determining the chemical enrichment. The
prospective application of this method to interpret large redshift galaxy
spectra is briefly discussed.

1. INTRODUCTION

So far population synthesis methods in galaxies have used libraries of
stellar spectra and in some cases libraries containing a mixture of stars and
globular clusters (Tinsley, 1980 ; Pagel and Edmunds, 1981 ; Pickles, 1985
and references therein). We have undertaken a different approach using
exclusively a library of integrated spectra of star clusters spanning wide
ranges in age and metallicity. A stellar library depends on T, g, Z whereas for
star clusters it is only a two parameter analysis : age and metallicity (Bica
and Alloin, 1986a ; hereafter we adopt the following notation for this series
of papers : BA 86a). Using a stellar library involves additional degrees of
freedom which are the slope and the lower and upper mass limits of the Initial
Mass Function (IMF). Using a library of star clusters avoids any assumption to
be made about the IMF which is implicit. The present method is an
observational approach, complementary to theoretical ones. We have observed
63 star clusters and 164 galaxy nuclei at the ESO 1.52m telescope using an IDS
detector in the visible range (BA 86a ; BA 87a). We have used the ESO 2.2m
telescope and a CCD detector for the near—infrared range, observing a
subsample of the previous set of objects which however still spans the same
variety of star cluster and galaxy types (BA 87b). In the latter paper we have
also developed a particularly powerful method for eliminating interference
fringes. A spectral resolution of ~ 12 Å was chosen, matching the stellar
velocity dispersion in galaxy nuclei. We are presently extending our method
to the ultraviolet range using the IUE data bank and new observations in the
near—ultraviolet with a coated CCD detector.

R. G. Kron and A. Renzini (eds.), Towards Understanding Galaxies at Large Redshift, 77–84.
© 1988 by Kluwer Academic Publishers.

Figure 1. Globular cluster spectra, a metallicity sequence ($[Z/Z_{\odot}]$ is given on the right). The flux F_{λ} is normalized to $F(5870\ \overset{\circ}{A}) = 10$.

Figure 2. Open cluster spectra, an age sequence (the age is given on the right). Same unit as in Figure 1.

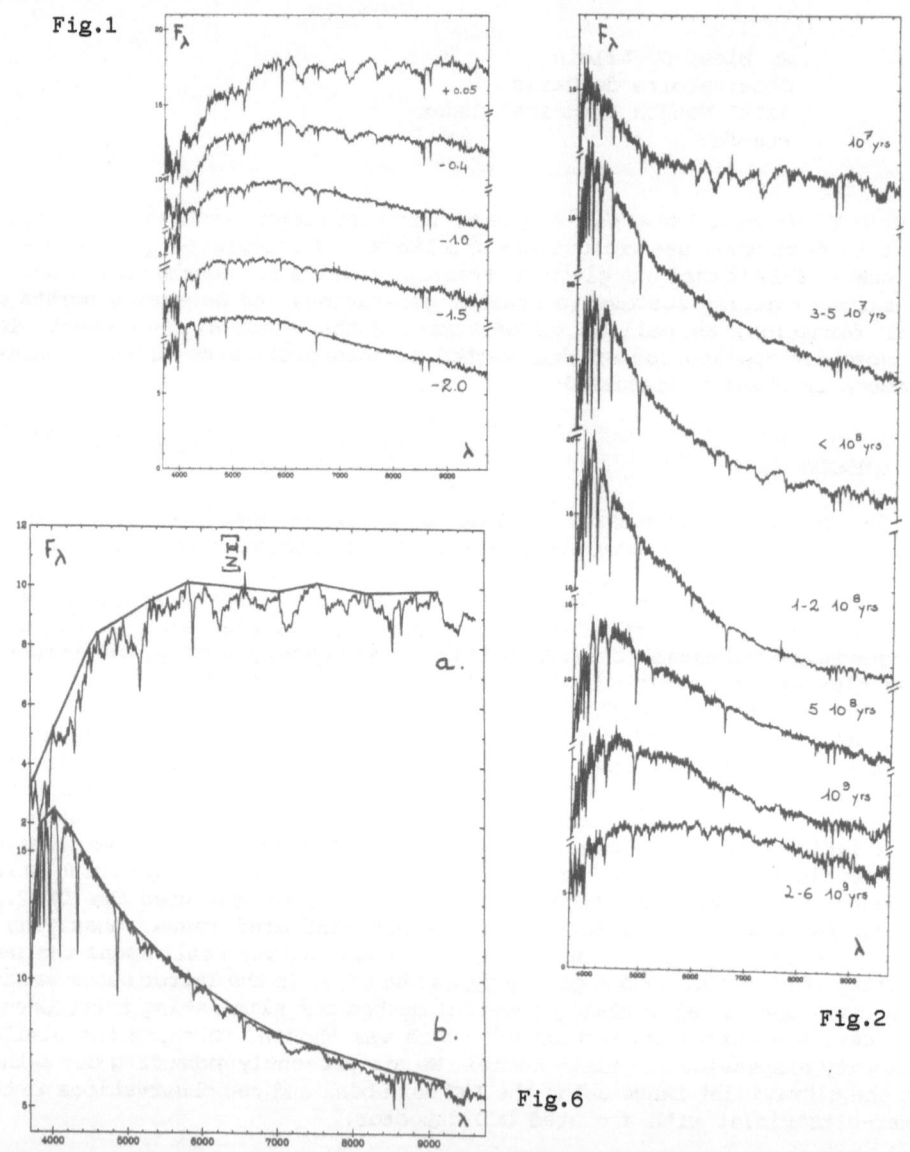

Figure 6. a. and b. Continuum tracing for a typical red and blue population respectively.

2. THE STAR CLUSTER SAMPLE

The cluster sample consists of Galactic globular and open clusters and LMC and SMC clusters. Thus we span ages from less than 10^7 to more than 10^{10} yr, metallicities $- 2 \leqslant [Z/Z_\odot] \leqslant + 0.1$(BA 86a). Galactic globular clusters form metallicity groups for which we show the visible and near-infrared spectral properties in Figure 1. The group with $[Z/Z_\odot]$ slightly above solar consists of the very strong-lined Globular Clusters (GC) NGC 6440, 6528 and 6553. It has been fundamental in the present synthesis method, because these clusters exhibit metallic features comparable to those observed in giant galaxy spectra. Globular clusters usually called metal rich like NGC 104 (47 Tuc) would clearly not be suitable because they hardly present an overlap of spectral properties with respect to the massive galaxies ; 47 Tuc is one of the weakest-lined spectra in the GC group centered at $[Z/Z_\odot] = -0.4$. Intermediate and young age groups are shown in Figure 2. The spectra are from LMC and Galactic disc clusters (BA 86a ; BA 87b). SMC clusters have been excluded from this figure because of their low metallicity. An interesting result is that some blue clusters are dominated by red stars in the near-infrared range : NGC 2004, at 10^7 yr, contains red supergiants and NGC 1866 at $t \simeq 10^8$ yr contains possibly AGB stars of spectral type M. Further studies of these red phases in young clusters will be very important for dating recent bursts of star formation in galaxies (BA 87b). Figure 2 also shows how Balmer lines and continuum distribution evolve with age.

Because these star clusters had previously known age, metallicity and reddening it has been possible to study the equivalent widths (W) and the intrinsic continuum distribution directly as a function of age and metallicity (BA 86a ; BA 87b). The main results are as follows : (i) the W of metallic features in the blue like CaIIK break up into different age groups in a plot against $[Z/Z_\odot]$ (Figure 3a). This is caused by the presence of luminous hot stars in young clusters which dilute metallic features in the blue. On the contrary the age groups tend to merge for metallic features in the near infrared like CaII 8542 Å (Figure 3b). Thus metallicity is the dominant parameter towards long wavelengths whereas, in addition to metallicity, there is a strong age dependence towards short wavelengths. Absorption Balmer lines are maximum at around 5×10^8 yr, when A stars dominate the integrated spectrum, while the continuum slope increases steadily towards younger ages (BA 86a). From this analysis it has been possible to interpolate a grid of star cluster spectral properties at suitable steps in age and metallicity (BA 86b ; BA 87b). As well, it has been possible to extrapolate the spectral properties to Super Metal Rich (SMR) star clusters at $[Z/Z_\odot] = +0.6$. We emphasize that this extrapolation is straightforward due to the inclusion in our sample of very strong-lined globular clusters. We also show in Figure 3a histograms for the galaxy values. In fact only very strong-lined galaxies require the grid extrapolation to $[Z/Z_\odot] = 0.6$ in the synthesis computations (Bica, 1987).

3. THE GALAXIES

The sample of galaxy nuclei spans morphological types from E to Sc and luminosities $-23 \leqslant M_B \leqslant -16$ (BA 87a ; BA 87b). It is not complete in a

Figure 3. a. W(CaIIK, 3933 Å) in A°, versus metallicity [Z/Z☉]. On the right we show an histogram of the W values observed for spiral galaxies (hatched) and for E + S0 galaxies. b. Same diagram for W(CaII 8542 Å).

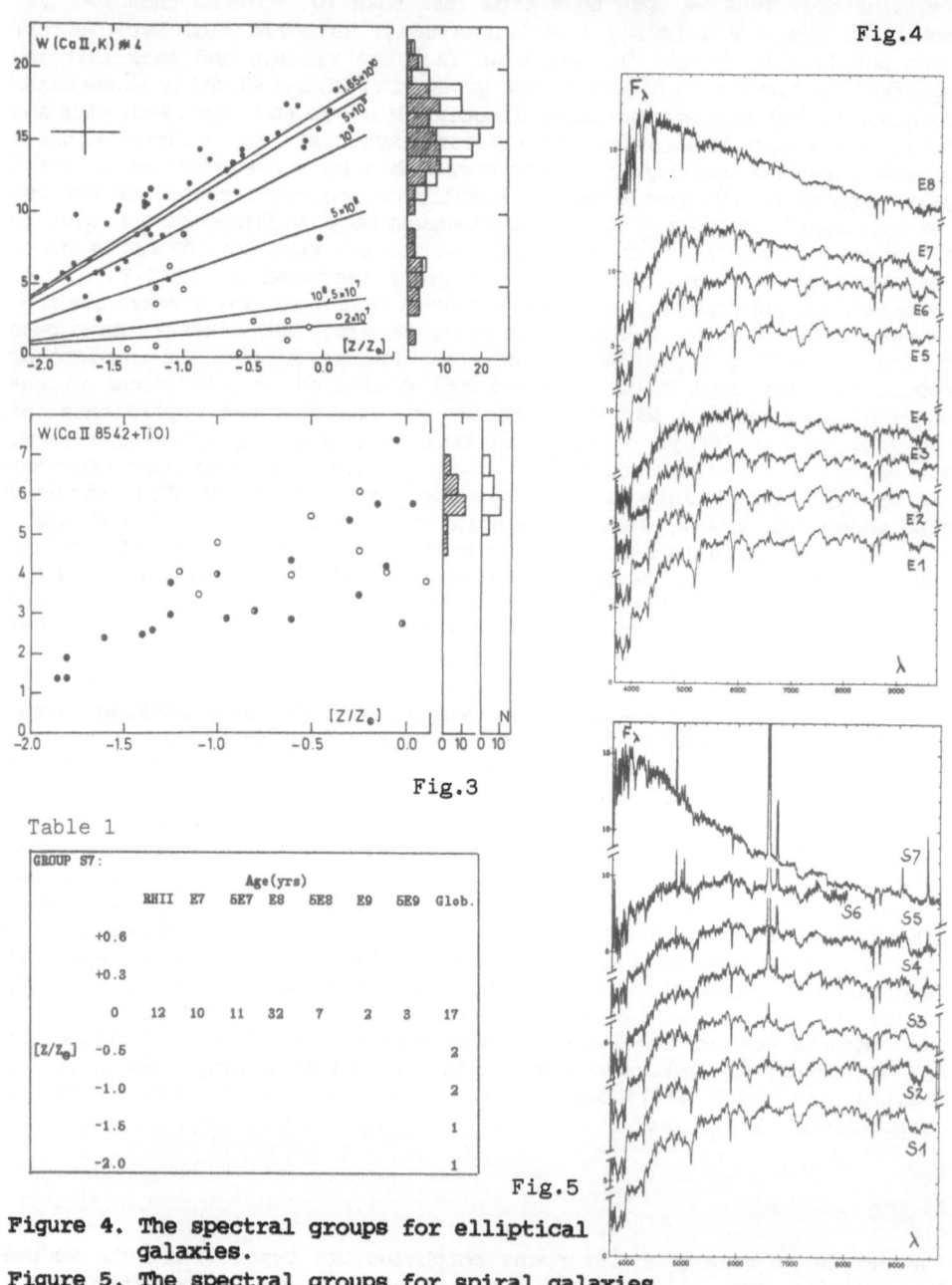

Fig.3

Fig.4

Table 1

Fig.5

Figure 4. The spectral groups for elliptical galaxies.

Figure 5. The spectral groups for spiral galaxies.

statistical sense but all types of normal nuclei occuring in nearby galaxies are well represented. In view of the population synthesis we have grouped the nuclei sharing the same spectral properties within error bars (Bica, 1987). Nine groups have been derived for early type galaxies (Figure 4). The red groups E1 to E4 follow the normal relationship between metal indicator and total magnitude (BA 87c). The very strong—lined E1 nuclei occur in giant E and S0 galaxies while the weak—lined E4 group occurs for $M_B \simeq -18$. The relatively less numerous bluer groups E5 to E8 correspond mostly to age effects with respect to groups E1 through E4. Evidence has been found that group E7, the spectrum of which shows a bump in the visible range and enhanced Balmer lines, has an important intermediate age component (BA 87a ; Bica, 1987). Group E8, which is made of the particular case NGC 5102, has a burst of star formation at $t \simeq 4 \times 10^8$ yr (BA 87a ; Bica, 1987). Group E9 is spectroscopically very similar to group E2, but occurs in less luminous galaxies, most of which show structural evidences of tidal stripping (BA 87c and references therein).

Spiral nuclei must be corrected for the reddening arising from inclination effects. Let us recall first how all sources of reddening are considered in the present synthesis method. For star clusters we have simply adopted the foreground reddening from the litterature (BA 86a). For galaxies, the foreground reddening is from a cosec law with reddening free polar caps (BA 87a). No correction for internal reddening was applied to the star clusters. This is an important guide—line of the method : reddening is important only for young clusters still associated with star forming regions. Therefore it must be kept in the input data because, whenever star forming regions are present in galaxy nuclei, the associated dust will certainly be localized in these regions, not necessarily affecting the older surrounding populations. Thus the reddening which is derived by the comparison of the computed synthetic spectra and the observed galaxy spectra should correspond to global diffuse dust occuring in the bulges and central regions and should not be direcly associated to star forming regions. In fact the synthesis computations indicate that this diffuse reddening is small $E(B-V) \simeq 0.04$ for all morphological types (Bica, 1987). Prior to the synthesis, the reddening due to inclination in spirals was corrected with the method developed in BA 87a. Our sample contains a large number of face—on and inclined spirals. Thus for each inclined galaxy, we search for a template with the same stellar population content among the nearby face—on galaxies. This is achieved when W of metallic features are the same all over the galaxy spectrum, based on the analysis of the star cluster spectra (Section 2). For some nearby edge—on spirals, the reddening due to inclination can be quite large ($E(B-V) \simeq 0.40-0.50$). Then, we also detect a strong NaID line excess (BA 86c). This is a well known interstellar line and thus we conclude that the excess arises from gas in the inclined discs. Obviously the NaID line is not suitable for population synthesis because of its interstellar contribution, which may be stronger than the stellar one in particular cases like edge—on spirals.

We obtained 7 groups of spiral spectra (Figure 5). Basically group S1 corresponds to the K type and group S7 to the A type in the early blue-violet spectral classification. The red groups S1 to S4 are similar to the E, S0 ones. The spectral properties of these groups are related to the bulge luminosity, but notice that group S4 with the weakest—lines is still strong-lined, very different from metal poor globular clusters (BA 87a ; Bica, 1987).

4. THE SYNTHESIS ALGORITHM

Input lines for the synthesis are the strongest metallic features which have a well-modeled behaviour as a function of age and metallicity in the star cluster analysis (BA 86a,b ; 87b). These are CaIIK 3933 Å, CN 4200 Å, CH 4300 Å (G-band) ; MgI + MgH 5175 Å and CaII 8542-8662 Å. Balmer lines are used only when they are emission-free. It is not necessary to test the whole plane age versus metallicity simultaneously. A chemical evolution argument, telling us that the nuclear volume is small enough for the gas to be chemically homogeneous at any time suggests that the solution must describe a continuous path in the plane age versus metallicity. Another astrophysical constraint is that the metal content cannot decrease towards younger ages. So we test all combinations of N grid points (star clusters) at a small step along a given path (chemical evolution). In a first phase, we consider N = 8 with a 10% step, and the algorithm selects all solutions which reproduce the galaxy equivalent widths within error bars. In a second phase we use more star clusters and a 5% step to explore the space of combinations around the solution well derived in the first phase. Typically we test 5 different paths. Previous population synthesis methods have in general used minimization procedures which looked for an optimal solution and it has always been very difficult, if not impossible at all, to estimate the uniqueness of the solution, because of the many parameters involved. The present algorithm allows to obtain a quantitative estimate of the uniqueness of the solution as the ratio of the number of solutions to the number of tested combinations. It turns out to be 1 part among 10 000 possibilities.

An example of the synthesis computations is given in Table 1 for the spectral group S7, which presents a burst of star formation at $t \simeq 10^8$ yr extending up to the present day. The solution is expressed in terms of the percentages of flux contribution at 5870 Å. Thus the method allows to date the successive stellar generations, to identify and to date eventual bursts of star formation, as well as to determine the chemical enrichment. Thus it is more than a simple population synthesis, providing for the first time a direct estimate of the chemical evolution in these nuclei. The spiral group S7 contains an important contribution of star forming regions which has been isolated as well. Detailed synthesis for our 15 types of galaxy nuclear spectra are provided in Bica, 1987. In the latter paper we have also created empirically, the spectrum of an SMR globular cluster at $[Z/Z_\odot]$ = +0.6, using the population synthesis results and a spectral decomposition of strong-lined galaxies. It would be very interesting to compare this empirical spectrum with synthetic spectra, when available.

A preliminary analysis from the ultraviolet to the near infrared is provided in BA 87d. We study the nature of the ultraviolet turnup in giant elliptical galaxies assuming that it arises either from a recent star formation event or from a metal poor population. We show that a real star cluster of age 10^7 yr explains the ultraviolet turnup in giant ellipticals quite well, without affecting significantly the visible and near infrared ranges, whereas a UV-bright metal poor globular cluster cannot account for it. This analysis also demonstrates the importance of astrophysical constraints provided by the frequence of occurence of particular types of stars with respect to the associated populations. Blue HB stars in past stellar synthesis have usually been considered as a free stellar group and consequently quite arbitrary contributions were allowed. Their contribution is in fact tied to metal poor giants and only integrated spectra of star clusters can indicate Nature's right proportions. Other types of UV-bright stars are also possible, like post-AGB stars in the planetary nebular phase (Bertola, 1986). The frequence of occurence of planetary nebula in star clusters of the Galaxy and Magellanic Clouds, as well as in the solar neighbourhood, Galactic and M31 bulges may provide contraints on this possibility. We also emphasize that our sample star clusters also show the relative importance of certain types of red stars which are difficult to include in other synthesis methods, like red supergiants and AGB stars (Section 2).

In view of future comparisons and use of the present synthesis method, we recall in Figure 6 high S/N spectra of typical blue and red populations and the criteria for continuum tracings. The method consists of measuring equivalent widths with fixed rest-frame window limits, which are listed in BA 86a or BA 87b respectively for the visible and near infrared ranges. These correspond to atomic and molecular lines from various absorbers. Next we shall provide measurements in other systems (e.g. Burstein et al., 1984).

5. SUMMARY

We have collected a library of star cluster and nearby galaxy nuclear spectra in the visible and near infrared ranges (BA 86a, 87a, 87b). We have derived a grid of star cluster equivalent widths and continuum distribution as a function of age and metallicity (BA 86b, 87b). The NaID doublet was studied in star clusters and galaxies and we have concluded that its use is not suitable for population synthesis, owing to an important interstellar contamination in many cases (BA 86c). The metallicity versus luminosity relationship in E and SO has been discussed in BA 87c. A population synthesis method using the library of star clusters was applied to nuclei of E to Sc galaxies of various luminosities (Bica, 1987). We emphasize that the computations are based on a complete grid of spectral properties as a function of age and metallicity, but the visualisation of the computations is also provided in terms of galaxy spectra decompositions *using the nearest star clusters available*. The method allows a direct estimate of the chemical evolution in the galaxy nuclei. In a preliminary study using the ultraviolet together with the visible and near infrared ranges, we have studied the nature of the ultraviolet turnup in giant ellipticals (BA 87d).

A detailed extension of the present method to the near ultraviolet is in progress. This will be important for more precise studies of the hot component contributions. It will also be useful to implement the grid of cluster properties with more observations, particularly of young and intermediate age clusters in M31, M33, LMC, SMC and the Galactic disc. The detailed analysis of star clusters in terms of synthetic spectra is also fundamental. Synthesis methods based on stellar libraries and the present method are complementary. Both techniques should be further developed.

6. PROSPECTIVE APPLICATIONS TO INTERMEDIATE AND LARGE REDSHIFT GALAXIES

The present method, which has already given interesting results for nearby galaxy nuclei, will certainly be an important tool in the study of distant will be included in the slit for intermediate redshift galaxies : E and SO galaxies will be directly comparable to global galaxy chemical evolution models. For spirals, it will be interesting to explore the effect of the disc contribution. We emphasize that in the latter case the synthesis solution should not necessarily describe a path in the age versus [$Z/Z_\odot$] diagram. At these intermediate redshifts it will also be possible to compare directly the integrated spectra of mergers and star burst galaxies to those of integrated spirals.

Another problem which intermediate redshift galaxies allow to study is the effect of disc inclination on the *total* integrated spectrum of spiral galaxies. Our previous analysis of nearby spirals has shown that both an enhancement of the reddening and of the interstellar NaID contribution in the *nuclear* spiral spectra arise from this effect.

Some modelisation can be performed with the present library of star clusters and nearby nuclei : (i) following chemical evolution models the effect of the extra nuclear metal poor components can be simulated using real E nuclei and metal poor globular clusters, (ii) young and intermediate age bursts of star formation can be empirically added, with different relative intensities, to the old underlying population, (iii) the star cluster spectra can be likewise combined according to different scenarios for the star formation rate evolution in galaxies, including stochastic burst events, and the results can be compared to large redshift galaxies. Using this method, metallicity effects can be easily taken into account. Combinations of real star cluster and galaxy spectra may provide more insight into the nature of the E + A galaxies (Dressler, 1986) ; such analysis will be presented in a forthcoming paper.

ACKNOWLEDGMENT : We thank Dr. D. Burstein for interesting discussions. E.B. thanks the Brazilian Institution CNPq for a fellowship.

REFERENCES

Bertola, F., 1986, 'Spectral Evolution of Galaxies', Eds. C. Chiosi and A. Renzini, *Reidel*, p.363.
Bica, E., Alloin, D., 1986a, *Astron. Astrophys.* **162**, 21.
Bica, E., Alloin, D., 1986b, *Astron. Astrophys. Suppl. Ser* **66**, 171.
Bica, E., Alloin, D., 1986c, *Astron. Astrophys.* **166**, 83.
Bica, E., Alloin, D., 1987a, *Astron. Astrophys. Suppl. Ser.*, in press.
Bica, E., Alloin, D., 1987b, *Astron. Astrophys.*, in press.
Bica, E., Alloin, D., 1987c, *Astron. Astrophys.*, in press.
Bica, E., Alloin, D., 1987d, *Astron. Astrophys.*, submitted.
Bica, E., 1987, *Astron. Astrophys.*, submitted.
Burstein, D., Faber, S., Gaskell, C., Krumm, N., 1984, *Astrophys. J.* **287**, 586.
Dressler, A., 1986, 'Spectral Evolution of Galaxies', Eds. C. Chiosi and A. Renzini, *Reidel* p.375
Pagel, B., Edmunds, M., 1981, *Ann. Rev. Astron. Astrophys.* **19**, 77.
Pickles, A., 1985, *Astrophys. J.* **296**, 340.
Tinsley, B., 1980, *Fund. of Cosmic Physics* **5**, 287.

GALACTIC WINDS IN ELLIPTICALS: CONSEQUENCES FOR GALACTIC AND
INTERGALACTIC ENRICHMENT

F. Matteucci[1], A. Tornambè[2] and P. Vettolani[3]

[1] European Southern Observatory, Karl-Schwarzschild-Str. 2,
   D-8046 Garching bei München, Federal Republic of Germany
[2] Istituto di Astrofisica Spaziale, Frascati, Italy
[3] Istituto di Radioastronomia, Bologna, Italy

ABSTRACT. Models of chemical evolution of elliptical galaxies, where
galactic winds powered by SN explosions (of type I and II), as well as
detailed nucleosynthesis prescriptions for single elements, are taken
into account, predict the following results:
 i) There are different mass-metallicity relations for different
    heavy elements (O, Mg, Si and Fe), the one for iron being the
    steepest one.
 ii) In spite of the fact that star formation stops at early epochs
    (< 1 Gyr), explosions of SNe of type I at the present time are
    predicted, in agreement with observations. Type I SN rates higher
    by a factor of ~ 20 with respect to the current ones are
    predicted for ellipticals at high redshift (> 3).
iii) The contribution to the enrichment of Fe and Si of the
    intergalactic medium from ellipticals is computed. Good agreement
    is obtained with the observed abundances from X-rays, and it is
    claimed that the majority of the gas in clusters should be of
    primordial origin.

1. INTRODUCTION

The study of the evolution of elliptical galaxies is very important to
understand the behaviour of galaxies at high redshift as well as the
chemical enrichment of the intergalactic medium (IGM) in galaxy
clusters. In fact, it is generally believed that elliptical galaxies
have undergone active star formation only at early epochs due to the
fact that they possess small quantities of gas, if compared with their
total mass, and that no type II SNe, generally associated with active
regions of star formation, have ever been seen in them. Galactic winds
have often been claimed to explain the absence of gas in ellipticals
as well as to reproduce the well-known observed mass-metallicity relation (Faber 1977; Terlevich et al. 1981; Dressler 1984; Vader 1986).
It has been shown, in fact, that supernova driven winds could lead to
a mass-metallicity relation (Larson 1974; Tinsley 1980; Hartwick 1980;
Arimoto and Yoshi 1987, hereafter referred to as AY; and Matteucci and

85

R. G. Kron and A. Renzini (eds.), Towards Understanding Galaxies at Large Redshift, 85–92.
© 1988 by Kluwer Academic Publishers.

Tornambé 1987, hereafter referred to as MT). Although the existence of galactic winds is not yet proved and other possible explanations for the mass-metallicity relation can be found (Tinsley and Larson 1979), their existence could provide an explanation for the abundances of Fe and other heavy elements measured in the IGM of galaxy clusters (see Rothenflug and Arnaud 1985 and references therein). We will present here a model for chemical evolution of elliptical galaxies, where galactic winds, powered by SN explosions, are taken into account.

The model predicts the behaviour of the abundances, relative to the sun, of several chemical elements (O, Mg, Si and Fe) in the gas. The average galactic abundances of Mg and Fe, derived from the abundances of the stars still alive at the present time, is also computed. These average abundances are those which, in principle, should be compared with the observational estimates. Unfortunately, the metal content of elliptical galaxies is known only through metallicity indicators, and it is not clear how they relate to real abundances. Therefore a quantitative comparison between predictions and observations is not very meaningful.

The most important new feature of this model, with respect to the previous ones, is that the different rôles of type I and II SNe in the chemical enrichment are taken into account, and predictions of the current type I SN rate in ellipticals are given. The assumed model for type I SN progenitors is that of Whelan and Iben (1973), where type I SNe originate from white dwarfs, exploding by C-deflagration after accretion of matter from a companion in a binary system. In this framework, type I SNe turn out to be the most important contributors to iron enrichment in galaxies. Under this assumption the trends in the ratios of oxygen and some α-elements (Mg and Si), which are believed to originate from massive stars, with respect to iron observed in solar neighbourhood dwarfs (see Gustafsson 1987 for a review on this subject), can be very well reproduced (Matteucci and Greggio 1986; Matteucci 1986).

2. THE CHEMICAL EVOLUTION MODEL

The basic equations of the model as well as the detailed prescriptions can be found in MT.

Galactic winds are assumed to occur when the energy input from type I and II SNe equals the binding energy of the gas (Larson 1974).

The computation of the energy input from SNe is done separately for the two SN types, since their rates are computed in different ways (see MT). The thermal energy content in the interior of a SN remnant of both types is computed according to Cox (1972) and the binding energy of the gas according to AY.

The assumed nucleosynthesis prescriptions are the most up to date and are the same as those used in successful models of chemical evolution of the solar neighbourhood (Matteucci 1986). In particular, for massive stars (M > 8 $M_\odot$) we take the results of Woosley and Weaver (1986), obtained with a new rate for the $^{12}C(\alpha,\gamma)^{16}O$ reaction. For low and intermediate mass stars (0.8 < $M/M_\odot$ < 8) we take Renzini and Voli's (1981) results for η = 0.33 and α = 1.5. For the nucleosynthesis in type I SNe, namely C-deflagrating C-O white dwarfs, we assume the computations of Nomoto et al. (1984). In this framework type I SNe are the most important contributors to iron production in galaxies, whereas type II SNe are responsible for O, Mg and Si production. These assumptions are fundamental for understanding the results which will be discussed in the next sections. Two different prescriptions for the IMF have been adopted: Salpeter (1955) and AY.

3. OBSERVATIONAL CONSTRAINTS

From the point of view of chemical evolution the most relevant constraints on elliptical galaxies are: a) the mass-metallicity relation; b) the amount of gas at the present time and c) the current rate of type I SNe.

Concerning point a), as already mentioned in the introduction, estimates of abundances in elliptical galaxies are quite uncertain since they are based on spectrophotometric indices, whose relation with the real abundances is not well known. Therefore, a detailed quantitative comparison with predictions of abundances is not yet possible.

One of the most commonly used metallicity indicators is the Mg_2 index at 5175 Å, as defined by Faber et al. (1977). Some recent data from Vader (1986) give the following mass-metallicity relations for galaxies in the field (F), Coma (C) and Virgo (V), respectively:

$$Mg_2 = 0.04 \; \log M_G - 0.116 \quad\quad F$$
$$Mg_2 = 0.037 \log M_G - 0.118 \quad\quad C \quad\quad\quad (1)$$
$$Mg_2 = 0.051 \log M_G - 0.278 \quad\quad V$$

Burstein (1979) made an attempt to transform the index Mg_2 into the abundance of iron relative to the sun, $[Fe/H]$. To obtain this he took results from Mould (1978), who modelled the behaviour of Mg_2 by means of spectral synthesis in the high metallicity range (the one of interest for ellipticals), and data on globular clusters in the low metallicity range.

The resulting transformation between $[Fe/H]$ and Mg_2 is

$$[Fe/H] = 3.9 \; Mg_2 - 0.9 \; , \quad\quad\quad\quad\quad\quad\quad (2)$$

where Mg_2 is expressed in magnitudes. If relation (2) is applied to (1), one obtains:

$$[Fe/H] = 0.14 \quad \log M_G - 1.35 \quad\quad F$$
$$[Fe/H] = 0.14 \quad \log M_G - 1.36 \quad\quad C \quad\quad\quad (3)$$
$$[Fe/H] = 0.199 \log M_G - 1.978 \quad\quad V$$

However, eq. (3) should be taken only as indicative, given the uncertainties present in Burstein's transformation. In fact, one could argue that it is not possible to compare globular clusters and elliptical galaxies, whose chemical histories could have been very different. In addition, the theoretical models of Mould, where the metallicity parameter is Z and not $[Fe/H]$, assume implicitly that these two quantities are proportional, whereas abundance data on dwarf stars in the solar neighbourhood (Gustafsson 1987) indicate that different heavy elements behave quite differently from one to another, and in particular iron. Therefore iron, among the other heavy elements, is the least indicated to represent Z. In the light of this, any quantitative comparison is very dangerous.

Concerning point c), most elliptical galaxies possess very small quantities of gas (several times 10^8 $M_\odot$) and some of them have not even been detected in the HI line. The origin of this gas is still controversial; in a very recent paper, Knapp (1987) argues that the HI content in ellipticals has an external origin and therefore is not produced by stellar mass loss.

Finally, the current type I SN rate in ellipticals (point d)) can represent a very important constraint on the past star formation rate and theories of SN progenitors. Type II SNe, which are believed to originate from stars more massive than 8 $M_\odot$, have never been observed in ellipticals. Estimates of the current type I SN rate in ellipticals (Tammann 1974) indicate a value of ~ 0.22 SNu (SNu = 10^{-10} $L_{B\odot}$ 100 yr^{-1}). However, Tammann's rate could be overestimated, as has been shown for his estimate of SN rates in spirals by van den Bergh et al. (1987), by a factor of ~ 3.

Since type I SNe are believed to have progenitors in the range of low and intermediate mass stars, in particular white dwarfs in binary systems, the study of the type I SN rate in ellipticals can be useful to put constraints on the assumed SN progenitors. It has already been shown (Greggio and Renzini 1983; MT; Tornambè and Matteucci 1987), that the white dwarf model for type I SN progenitors predicts the occurrence of SN explosions at the present time and at the observed rate, even in systems where star formation was assumed to have stopped several billion years ago.

4. THEORETICAL PREDICTIONS

4.1. Abundances in the Gas

The evolution of O, Mg and Fe in the gas in ellipticals of various initial masses has been predicted in detail up to the onset of galactic winds. The occurrence of these winds is a function of the initial galactic mass, in the sense that winds occur later for more massive objects. In particular, the results we have obtained are similar to those of AY.

The variations of the absolute abundances in the gas, as functions of the galactic mass, $\Delta[X/H]/\Delta\log M_G(0)$, are of the order of 0.054 for O and 0.27 for Fe when AY's IMF is used, and -0.037 for O and 0.28 for Fe, when Salpeter's IMF is adopted. The other studied elements show variations which are betweeen those of O and Fe.

The fact that the predicted variation of Fe with respect to the galactic mass is the biggest one among all the other elements, for both the IMFs, is again a consequence of the assumed iron nucleosynthesis and type I SN progenitors. In fact, the general variation of the abundances of all the elements as functions of the galactic mass is the effect of galactic winds occurring at different times in galaxies of different masses, but the absolute value of this variation is a function of the assumed progenitors and nucleosynthesis of the studied elements. The galactic wind effect is therefore enhanced on iron by the fact that iron is mostly produced by long living progenitors which, in small galaxies, where winds start earlier, do not have time to die in a substantial proportion before them, whereas in big galaxies they do.

4.2. Abundances in Stars

We have computed the average abundances in ellipticals of different masses by weighting the abundances of dwarfs and giants, still alive at the present time, on the stellar visual luminosities, with a procedure very similar to that adopted by AY. The resulting mass-metallicity relations for Fe and Mg, which are the two elements most accessible from the observational point of view, are:

$$<[Fe/H]> = 0.28 \log M_G - 3.27$$
$$<[Mg/H]> = 0.05 \log M_G - 0.87$$
(4)

for AY's IMF and:

$$<[Fe/H]> = 0.14 \log M_G - 1.74$$
$$<[Mg/H]> = 0.05 \log M_G - 0.84$$
(5)

for Salpeter's IMF.

Again, the relation for iron looks flatter and its interpretation is the same as given before. From an observational point of view it would be interesting to check if this is true. Unfortunately, a direct comparison of (4) and (5) with (1) and (3) could be meaningless, for the reasons discussed in the previous section, even if, in this case, predictions seem to agree with observations. On the other hand, measurements of iron lines and Mg_2 in ellipticals by Burstein et al. (1984) seem to show the opposite trend for Mg and Fe than the one predicted here. However, also in this case one should remember that metallicity indicators might not vary in the same way as real abundances. Therefore, eqs. (4) and (5) should be considered as pure theoretical predictions which will be tested when a better knowledge of real abundances in ellipticals will be achieved.

4.3. SN Rates

In Fig. 1 is shown the behaviour of type I and II SN rates until the onset of galactic winds in a typical elliptical of initial mass 10^{12} $M_\odot$ and for AY's IMF. The qualitative behaviour is the same for a Salpeter IMF. It is clear from this figure that type II SNe exist in elliptical galaxies only for a very short length of time. On the other hand, despite the lack of star formation after a time smaller than 1 Gyr from galaxy formation, type I SNe, as shown in Fig. 2, continue to exist until the present time, in agreement with the observations. The two curves in Fig. 2 refer to the two choices for the IMF, as indicated on each curve. In both cases the type I SN rate has a

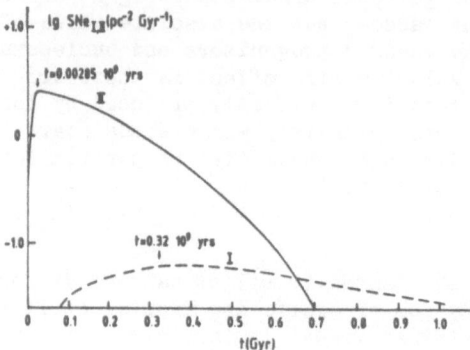

Figure 1. The logarithm of the type I and II SN rates, for a galaxy of initial mass $M_G(0) = 10^{12}$ $M_{\odot}$ and AY's IMF, as a function of time, during the first billion years after galaxy formation. The numbers reported above each curve represent the time at which the maximum rate occurred.

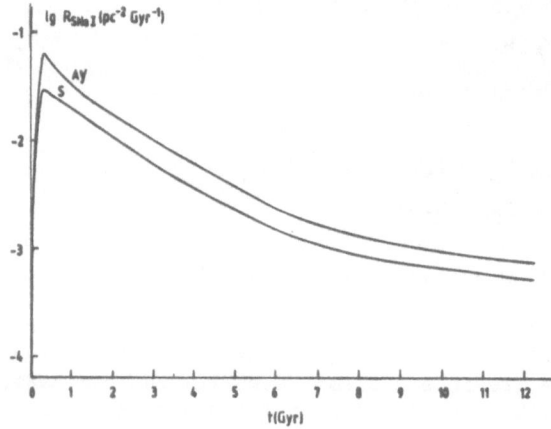

Figure 2. The logarithm of the type I SN rate as a function of the galactic age for a galaxy with initial mass $M_G(0) = 10^{12}$ $M_{\odot}$, and for two different IMFs (see text).

maximum between 0.3 and 0.5 Gyr from galaxy formation. At ~ 1.5 Gyr, corresponding to redshift ~ 3, the SN rate is ~ 300 and ~ 100 SNe 100 yr^{-1}, for AY's and S's IMF, respectively. The predicted absolute values of the current type I SN rate for the two cases are ~ 0.25 SNU for S's IMF and ~ 0.42 SNU for AY's IMF. These rates are derived by assuming a M/L_B ratio of ~ 10.

4.4. Enrichment of the IGM

We have also computed the expected abundances in the IGM, under the assumption that elliptical galaxies eject their enriched residual gas at the onset of galactic winds at early epochs, as well as, after this stage, the gas restored by dying stars in several smaller events. This latter is assumed to be lost because the energetic balance between SN energy input and binding energy of the gas always favours the SN energy, due to the continuous type I SN explosions. The contribution to the IGM enrichment from spirals and irregulars should be negligible if compared with ellipticals, since the former retain most of their gas inside. The detailed prescriptions for the computation of the abundances of Fe and Si in the IGM can be found in Matteucci and Vettolani (1987). Here we will recall only the main assumptions and present a single case. In summary, we have derived the relations between the ejected masses in the form of single elements and total gas mass as functions of the initial galactic mass from MT, and integrated them over the mass function distribution of a cluster, starting from Schechter's (1976) luminosity function. The predicted total masses in form of Fe, Si and gas mass for the Coma cluster are: $M_{Fe} = 2.5 \cdot 10^{11}$ $M_{\odot}$, $M_{Si} = 10^{11}$ $M_{\odot}$, $M_{gas} = 2 \cdot 10^{13}$ $M_{\odot}$. Since the observed mass of gas in Coma is $\sim 4 \cdot 10^{14}$ $M_{\odot}$ (Rothenflug and Arnaud 1985), the abundances of Fe and Si predicted for this gas are: $X_{Fe} = 6.25$ (-4), $X_{Si} = 2.5$ (-4), which are roughly ½ of the corresponding solar abundances (Cameron 1982). This is in very good agreement with the abundances of Fe and Si deduced from X-ray spectroscopic observations (see Rothenflug and Arnaud 1985 and references therein). Since the total ejected mass of gas is smaller by about an order of magnitude than the observed gas mass in Coma, the suggestion arises that most of the gas present in the IGM should have a primordial origin. However these results are only preliminary, and the interested reader will find more details and a critical discussion of them in Matteucci and Vettolani (1987).

5. CONCLUSIONS

The main conclusions can be summarized as follows:

i) It is confirmed that the existence of galactic winds in ellipticals can account for the observed mass-metallicity relation. Different chemical elements have different mass-abundance relations. This result is new and is due to the assumptions made on nucleosynthesis and progenitors of SNe of type I and II. In particular, the iron - mass relation is predicted to be steeper by about a factor of 3 than the magnesium - mass relation.

ii) The predicted current type I SN rates in ellipticals are in agreement with the observational estimates. This confirms the validity of the assumed model for type I SN progenitors.

Rates of the order of hundreds of SNe of type I every 100 years are predicted for typical ellipticals (e.g. with initial masses 10^{11}-10^{12} $M_{\odot}$) at a redshift of ~ 3. These rates are greater than almost a factor of 20 than the current ones and therefore they could be observable at high redshift.

iii) The existence of galactic winds in ellipticals can also explain the presence of iron and other heavy elements in the IGM in galaxy clusters. Preliminary calculations of the integrated ejected masses in form of heavy elements (Fe and Si) in the IGM show good agreement with the observations and indicate that the major part of gas in clusters should have a primordial origin.

REFERENCES

Arimoto, N., Yoshi, Y.: 1987, Astron. Astrophys. **173**, 23. (AY)
Burstein, D.: 1979, Astrophys. J. **224**, 768.
Burstein, D., Faber, S.M., Gaskell, C.M., Krumm, N.: 1984,
    Astrophys. J. **287**, 586.
Cameron, A.G.W.: 1982, in "Essays in Nuclear Astrophysics" eds. C.A.
    Barnes, D.D. Clayton and D.N. Schramm, Cambridge University
    Press, p. 377.
Cox, D.P.: 1972, Astrophys. J. **178**, 159.
Dressler, A.: 1984, Astrophys. J. **181**, 512.
Faber, S.M.: 1977, in "The Evolution of Galaxies and Stellar
    Populations", eds. B.M. Tinsley, R.B. Larson, Yale University
    Observatory, p. 157.
Faber, S.M., Burstein, D., Dressler, A.: 1977, Astron. J. **82**, 941.
Greggio, L., Renzini, A.: 1983, Astron. Astrophys. **118**, 217.
Gustafsson, B.: 1987, ESO Workshop on "Stellar Evolution and Dynamics
    in the Outer Halo of the Galaxy", eds. M. Azzopardi,
    F. Matteucci, K. Kjär, in press.
Hartwick, F.D.A.: 1980, Astrophys. J. **236**, 754.
Knapp, G.R.: 1987, IAU Symp. 127.
Larson, R.B.: 1974, Mon. Not. R. astr. Soc. **166**, 585.
Matteucci, F.: 1986, Astrophys. J. Lett. **305**, L81.
Matteucci, F., Greggio, L.: 1986, Astron. Astrophys. **154**, 279.
Matteucci, F., Tornambé, A.: 1987, Astron. Astrophys. in press. (MT)
Matteucci, F., Vettolani, P.: 1987, in preparation.
Mould, J.R.: 1978, Astrophys. J. **220**, 434.
Nomoto, K., Thielemann, F.K., Yokoi, K.: 1984, Astrophys. J. **286**, 644.
Renzini, A., Voli, M.: 1981, Astron. Astrophys. **94**, 175.
Rothenflug, R., Arnaud, M.: 1985, Astron. Astrophys. **144**, 431.
Salpeter, E.E.: 1955, Astrophys. J. **121**, 161.
Schechter, P.: 1976, Astrophys. J. **203**, 297.
Tammann, G.A.: 1974, in "Supernovae and Supernova Remnants", ed. C.B.
    Cosmovici (Dordrecht: Reidel), p. 155.
Terlevich, R., Davis, R.L., Faber, S.M., Burstein, D.: 1981, Mon. Not.
    R. astr. Soc. **186**, 381.
Tinsley, B.M.: 1980, in Fund. of Cosmic Phys. Vol. 5, p. 287.
Tinsley, B.M., Larson, R.B.: 1979, Mon. Not. R. astr. Soc. **186**, 503.
Tornambé, A., Matteucci, F.: 1987, Astrophys. J. Lett. **318**, L25.
Vader, J.P.: 1986, Astrophys. J. **306**, 390.
van den Bergh, S., McClure, R.D., Evans, R.: 1987, Jan. 1987 preprint
    of Dominion Astrophysical Observatory.
Whelan, J.C., Iben, I. Jr.: 1973, Astrophys. J. **186**, 1007.
Woosley, S.E., Weaver, T.A.: 1986, IAU Coll. 89, "Radiation
    Hydrodynamics in Stars and Compact Objects", eds. D. Mihalas and
    K.H. Winkler, p. 91.

GLOBAL PROPERTIES AND STELLAR POPULATIONS OF SPIRAL GALAXIES

David Burstein
Department of Physics
Arizona State University
Tempe, AZ, 85287   U.S.A.

ABSTRACT. The role of the 'stellar population' of a sprial galaxy in determining the global properties of spirals is often overlooked. This paper uses two different measures of stellar population to reconsider a) the physical meaning of the near-IR Tully-Fisher relationship for spirals and b) the relationship between the old and young components of the stellar populations in spirals.

1. INTRODUCTION

To understand how a sample of spiral galaxies would appear at an arbitrary look-back time, one must first understand the physical source (or sources) of the global properties of present-day spirals. This, in turn, requires that the evolutionary histories of these stellar populations of spirals be reasonably understood. In order to avoid confusion in the discussion that follows, define the term 'global stellar population' in reference to spirals to be the time-averaged, luminosity-weighted (TALW) source of stellar light.

From our current knowledge of the kinds of stellar populations in our own spiral galaxy, one can predict that the sources of TALW light will vary with wavelength: X-ray flux comes both from old stars (binary stars) and from young stars (supernova remnants); ultraviolet flux comes from hot stars, both young (main sequence) and old (post-asymptotic giant branch); both optical flux __and__ near-infrared flux comes from both old and young stars, with the percentage varying with wavelength (cf. Renzini 1981); far-infrared flux comes from heated dust, which in turn comes from both old and young stars; radio continuum comes primarily from free-free and synchrotron emission, which can originate in both old (M giants?) and young stars; radio HI emission comes from gas that forms young stars, as well as gas that is expelled from old stars; the mass-to-light ratio is a function of the TALW luminosity at a given wavelength, as well as of the mass of the dark matter.

To further complicate matters, it is well-known that the TALW light at all wavelengths can be affected by variations in hard-to-determine physical properties such as the initial mass function of stars, the mean metallicity (Z) as a function of time and elemental abundance (Z(Fe), Z(C), Z(N), Z(O), ...). In addition, it is likely that both the kind

93

R. G. Kron and A. Renzini (eds.), Towards Understanding Galaxies at Large Redshift, 93–102.
© 1988 by Kluwer Academic Publishers.

and the amount of dust mixed together with the stars is dependent on the IMF and metallicity history. Finally, the known 'disk-halo' conspiracy of spiral galaxies closely links the amount of dark matter in a spiral galaxy to the amount of TALW light in the optical and near-infrared (Burstein and Rubin 1985; Bahcall and Casertano 1985).

The wealth of <u>possible</u> variations in global stellar populations among spiral galaxies is so great that it would appear prudent to use the observed range of stellar population properties to constrain the theoretical interpretations. This paper explores the systematic properties of two different measures of the global stellar populations of spirals - near-infrared H luminosity and 20 cm radio continuum emission - employing absolute magnitudes from distances based on radial velocities and the velocity field model of Lynden-Bell et al. (1988).

2. DATA: OBSERVED AND ABSOLUTE VALUES

It is fortuitous that the physical properties of spiral galaxies have been used for the past ten years to predict distances in the universe (the Tully-Fisher (1977) relationship). As a result, a significant amount of accurate data relating to the global stellar populations of nearby galaxies has been collected at optical and near-infrared wavelengths (Aaronson et al. 1982), together with data related to their mass properties (i.e., rotation velocities). In order to further study the stellar populations of these spirals, 20 cm radio continuum VLA observations have recently been made by Condon et al. (1988) for 161 of the galaxies in the Aaronson et al. sample.

The relationships among the near-IR magnitudes ($H_{-0.5}$), optical diameters (D_{25}), 20 cm continuum data, rotation velocities ($\Delta V/2$) and Hubble types for these spirals were analyzed in a manner that is distance-independent by Burstein et al. (1987). However, the success of the 'Great Attractor' velocity field model of Lynden-Bell et al. (1988) in predicting distances of galaxies within 4000 km/s of the Local Group (Burstein et al. 1988a,b; see also below) results in more reliable determinations of the absolute global properties of spiral galaxies. As such, this paper will use the distant-dependent relationships of the data analyzed in other ways by Burstein et al. (1987).

The Great Attractor velocity field model incorporates two systematic motions in the nearby universe, over and above motion due to the expansion of the universe: 1) A streaming motion towards a centralized mass that is centered at a distance of 87±6 Mpc (H_o = 50 km/s/Mpc) towards the direction $l=307°$, $b=9°$. The streaming motion varies as 1/R from this center, and this mass (the 'Great Attractor') induces a streaming velocity at the position of the Local Group of 570 ± 60 km/s. 2) A motion towards the center of the Virgo cluster, giving an infall velocity at the position of the Local Group of 200 ± 30 km/s.

The errors quoted here for the distant-dependent parameters are due to errors in measurement of fluxes, sizes and distances. Errors in distance can be estimated from the near-IR Tully-Fisher relation (Fig. 1(a)); errors in fluxes and sizes come from Aaronson et al. (1982), Burstein (1982) and Condon et al. (1988). The overall error estimated for each of the absolute quantities is given in the captions

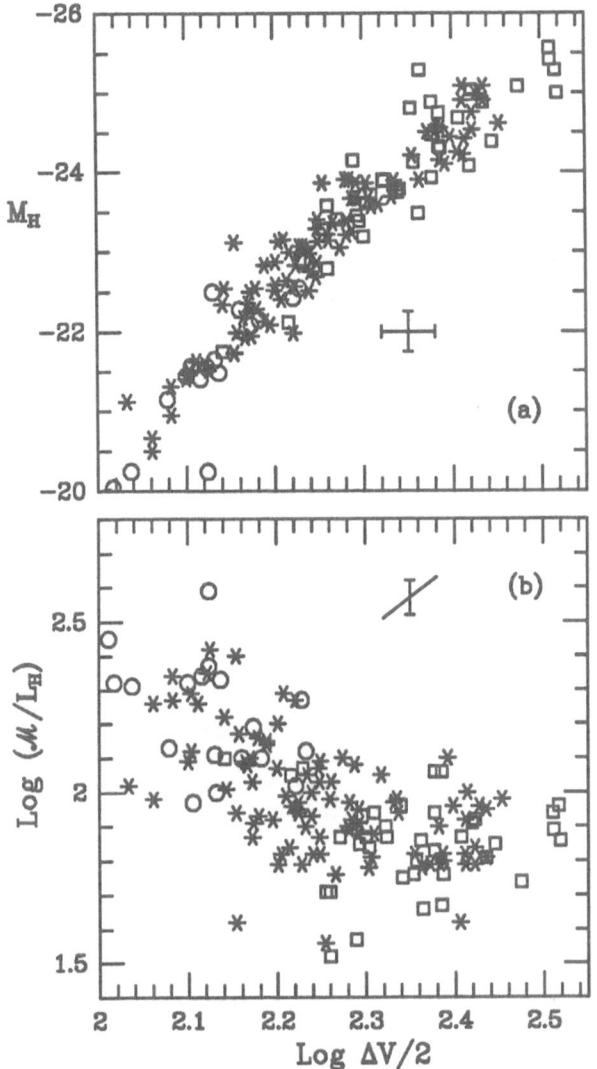

Fig. 1. (a) The near–IR
Tully–Fisher relationship for
149 galaxies observed by
Aaronson et al. (1982) that
have both accurate blue
diameters and 20 cm radio
continuum fluxes from
Condon et al. (1987).
Distances predicted using
'Great Attractor' model of
Lynden–Bell et al. (1988;
see text). Error bars of
0.04 dex in Log $\Delta V/2$ and
0.25 mag in M_H are shown.
The symbols in these
figures represent three
ranges of Hubble types:
squares – Sa, Sab, and Sb;
stars – Sbc, Sc and Scd.
circles – Sd and later
(cf. Burstein et al. 1987).
(b) The mass–to–near–IR
luminosity ratio, $\mathcal{M}/L_H$,
versus Log $\Delta V/2$ for the
same galaxies. The error
bars represent an error
of 0.05 dex in Log $\mathcal{M}/L_H$
alone, together with the
correlated effect in
both coordinates due to
an error of 0.03 dex in
Log $\Delta V/2$.

to Figures 1 and 2, and represented by the error bars in each figure.
Absolute H magnitude, M_H is defined in the standard manner from $H_{-0.5}$,
while the absolute 20 cm magnitude, M_{20cm} is defined here to be
$-2.5\log$(absolute 20 cm flux, in Janskys). The mass–to–near–infrared
luminosity ratio, M/L_H is a hybrid of optical and near–IR properties,
and is directly proportional to $(\Delta V/2)^2 D_{25}/L_H$, with an arbitrary zero
point. These absolute parameters are plotted against one another in
Figures 1 and 2 (above and on the following page).

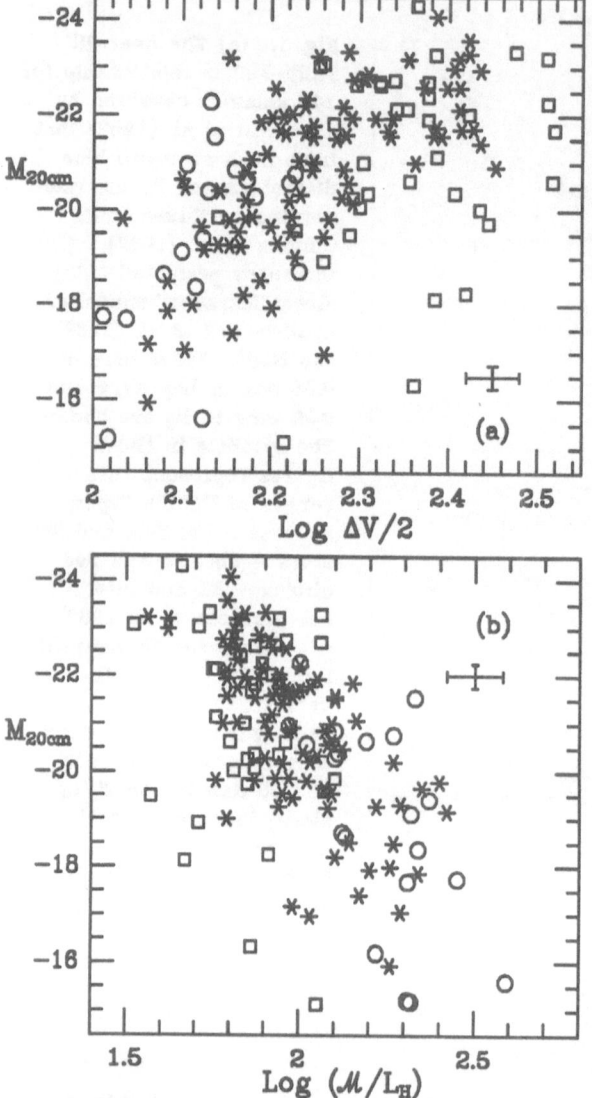

Fig. 2. The absolute radio 20 cm continuum flux, M₂₀cm, expressed in magnitudes, plotted versus: (a) The logarithm of rotation velocity, Log ΔV/2 and (b) The logarithm of mass-to-near-IR luminosity ratio, Log 𝓜/L_H. Error bars correspond to a nominal 0.25 mag error in 20 cm absolute mag, and a total of 0.09 dex error in 𝓜/L_H. The symbols are the same as in Figure 1, and distances are predicted with the same velocity field model as in Figure 1.

3. CORRELATIONS AMONG ABSOLUTE GLOBAL PROPERTIES

Fig. 1(a) is the IR Tully-Fisher relationship that is predicted by the Great Attractor model. As discussed in Burstein et al. (1988a,b), the errors in these data are consistent with an upper limit error of 0.25 mag in M_H combined with an error of 0.03 dex in $\Delta V/2$. The total scatter in this relationship, including remaining distance errors, is 0.32 mag, significantly less than the error (0.45 mag) originally estimated by Aaronson et al. (1982). The smaller scatter found here is the result of using only those galaxies in the Aaronson et al. survey that have values of D_{25} given by de Vaucouelurs et al. (1976, RC2; see discussion in Burstein et al. 1988b). Moreover, the scatter seen in Fig. 1(a) is an upper limit to the intrinsic scatter in the IR T-F relation: much of the apparent scatter is due to remaining errors in distance.

The relationship in Fig. 1(a) is consistent with the curved form of the IR Tully-Fisher relation as defined by Aaronson et al. (1986).

In addition, early-type, high-luminosity galaxies appear to scatter significantly more in the IR T-F relation than do late-type, high-luminosity galaxies, a point which is discussed further in Burstein et al. (1988b).

On the other hand, the relationships between M/L_H and $\Delta V/2$ (Fig. 1(b)) and between M_{20cm} and $\Delta V/2$ (Fig. 2(a)) show significant intrinsic scatter, both with $\Delta V/2$ (as shown) and with absolute H magnitude (as inferred from Fig. 1(a)). Both M_{20cm} and M/L_H vary systematically with absolute H magnitude: M/L_{20cm} becomes larger as the luminosity becomes fainter (cf. Burstein 1982); radio continuum emission decreases as luminosity decreases. In both relationships, early-type galaxies define one end of the correlation, late-type galaxies the other end. However, the absolute 20 cm flux varies by at least a factor of 40, and M/L_H varies by at least a factor of two for galaxies of the same absolute H magnitude. The intrinsic range in M/L_H at a given L_H must be the result of an intrinsic range in D_{25} at a given L_H: galaxies with blue diameters that differ by a factor of two can have the same H luminosity.

The existence of a direct relationship between M/L_H and M_{20cm} (Fig. 2(b)) was inferred by Burstein et al. (1987). As with the other correlations, this relationship is also dependent on Hubble type: early-type galaxies have both high 20 cm flux and low values of M/L_H; late-type galaxies have low values of 20 cm flux and high values of M/L_H. Significant intrinsic scatter also exists in this relationship, together with a small number of early-type galaxies with anomalously low 20 cm flux (cf. Condon et al. 1988).

4. HOW MANY SOURCES OF INTRINSIC SCATTER EXIST IN THESE RELATIONSHIPS?

It is remarkable that, among all the relationships combining stellar population-related parameters and mass-related parameters, the Tully-Fisher relation should be relatively tight, while the other relationships show considerable intrinsic scatter. Is the intrinsic scatter in each relationship responding to separate physical processes in spirals? If so, then the scatter in each relationship would not be correlated with that from another relationship.

Fig. 3(a) plots the residuals of M_H from the quadratic IR T-F relationship of Aaronson et al. (1986), δ(absolute H mag), versus the residuals of an assumed power-law relationship between M/L_H with $\Delta V/2$ (Log (M/L_H) = 1.5Log $(\Delta V/2)$ + 5.35). The direction of correlated errors due either to distance errors or to errors in $\Delta V/2$ are shown. Apart from a handful of galaxies with relatively large distance errors, there appears to be little or no correlation between the scatter in M/L_H and in L_H at a given rotation velocity. Thus, intrinsic variations in M/L_H do not appear to appreciably affect the relationship between L_H and rotation velocity!

Before comparing the residuals of the M_{20cm} - $\Delta V/2$ relationship with M/L_H residuals, it is useful to define another quantity related to M/L_H, but which is distance independent: HSB = $L_H/(D_{25})^2$. This hybrid IR/optical surface brightness has been used by Burstein (1982) and Burstein et al. (1987) to infer mass-to-light properties of

98

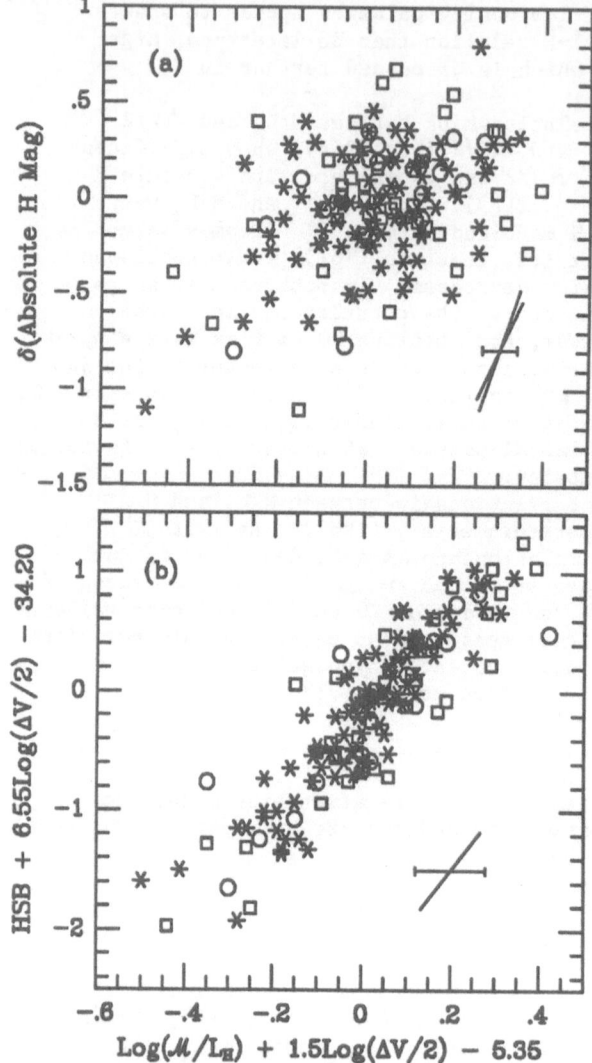

Fig. 3. (a) The residuals (in mag) from the Tully–Fisher relation versus the residuals (in dex) from the relationship between $\mathcal{M}/L_H$ and $\Delta V/2$. Note the relative lack of correlation for the majority of the sample. The effect of correlated errors in $\Delta V/2$ and H mag are shown, together with an error of 0.05 dex in blue diameter.
(b) The residuals in the relationship between the hybrid H mag surface brightness (HSB) and Log $\Delta V/2$, versus the residuals in the $\mathcal{M}/L_H$ – Log $\Delta V/2$ relationship. The effect of correlated errors in $\Delta V/2$, H mag and blue diameter are shown, along with the effect of a 20% error in distance (in $\mathcal{M}/L_H$ alone). The tight relation in this figure is a direct result of the tight relationship between absolute H mag and rotation velocity for these galaxies.

the Aaronson et al. spirals in a manner that is distance-independent. The correlation of HSB with $\Delta V/2$ for the present sample of spirals is given in Burstein et al. (1987), and is not repeated here. However, the residuals of that correlation, HSB + 6.55Log ($\Delta V/2$) – 34.20 are plotted versus the residuals of M/L$_H$ with $\Delta V/2$ in Fig. 3(b). As explained in Burstein (1982), the strong relationship between these two residual quantities follows directly from the tight IR T-F relationship and the scatter in M/L$_H$: the scatter in both quantities comes from the scatter of blue diameter with both L$_H$ and $\Delta V/2$. In the case of HSB, this scatter is measured in a distance-independent manner; in the case of M/L$_H$, the scatter is measured in a distance-dependent manner.

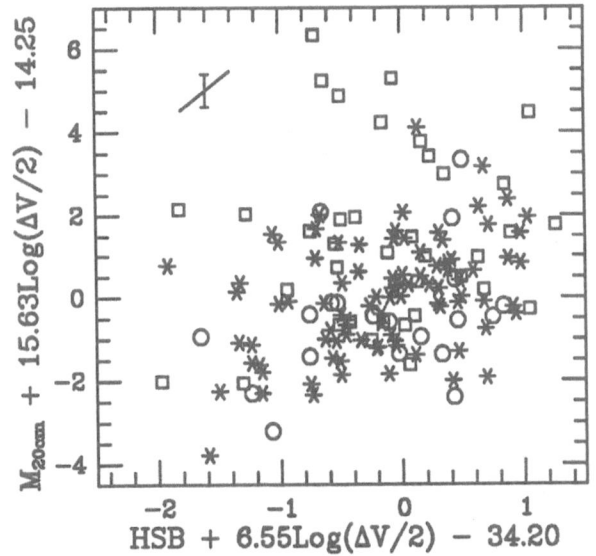

Fig. 4. The residuals (in mag) from the relationship between M_{20cm} and Log $\Delta V/2$, plotted versus the residuals of the relationship between HSB and Log $\Delta V/2$. The effect of correlated errors in $\Delta V/2$ are shown, together with the effect of a 20% distance error in M_{20cm} alone.

Thus, HSB can be substituted for M/L_H in any of the above analyses. In particular, distance errors have a proportionally larger effect for the correlation of M/L_H with $\Delta V/2$ than for the correlation of M_{20cm} with $\Delta V/2$ (cf. size of error bars in Fig. 1(b) to those in Fig. 2(a)). It is therefore advantageous to compare residuals of the $M_{20cm}-\Delta V/2$ relationship with the residuals of the HSB-$\Delta V/2$ relationship, to avoid additional complications due to distance errors. This is done in Fig. 4. In contrast to the residuals of the IR T-F relation, the scatter in the $M_{20cm}-\Delta V/2$ relation appears to be related to the scatter in the relationship between HSB (=M/L_H) and $\Delta V/2$: Galaxies of Hubble type Sbc and later show a general trend, in the sense that, at a given value of $\Delta V/2$, galaxies with a brighter HSB (= lower M/L_H) have higher absolute 20 cm fluxes. There appears to be little correlation in Fig. 4 for early-type galaxies, although those galaxies with anomalously low values of absolute 20 cm flux appear to follow a separate relationship, in which galaxies with fainter HSB have higher absolute 20 cm fluxes.

Despite signficant intrinsic scatter in the relationship between 20 cm radio continuum flux and M/L_H, the general correlation in Fig. 4 suggests that both stellar population properties are more strongly related to each other than either parameter is to the absolute size of the galaxy. This same conclusion was reached from the distance-independent analysis of these data by Burstein et al. (1987).

4. DISCUSSION

As discussed in the Introduction, the H magnitude flux from normal spiral galaxies is thought to be dominated by relatively old stars (Aaronson and Mould 1983), while the 20 cm flux is believed to originate primarily from relatively young stars (e.g. Biermann 1976). In contrast, the mass of the galaxy within D_{25} is determined by both luminous matter and dark matter (e.g. Burstein and Rubin 1985), and is related to the global stellar population in two ways: $[mass(total)]/L_H$ is the sum of $[mass(H) + mass(dark)]/L_H$; in addition, mass(total) is necessarily determined by the optical size of the galaxy, D_{25}, yet another TALW measure of global stellar population.

In this context, then, what are the physical relationships that are implied by the existence of a very strong IR Tully-Fisher relationship?

a) The IR Tully-Fisher relationship is the result of a direct relationship between a measure of the luminosity of a spiral galaxy and the maximum rotation velocity reached by that galaxy. In this context, it is worth noting that Tully-Fisher relationships also exist for passbands other than H magnitude.

b) The IR Tully-Fisher relationship is not a direct function of the global stellar population that produces the luminosity: Galaxies with the same luminosity have an intrinsic range in both M/L_H and in 20 cm radio continuum emission.

c) The IR Tully-Fisher relationship is not the result of a constant relationship between the observed mass of a galaxy and the observed luminosity of a galaxy. Average values of mass-to-light ratios correlate with absolute luminosity.

These conclusions regarding the IR Tully-Fisher relation are substantially the same as those reached in an earlier analysis with fewer data (Burstein 1982). The new information added here is that the IR Tully-Fisher relationship shows very little intrinsic scatter, while all other relationships involving luminosities and masses show considerable intrinsic scatter.

What are the physical implications of the correlation of M/L_H with M_{20cm}? As discussed in Burstein et al. (1987), this correlation implies a physical link between the relatively short-term stellar populations in spirals (perhaps over the last 100 million years) and their relatively long-term stellar populations (perhaps over the last 2-3 billion years). The connection is not perfect: intrinsic scatter exists in the relationship between M/L_H and M_{20cm}. Nonetheless, the fact that a link exists at all requires the existence of a physical mechanism that couples star formation processes over periods of time long compared to the lifetimes of massive stars. Furthermore, the present data also imply that the effect of this physical mechanism also differs intrinsically among spiral galaxies of similar absolute luminosity, Hubble type or observed mass.

In order to further understand the physical properties of present-day spirals, we are faced with a series of hard questions:

a) What is the physical origin of the Tully-Fisher relationship? It is demonstrably not a relationship between observed mass and observed luminosity.

b) What is this physical mechanism(s) that couples old and young
stellar populations, but which can vary among galaxies of
otherwise similar overall luminosity?

c) Is the intrinsic variation of M/L_H simply a function of the
stellar population alone, or is it possible that the percentage
of dark matter differs within the optical confines of these
galaxies?

d) Why is the IR Tully-Fisher relationship apparently unaffected
by the intrinsic range in global stellar population among
spirals?

It is also worth noting the apparent similarities, and differences,
in the systematic mass-luminosity properties of spiral galaxies and
elliptical galaxies (cf. the companion papers on ellipical galaxies in
these proceedings): Both elliptical galaxies and spiral galaxies
evidence an intrinsic range of mass-to-light ratio at a given
luminosity, and a trend of mass-to-light ratio with luminosity. In both
cases, these variations do not appear to affect the present-day
relationship between luminosity and mass-related parameter. However,
there is circumstantial evidence that the luminosity and the mass being
measured in elliptical galaxies are coming the same physical, stellar,
component. That is not the case for spiral galaxies. Moreover, the
luminosity that is measured in late-type spiral galaxies has a very
different evolutionary history than the luminosity that is measured in
typical elliptical galaxies.

As was the case with elliptical galaxies, the questions raised
above are directly relevent to the context of this conference. How do
the global stellar populations of spiral galaxies evolve with time? Are
the intrinsic differences among stellar populations amplified or muted
with time? Should we expect to see a much wider range in stellar
population properties at large look-back times, or a signficantly
smaller range?

It is possible that, just as with elliptical galaxies, one may have
to study the stellar populations of spirals at all evolutionary times,
in order to understand the physical parameters that govern the
present-day stellar populations in spirals.

5. REFERENCES

Aaronson, M., Huchra, J., Mould, J., Schechter, P.L. and Tully, R.B.
1982, Ap. J. 258, 64.
Aaronson, M., Bothun, G., Mould, J., Huchra, J., Schommer, R.A. and
Cornell, M.E. 1986, Ap. J. 302, 536.
Aaronson, M. and Mould, J. 1983, Ap. J. Suppl. 50, 241.
Bahcall, J.N. and Casertano, S. 1985, Ap. J. Lett. 293, L7.
Biermann, P. 1976, Astron. Ap. 53, 295.
Burstein, D. 1982, Ap. J. 253, 539.
Burstein, D., Condon, J.J. and Yin, Q.F. 1987, Ap. J. Lett. 315, L99.
Burstein, D. and Rubin, V.C. 1985, Ap. J. 297, 423.

Burstein, D., Davies, R.L., Dressler, A., Faber, S.M., Lynden-Bell, D., Terlevich, R.L, and Wegner, G.A. 1988a, in <u>Large Scale Structures of the Universe</u>, proceedings of IAU Symposium 130, ed. J. Audouze and A. Szalay (Reidel).

Burstein, D. et al. 1988b, in prepartion.

Condon, J.J., Yin, Q.F. and Burstein, D. 1988, Ap. J. Suppl., in press.

de Vaucouleurs, G., de Vaucouleurs, A. and Corwin, H. 1976, <u>Second Reference Catalogue of Bright Galaxies</u> (Univ. of Texas). (RC2)

Lynden-Bell, D., Faber, S.M., Burstein, D., Davies, R.L., Dressler, A., Terlevich, R.J. and Wegner, G. 1988, Ap. J., in press.

Tully, R.B. and Fisher, J.R. 1977, Astron. Ap. 54, 661.

THE FORMATION OF GALACTIC DISKS

S. Michael Fall
Space Telescope Science Institute
3700 San Martin Drive, Baltimore, MD 21218
and
Department of Physics and Astronomy
The Johns Hopkins University
Homewood Campus, Baltimore, MD 21218

The disk components of galaxies are usually assumed to form later than the spheroidal components but how much later is still an open question. The following argument, based on the spin-up of collapsing protogalaxies, suggests that the disk components formed recently (Fall and Efstathiou 1980, Fall 1985). Since the specific angular momentum of a disk today is likely to be nearly the same as that in the protogalaxy, the initial size can be calculated as a function of the initial rotation velocity. The initial size of the protogalaxy then determines the free-fall time and hence the maximum redshift of collapse. The initial rotation is most conveniently expressed in terms of the dimensionless spin parameter, $\lambda \equiv J|E|^{1/2}G^{-1}M^{-5/2}$, where J, E and M are respectively the total angular momentum, energy and mass of the protogalaxy, including any dark matter it may contain. For an exponential disk with a scale length α^{-1} embedded in an isothermal halo with a circular velocity v_c, one finds that the free-fall time is $\tau_{ff} = 1.9(\alpha v_c \lambda)^{-1}$. The values of α^{-1} and v_c can be derived from observation, and for all theories that invoke hierarchical clustering. the values of λ can be derived from N-body simulations. The distribution of spin parameters turns out to depend only weakly on the spectrum of perturbations at recombination and always has a median value near $\lambda \simeq 0.05$ (Barnes and Efstathiou 1987, Quinn, Salmon and Zurek 1987). I have used these results to show that the median redshift of disk formation is $z_d \leq 2$ for $q_0 = 1/2$ and $z_d \leq 7$ for $q_0 = 0$ (Fall 1987).

One of the standard arguments for late infall is based on the so-called "G-dwarf problem" in the solar neighborhood. Several years ago, Cedric Lacey and I computed a series of models for the chemical evolution of the disk of the Milky Way (Lacey and Fall 1983, 1985). We considered the infall of material from outside the disk and the radial flow of gas within the disk. The models were compared with the observed age-metallicity relation and metallicity distribution of stars in the solar neighborhood as well as the variations in the metallicity, star formation

R. G. Kron and A. Renzini (eds.), Towards Understanding Galaxies at Large Redshift, 103–104.
© *1988 by Kluwer Academic Publishers.*

rate and gas density with distance from the galactic center. We found that. to fit these and several other constraints, the time scale of the infall had to be in the range 2 Gyr $\lesssim \tau_{in} \lesssim$ 6 Gyr, i.e. the formation of the disk was a slow process. Similar conclusions have been reached by Tosi (this meeting). Another indication for late infall comes from the kinematical evolution of the galactic disk (Gunn 1982, Sellwood and Carlberg 1984). The higher velocities of the older stars were once thought to be the result of interactions with randomly distributed clouds in the disk but it now appears that perturbations on larger scales are required. The best candidates, spiral arms and spurs, involve gravitational instabilities in the disk. Since the stellar component always grows hotter and therefore more stable, the overall instability of the disk can only be maintained if the gas, the coldest component, is replenished by infall. Further discussion of the kinematical evolution of galactic disks can be found in the contribution by Athanassoula (this meeting).

REFERENCES

Barnes, J., and Efstathiou, G. 1987, *Ap. J.*, **319**, 575.
Fall, S. M. 1985, in *IAU Symposium 106, The Milky Way Galaxy*, eds. H. van Woerden, R. J. Allen, and W. B. Burton (Dordrecht: Reidel) p. 603.
Fall, S. M. 1987, in preparation.
Fall, S. M., and Efstathiou, G. 1980, *M.N.R.A.S.*, **193**, 189.
Gunn, J. E. 1982, in *Astrophysical Cosmology*, eds. H. A. Bruck, G. V. Coyne, and M. S. Longair (Vatican: Pontificia Academia Scientarium), p. 233.
Lacey, C. G., and Fall, S. M. 1983, *M.N.R.A.S.*, **204**, 791.
Lacey, C. G., and Fall, S. M. 1985, *Ap. J.*, **290**, 154.
Quinn, P. J., Salmon, J. K., and Zurek, W. H. 1987, in preparation.
Sellwood, J. A., and Carlberg, R. G. 1984, *Ap. J.*, **282**, 61.

SPIRAL STRUCTURE AND GALAXY EVOLUTION.

Preben Grosbøl
European Southern Observatory,
Karl-Schwarzschild Straße 2,
D-8046, Garching, Fed. Rep. of Germany.

ABSTRACT. The paper reviews the main mechanisms for spiral structure in galaxies with emphasis on normal spirals. The general distribution of the different spiral types are given and new data on spiral parameters are presented. The possible evolutionary effects caused by the spiral structure in galaxies are discussed. It is concluded that the spiral structure is unlikely to affect the global evolution of a galaxy significantly because the typical dynamical time scale is too long. Although the star formation rate may be altered by the pattern other effects are as important.

1. INTRODUCTION

Spiral galaxies can be distinguished from ellipticals out to relative large redshift due to their morphological appearance or bluer colors. For cosmological studies it is important to estimate possible evolutionary effects for different types of galaxies in order to correct for them. The grand design spiral structure in galaxies can be explained by the density wave theory by Lin and Shu (1964). This makes it possible to estimate the evolutionary effects due to the presence of a spiral pattern in a galaxy.

A linear resolution of at least 1 kpc is needed to observed the detailed spiral structure in a galaxy. Even with an angular resolution of 0.1 arcsec, it is only possible to study the morphology of spirals up to a redshift of $z \approx 0$ 1 corresponding to a relatively short look-back time.

The present paper reviews the theory of a spiral structure in galaxies and summaries the distribution of different spiral types. Finally, the possible evolutionary effects are discussed.

2. SPIRAL STRUCTURE

The spiral structure in galaxies can be divided into two general groups based on their dynamical properties, namely : material arms or density waves. Due to the differential rotation in typical disk galaxies material arms will be wound up and destroyed in a few rotation periods while spiral structures supported by a density wave can exist over much longer

R. G. Kron and A. Renzini (eds.), Towards Understanding Galaxies at Large Redshift, 105–110.
© 1988 by Kluwer Academic Publishers.

time. The latter will therefore play a more important role for the long
term evolutionary effects. These two categories are discussed below in
more details.

2.1. Material Arms

Material arms are effected by the differential rotation and are therefore
in general transient phenomena. Their are formed by three main processes,
namely:

a) By interactions or tidal encounters between galaxies.
b) By ejection of matter in the galaxies typically from the center.
c) By Stochastic self-propagating star formation suggested by Gerola
   and Seiden (1978).

Whereas the two former processes produce short lived arm structures the
latter can maintain itself longer time by regenerating new arm segments
through propagation of star forming regions.

In all cases, however, material arms are secondary effects due to
other phenomena such as tidal effects, explosions or star formation. As
such they will not effect the evolution of the parent galaxy.

2.2. Density waves

The existence of grand design spiral structures in disk galaxies can be
explained by the density wave theory (Lin and Shu, 1964). A general review of the theory can be found in Lin and Lau (1979). The theory is based
on the quasi-stationary spiral structure (QSSS) hypothesis which assumes
that time dependent terms can be disregarded. Density waves can be triggered by general instability or tidal forces and then amplified near the
corotation region by stimulated emission (Mark, 1976; Lau et al., 1976).

The spiral pattern in a galaxy is determined by its three basic distribution functions, namely: a) the angular velocity $\Omega(r)$, b) the active
surface density $\sigma(r)$, and c) the velocity dispersion $a(r)$. The actual pattern, which rotates with constant angular speed, can be obtained from the
local dispersion relation as derived by Lau and Bertin (1978) :

$$\frac{Q^2}{4} = \frac{1}{K} - \frac{1 - \nu^2}{K^2 - J^2/(1 - \nu^2)} \tag{1}$$

where Q, J, K and ν are the dimensionless velocity dispersion, surface density, wave number, and frequency, respectively (for exact definitions see
e.g. Bertin et al. (1984)). Several modes can exist with different multiplicity m e.g. one, two or three armed patterns with $m = 1, 2, 3$. The modes,
actual present in a galaxy, are determined by their growth rate (Haass et
al., 1982). By mapping growth rates as functions of surface density J and
velocity dispersion Q, Bertin et al. (1984) determined typical regions in
$J - Q$ plane where normal spirals or barred galaxies would be formed.

3. DISTRIBUTION OF SPIRAL TYPES

A large survey of spiral structure in galaxies was made by Elmegreen and
Elmegreen (1982) who classified spiral arms in 305 galaxies on the Palo-

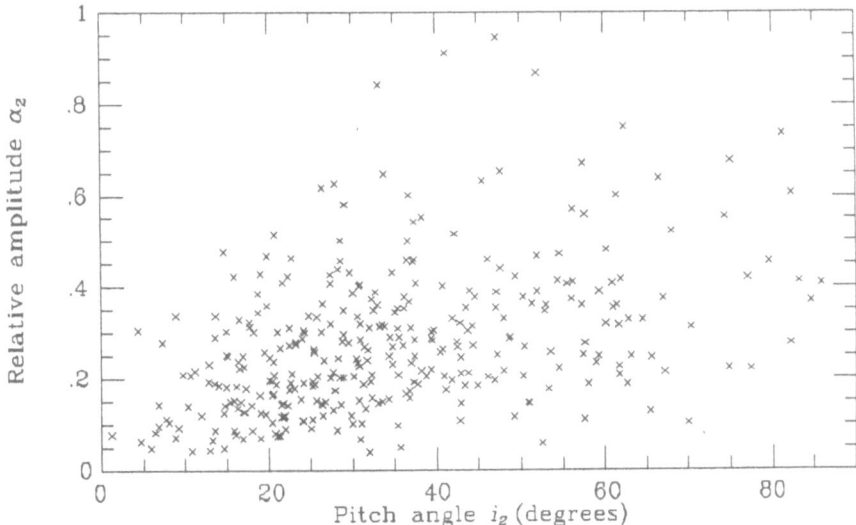

Figure 1. The distribution of relative amplitude α_2 as function of pitch angle i_2 for the second harmonics of the spiral pattern.

mar Observatory Sky Survey (POSS) plates in 12 classes from flocculent to grand design. They found that approx. 70% of the sample showed grand design spiral amrs while the rest were flocculent.

3.1. Environment

By dividing their sample into three groups depending on the environment (*i.e.* field, binary, and cluster galaxies) they detected a significant increase of the fraction of normal galaxies with grand design spirals in binary and cluster environment compared to that for field galaxies. A statistical analysis of the data (Elmegreen and Elmegreen, 1983) showed that either the life time of such patterns was at least 15 rotational periods or the effective cross section for interaction was as large as 45 times the galaxy radius R_{25}. This indicated that tidal forces are likely to induce spiral structures in disk galaxies whereas density waves are responsible for their preservation.

3.2. Distribution of Spiral Parameters

A sample of 605 spiral galaxies (Grosbøl, 1985) scanned on the red POSS plates has analyzed and parameters for spiral patterns in them were extracted. For each galaxy azimuthal intensity profiles were Fourier transformed and pitch angles i_m and relative amplitudes α_m for logarithmic spirals of mode $m=1,2,3$ and 4 were determined. The amplitudes are slightly overestimated due to H_α emission. The distribution of the relative amplitude α_2 as function of their pitch angle i_2 is given in Figure 1 for a subset of 337 galaxies showing grand design structures. The deficiency of galaxies with very small amplitudes or very tight arms is caused by selection effects in the analysis. It can be seen that open spiral generally have larger amplitudes reflecting the $\tan(i)$ term in the perturbing force. Both pitch angle and amplitude show smooth distributions which start to decrease significantly around $i_2 \approx 40°$ and $\alpha_2 \approx 0.4$.

The relative importance of different spiral modes can be estimated using the fraction

$$f_m = \frac{\alpha_m}{\alpha_1 + \alpha_2 + \alpha_3 + \alpha_4} \qquad (2)$$

of a given component over the sum. A mode is defined as dominant if its f_m is larger than 0.5. The fractions of galaxies in the total sample with different spiral modes are given in Table I.

Table I. Fraction of galaxies with different spiral modes.

Spiral mode	$m = 1$	$m = 2$	$m = 3$	$m = 4$	$m = 2+4$
Detected mode : $f_m > 0.0$	47%	56%	55%	51%	34%
Dominant mode : $f_m > 0.5$	16%	25%	11%	11%	27%
Single mode : $f_m = 1.0$	7%	5%	6%	4%	5%

Dominant spirals with two arms are, as expected, more frequent that those with other modes. However, a significant fraction of galaxies has more than one detectable spiral mode. Since the strong two armed spirals often have a fourth harmonic component the per cent of galaxies where the sum of these components is dominant is also given. Of special interest is the three armed patterns because they are difficult to excite through tidal interaction. This indicate that spirals not only are excited by encounters but also through instability and growth of spiral mode as described by Haass et al. (1982).

4.   EVOLUTION

The density wave theory uses the QSSS assumption and can therefore not be applied to the collapse phase of a galaxy. When the galaxy has reached in stable mass distribution and a pattern is excited the following estimated can be given for the evolutionary effects introduced by it.

4.1.   Global changes

The wave amplification by stimulated emission (Mark, 1976) has a time scale of less than a Gyear for typical galaxies. The pattern will therefore soon after its excitation reach the amplitude where non-linear effects limit its growth. The fast growth rate justifies the use of the QSSS assumption. Thus, a significant variation of the relative amplitude of spiral as function of look-back time cannot be expected.

The transfer of angular momentum by the wave has estimated by Bertin (1983) and found to have a time scale longer that the Hubble time for a standard model. Thus, global changes in the galactic disk are not either likely for the majority of disk galaxies.

4.2. Stellar Formation rate

The spiral density wave potential can cause a shock in the interstellar gas (Roberts, 1976) and thereby stimulate the star formation rate. Since the spiral pattern rotates with constant angular speed, the interstellar medium will pass through such shocks more frequent in the inner parts of a galaxy than in its outer parts. This would tend to increase the global star formation in the inner parts of galaxies and thereby produce a metallicity gradient. However, a study by Elmegreen and Elmegreen (1986) has shown that the total star formation rate in a galaxy is only marginally correlated with the presents of a grand design spiral. Thus, the spiral pattern is not the major factor which determines the total star formation rate in a galaxy.

5. CONCLUSIONS

Material spiral arms will not affect the evolution of a galaxies since they are only relatively short lived features formed by other phenomena such as tidal encounters, ejection of matter or propagating star formation. Whereas the events, which create the material arms, have a significant impact on the evolution the arm structures will not change it.

Spiral structures caused by density waves can exists in disk galaxies over long periods of time. They will change the dynamics of the galaxy and may therefore cause global changes in its disk due to angular momentum transfer. The estimate of the time scale for such changes suggests that they for most normal galaxies are insignificant during a Hubble time.

The spiral potential of a density wave may also create a shock in the interstellar gas and thereby change the star formation rate as a function of radius. However, recent studies indicate that less that half the star formation in a typical spiral galaxy is triggered by density wave induced shocks. Although some evolutionary changes may be caused by density waves the evolution of spiral galaxies depends mostly on other effects.

REFERENCES

Bertin,G.: 1983, *Astron. Astrophys.* **127**, 145.
Bertin,G., Lin,C.C. and Lowe,S.A.: 1984, *Proc. of Course and Workshop on Plasma Astrophysics* ESA SP-207, 115.
Elmegreen,D.M. and Elmegreen,B.G.: 1982, *Mon. Not. R. astr. Soc.* **201**, 1021.
Elmegreen,B.G. and Elmegreen,D.M.: 1983, *Astrophys. J.* **267**, 31.
Elmegreen,B.G. and Elmegreen,D.M.: 1986, *Astrophys. J.* **331**, 554.
Gerola,H. and Seiden,P.E.: 1978, *Astrophys. J.* **223**, 129.
Grosbøl,P.: 1985, *Astron. Astrophys. Suppl.* **60**, 261.
Haass,J., Bertin,G. and Lin,C.C.: 1982, *Proc. Natl. Acad. Sci. USA* **79**, 3908.
Lau,Y.Y., and Bertin,G.: 1978, *Astrophys. J.* **226**, 508.
Lau,Y.Y., Lin,C.C. and Mark,J.W.-K.: 1976, *Proc. Natl. Acad. Sci. USA* **73**, 1379.

Lin,C.C. and Shu,F.H.: 1964, *Astrophys. J.* **140**, 646.
Lin,C.C. and Lau,Y.Y.: 1979, *Studies in Appl. Math.* **60**, 97.
Lynden-Bell,D. and Kalnajs,A.J.: 1972, *Mon. Not. R. astr. Soc.* **157**, 1.
Mark,J.W.-K.: 1976, *Astrophys. J.* **205**, 363.
Roberts,W.W.,Jr.: 1969, *Astrophys. J.* **158**, 123.

SOME RECENT RESULTS ON THE DYNAMICAL EVOLUTION OF GALACTIC STRUCTURES

E. Athanassoula
Observatoire de Marseille
2 Place Le Verrier
13248 Marseille Cedex 4
France

ABSTRACT.   In this paper I discuss schematically a few aspects of the dynamical evolution of spiral galaxies.

Galactic structures, like spirals, bars etc., show clear signs of dynamical evolution within a few galactic rotation periods. Furthermore they will generate secular evolution of the axisymmetric background. Thus stationary solutions can only be considered as a first approximation.

I. THE NUMBER OF SPIRAL ARMS

The theory of the swing amplifier (Toomre, 1981) leads to very definite predictions of how various properties of the galaxy influence the spiral arms that can be amplified in it. The amplification rate depends on the values of the following three parameters in the region where the amplification takes place:
1) Shear. This is given by the parameter $\Gamma = -d\ln\Omega / d\ln r$, where $\Omega = V/r$ and V is the circular velocity in the disk. Γ is small in the central parts of galaxies, where rotation curves rise steeply, and becomes of order unity in the parts where rotation curves are flat. Since the amplification rate increases with Γ, it depends on the form of the rotation curve (Athanassoula, 1984, Section 9 and Fig. 26).
2) Q, or the ratio of radial velocity dispersion to the minimum amount necessary for axisymmetric stability (Toomre, 1964). Larger values of Q imply smaller amplification rates.
3) $X = \lambda_y / \lambda_{crit}$, where λ_y is the wavelength in the tangential direction $\lambda_y = 2\pi r/m$, λ_{crit} the critical wavelength $\lambda_{crit} = 4\pi^2 G\Sigma/\kappa^2$, Σ the surface density and κ the epicyclic frequency. Thus X gives us a measure of the disk to total mass in the galaxy and depends also on m, the number of arms (or spiral multiplicity).
    To allow amplification, X must not exceed a critical value X_{crit}, which is an increasing function of Γ, with $X_{crit} \simeq 3$ for $\Gamma = 1$. This allows one to set some lower limits on the ratio of disk to total mass,

R. G. Kron and A. Renzini (eds.), Towards Understanding Galaxies at Large Redshift, 111–116.
© 1988 by Kluwer Academic Publishers.

provided one can see a spiral structure of a given multiplicity
(Athanassoula et al., 1987). Likewise one can predict that in the inner
parts of at least some late type galaxies, with small or inexistant
bulges, the disk to halo mass ratio is high enough to permit $m = 1$
structures. Such asymmetries have indeed been often observed. The most
dramatic examples are the displaced bars (de Vaucouleurs and Freeman,
1972), like NGC 4027 (de Vaucouleurs et al., 1968) NGC 4625 (Bertola,
1967) and NGC 1313 (Marcelin and Athanassoula, 1982), while lesser
asymmetries seem to be more the rule than the exception.

Similar arguments can be made about the time evolution of the
spiral multiplicity in the case of a formation scenario where the disk
material gradually rains in in a dark matter potential well (White and
Rees, 1978; Gunn, 1982; etc..). Several numerical simulations with
different infall laws and different disk and halo mass models (Sellwood
and Carlberg, 1984; Villumsen et al., 1987) have shown that the number
of spiral arms decreases with time. All these simulations had rigid
haloes. Yet this problem is more complicated since, as the disk is
gradually formed, it pulls in the halo considerably (Blumenthal et al.,
1986; Van Albada and Sancisi, 1986; Barnes, 1987; Blumenthal, 1987;
Ryden and Gunn, 1987), thus enhancing locally the halo-to-disk mass
ratio and changing the form of the rotation curve and the shear.

I have made a few simple calculations, assuming that the disk
formation is slow enough for some adiabatic invariants to exist.
Following Blumenthal et al. (1986) I calculated the contraction of the
halo, the modified rotation curve and the values of the other relevant
parameters of the axisymmetric component. I then calculated the
amplification factors for different multiplicities as outlined by Toomre
(1981; see also Athanassoula, 1984 for more details). In some cases I
assumed forcings whose values did not depend on m, while in others not
all multiplicities were forced equally, as is the case in typical
astrophysical situations (e.g. companions). Then at each radius I found
which multiplicity has the highest response amplitude.

One example is given in Fig. 1. A fraction F (of 0.05, 0.1 and 0.2
respectively) of an initially isothermal sphere of core radius $r_c = 0.8$
is assumed to dissipate and form an exponential disk of scale length b =
0.06. No dissipation is assumed to occur beyond $r = 1$. The amplification
factors were calculated using Eqs. (18) and (19) of Toomre (1981), i.e.
the Goldreich-Lynden-Bell shear effect and the Lin-Shu-Kalnajs reduction
factor. Figure 1 shows, as a function of radius, which multiplicity
gives the maximum response. The radius range where a given multiplicity
dominates is not the same as the possible extent of the corresponding
spiral structure, since the latter can be about twice as large as the
former. The multiplicities are obviously integer-valued, but in order to
avoid overlaps I shifted the line corresponding to $F = 0.05$ upwards and
that corresponding to $F = 0.2$ downwards by a few percent. Only values of
m between 2 and 6 have been compared. Note that for low F values the
extent of the $m = 2$ will be small, so that the disk will be dominated by
higher multiplicities. Conversely for large F values bisymmetic spirals
will prevail.

Although these calculations are admittedly simple and should be

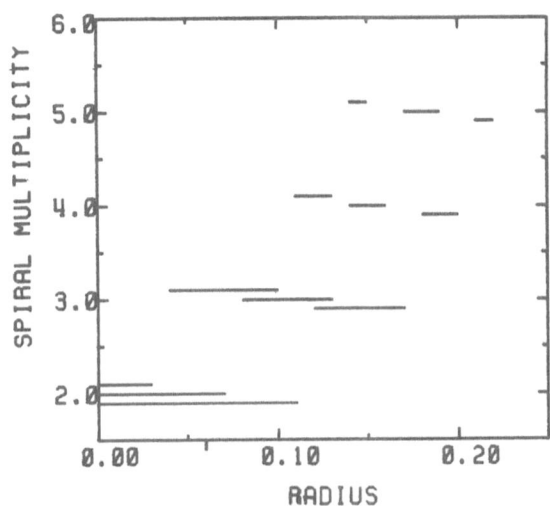

Figure 1. The spiral multiplicity m which has the maximum response, as a function of radius, for F = 0.05 (horizontal lines shifted a bit upwards), F = 0.1, and F = 0.2 (horizontal lines shifted a bit downwards).

followed up by numerical simulations, they clearly indicate that in models with smaller F fractions one can find spirals of larger multiplicities, despite the lesser inward contraction of the halo from the disk material. This suggests that galaxies had higher multiplicities in the past. It might be possible to test this with the space telescope by comparing the number of arms in galaxies of a suitably chosen far away sample with that of a comparable nearby one, observed from the ground.

II. SECULAR EVOLUTION IN BARRED GALAXIES

Lynden-Bell and Kalnajs (1972) showed that spiral waves transport angular momentum outwards and, in doing so, allow the disk to increase its central concentration. Carlberg (1987) has given an estimate of the corresponding rate of change of the surface density of the order of $dlnb/dt = -0.0055$ per rotation period at the half mass radius, where b is the scale length of the exponential disk which best represents the middle range of the disk density. Performing similar measurements on the numerical simulations of Athanassoula and Sellwood (1986) I found a wide range of values, reaching up to several times the above mentioned one, during the evolution of the more unstable disks.

In the case of gas there is one more mechanism that can drive radial flows, namely viscosity. Lacey and Fall (1985) have given for our Galaxy an estimate of the radial velocity $|u_r| \approx 0.1$ km/s at $r \approx 10$ kpc. In the more complicated case of a barred galaxy the radial velocity

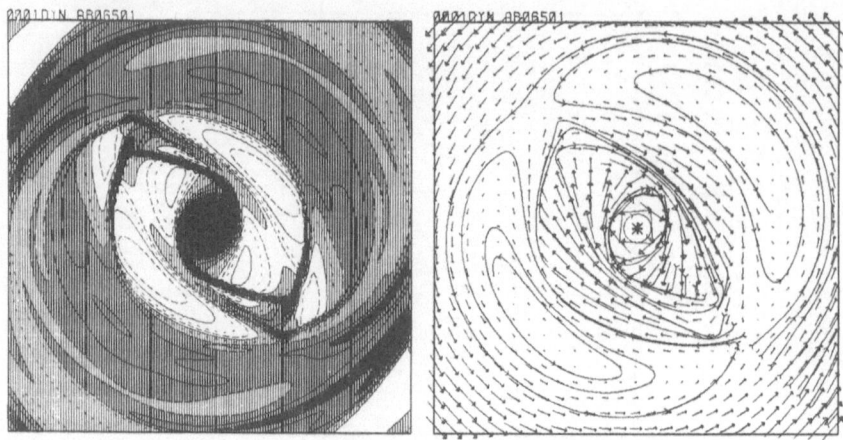

Figure 2. The left panel shows the density response to a typical barred galaxy potential. The bar is at 45°, measured clockwise, from the horizontal, has a semi major axis equal to 5/16 of the size of the outer box and rotates clockwise. Note the shocks at the leading edges of the bar and how their shape resembles that of observed dust lanes. The right panel shows, on the same scale, the flow lines and velocity vectors for the same case. Note the important inflow and the change of orientation of the flow lines near the center.

depends heavily on the properties of the gas flow. I have made a number of simulations with different galaxy potentials, using a gas code kindly proveded by G. D. van Albada. Within the bar region, in cases where there are no shocks, I find a very small inflow of the order of a few percent of the bar length per bar rotation period. On the other hand, in cases where there are strong shocks along the leading edges of the bar, as e.g. in Fig. 2, the gas may move inwards by about half a bar semi-major axis in that time. When radial motion is mainly a monotonic inflow it will contribute to the formation of metalicity gradients, as proposed initially by Tinsley and Larson (1978). However when the radial oscillations superposed on the mean inflow motion are large , they may, on the contrary, tend to smooth out abundance gradients. It is interesting to recall that Alloin et al. (1981) have reported small abundance gradients in the barred galaxy NGC 1365.

Thus spirals and bars cause radial flows which enhance the central concentration of the galaxy. The disk, and therefore the total, rotation curve becomes steeper in the central part, producing also a steeper Ω - κ / 2 curve. This can contribute to the formation of an inner Lindblad resonance (ILR) which will be fatal to linear modes, but not to recurrent transient spirals, nor, necessarily, to nonlinear modes (Tagger et al., 1987 and in preparation).

In order for the locii of shocks along the leading edges of the bars to be of the shape and type of the observed dust lanes, the galaxy must have an ILR, in the sense that the family of orbits oriented perpendicular to the bar must be present (Athanassoula, in preparation). This is true both for strong bars and for ovals. Assuming that the bar grew out of some initially low-amplitude mode(s), for which the linear theory is applicable, it could not initially have had an ILR. Thus the pattern should have acquired an ILR during its evolution. Indeed, as I have argued above, material will be pushed inwards, leading to a steeper $\Omega - \kappa / 2$ curve. At the same time dynamical friction slows down the bar (Sellwood, 1981; Weinberg, 1985; etc..) The combined effect of the two will be most favourable to the formation of an ILR. A more quantitative discussion of the results in this section will be given elsewhere (Athanassoula, in preparation).

REFERENCES

Alloin, D., Edmunds, M.G., Lindblad, P.O. and Pagel, B.E.: 1981, Astron. Astrophys., 101, 377
Athanassoula, E.: 1984, Physics Reports 114, 319
Athanassoula, E., Bosma, A. and Papaioanou, S.: 1987, Astron. Astrophys., 179, 23
Athanassoula, E. and Sellwood, J.A.: 1986, M.N.R.A.S., 221, 213
Barnes, J.: 1987, in "Nearly Normal Galaxies: From Planck Time to the Present", ed. S.M. Faber, Springer, p. 154
Bertola, F.: 1967, Cont. Obs. Astrofis. Univ. Padova No 197
Blumenthal, G.R.: 1987, in "Nearly Normal Galaxies: From Planck Time to the Present", ed. S.M. Faber, Springer, p. 401
Blumenthal, G.R., Faber, S.M., Flores, R. and Primack, J.R.: 1986, Astrophys. J., 301, 27
Carlberg, R.G.: 1987, in "Nearly Normal Galaxies: From Planck Time to the Present", ed. S.M. Faber, Springer, p. 129
de Vaucouleurs, G., de Vaucouleurs, A. and Freeman, K.C.: 1968, M.N.R.A.S., 139, 425
de Vaucouleurs, G. and Freeman, K.C.: 1972, Vistas in Astronomy, 14, 163
Gunn, J.E.: 1982, in "Astrophysical Cosmology", eds. H.A. Bruck, G.V. Coyne and M.S. Longair, Pontificia Academia Scientarium, Vatican, p. 233
Lacey, C.G. and Fall, S.M.: 1985, Ap. J., 290, 154
Lynden-Bell, D. and Kalnajs, A.J.: 1972, M.N.R.A.S., 157, 1
Marcelin, M. and Athanassoula, E.: 1982, Astron. Astrophys., 105, 76
Ryden, B. and Gunn, J.E.: 1987, Astrophys. J., 318, 15.
Sellwood, J.A.: 1981, Astron. Astrophys., 99, 362
Sellwood, J.A. and Carlberg, R.: 1984, Astrophys. J., 282, 61
Tagger, M., Sygnet, J.F., Athanassoula, E. and Pellat, R.: 1987, Ap. J. Let. 318, L43
Tinsley, B.M. and Larson, R.B.: 1978, Ap. J., 221, 554
Toomre, A.: 1964, Astrophys. J., 139, 1217
Toomre, A.: 1981, in "The Structure and Evolution of Normal Galaxies", eds. S.M. Fall and D. Lynden-Bell, Cambridge Univ. Press, p. 111
Van Albada, T.S. and Sancisi, R.: 1986, Phil. Trans. R. Soc. Lond., 320, 15

116

Villumsen, J.V., Gunn, J.E. and Casertano, S.: 1987, Astron. J. (sub.)
Weinberg, M.D.: 1985, M.N.R.A.S., 213, 451
White, S. and Rees, M.: 1978, M.N.R.A.S., 183, 341

STARBURST GALAXIES

Jorge Melnick
European Southern Observatory, Garching b. München, FRG

1. INTRODUCTION

1.1 Starbursts and starburst galaxies

There so is much confusion in the literature concerning the definition of *starbursts* (e.g. Weedman, 1987) that, at the risk of confusing things even more, I will introduce here a new definition. I will call *starbursts* objects, or parts thereof, that have extremely large *local* star formation rates (i.e. the star formation rate *per unit area*). The *global* SFRs of *starbursts* are on the average similar to those of "normal" star formation regions (i.e. 1-100 $M_\odot$ yr^{-1}) but the *local* SFRs of *starbursts* are 10^6 to 10^8 times larger than those of typical spiral galaxies (of course there are extreme objects, ARP 220 for example, where the *global* SFRs are very large but, even among *starbursts*, objects like this are rare). The term *violent star formation* is often used in the literature to characterize "bursting" processes. Thus, *starbursts* are also referred to as *Violent Star Formation Regions (VSFRs)*.

The definition of *starburst galaxies* is also confused and here I will call *starburst galaxies* to those where the (present) global SFR is dominated by VSFRs.

1.2 HII galaxies

In this contribution I would like to review the properties of a class of *starburst galaxies* characterized by having low masses, low metal abundances (corresponding to the low masses) and overall observable properties dominated by very young VSFRs. In fact, the integrated spectra of these objects, called *HII galaxies*, are essentially identical to those of high excitation giant HII regions (i.e. strong narrow emission lines and very weak blue continua). Thus HII galaxies are probably ideal targets to look for young galaxies and to study the collective properties of objects with very large young stellar populations. But, because of their large SFRs and in spite of their low masses, HII galaxies can be very luminous and therefore their properties may also be of interest to people working on galaxy counts out to large redshifts. During this workshop there was much discussion about the so-called "E+A" galaxies (see papers by Gunn and by Ku in this volume). This has led me to include a brief discussion of the properties of metal-rich *starbursts*. Although metal rich objects may not be directly relevant to understanding young galaxies,

R. G. Kron and A. Renzini (eds.), Towards Understanding Galaxies at Large Redshift, 117–126.
© *1988 by Kluwer Academic Publishers.*

118

outside galactic nuclei, "E+A" galaxies may be the only places where very massive super metal-rich *starbursts* might exist. Thus, the properties of metal rich VSFRs may be important to understand the evolution of large galaxies.

2. THE EVOLUTION OF MASSIVE STARS: WARMERS

The mass loss rate, $\dot{M}$, is a dominant parameter in the evolution of very massive stars. This is schematically illustrated in figure 1 where the evolution of a metal poor star ($\dot{M}=0$) is compared to that of a metal-rich star with strong mass loss. If $\dot{M}$ is sufficiently large, massive stars will lose their envelopes shortly after leaving the H-ZAMS and will spend most of their He-burning lives near the He-ZAMS where they will reach effective temperatures larger than 10^5K. These luminous hot stars have a dramatic influence on the ionisation of gas surrounding young VSFRs and this effect led Roberto Terlevich and myself (1985) to introduce the term WARMERS to characterize massive stars near the He-ZAMS.

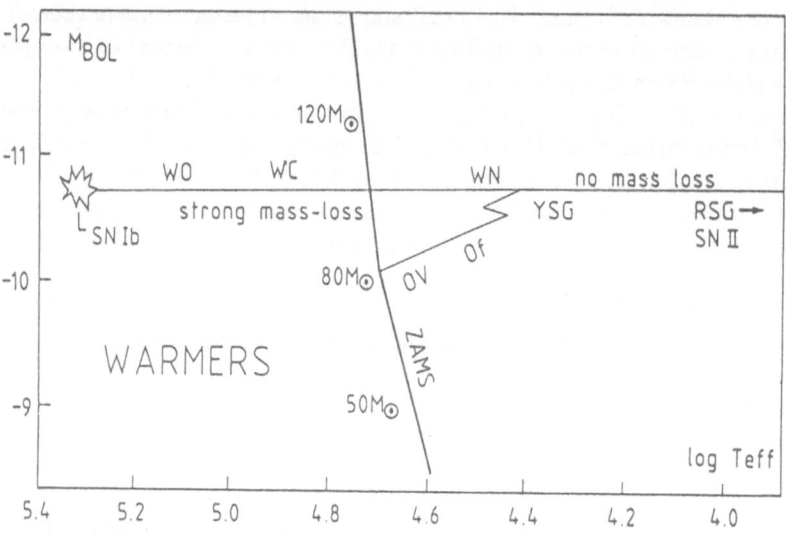

Fig. 1. Schematic representation of the evolution of a $60M_\odot$ star with and without mass loss. Adapted from Maeder (1981).

Since, besides mass and luminosity, $\dot{M}$ depends on chemical composition, the existence of a WARMER phase depends rather critically on metallicity. Thus, chemical composition plays a fundamental rôle in determining the *observed* properties of very young VSFRs. Empirically it is found that the WARMER phase has a substantial impact on the observable properties of VSFRs only for abundances larger than solar (Terlevich, Melnick and Moles, 1987). For lower abundances, the WARMER phase is either absent (massive stars explode before becoming WARMERS), or is too short (most of the He-burning time is spent in the RSG phase) to have a significant influence on the observed spectrum. Therefore I will discuss separately the properties of *starbursts* of abundances lower and larger than solar.

3. METAL POOR STARBURSTS: statistical properties of HII galaxies

3.1 Global parameters

We have compiled a catalogue containing spectrophotometric observations of HII galaxies discovered in objective prism surveys mostly of southern skies. The catalogue, called the *Spectrophotometric Catalogue of HII Galaxies* (Terlevich et al., 1987, hereafter the SCHG) contains positions, redshifts, line intensities and equivalent widths for over 400 HII galaxies out of which close to 100 have been observed with sufficient accuracy to detect the [OIII]λ4363 line and therefore to directly measure electronic temperatures and thus elemental abundances. In addition, for $\sim$ 70 of these galaxies we have obtained Echelle spectra that have allowed us to study their internal dynamics. The analysis of the abundances, including a new determination of the primordial He abundance, and of the dynamics and the correlations between global parameters have been published elsewhere (Melnick et al., 1987a,b; Campbell et al., 1986; Pagel et al., 1986) and I refer to these papers for further details. Here I will present the distributions of absolute luminosities and abundances that are most relevant to the subject of this workshop.

Figure 2 presents the distribution of oxygen abundances among HII galaxies. There is a strong selection bias in this plot because it includes only galaxies with "measurable" electronic temperatures and this restricts the sample to abundances lower that $12 + log(O/H) \sim 8.5$. Thus, it is not clear if there is a high abundance cutoff but there seems to be a low abundance cutoff. However, out of $\sim$ 400 objects, we have found none with oxygen abundances lower than about 0.03 times solar while the most common oxygen abundance is close to one tenth of solar. Figure 3 presents the distribution of absolute blue magnitudes of the sample calculated as,

$$M_B = -2.5log[L(H_\beta)/W(H_\beta)] + 79.4$$

where L(H_β) is the H_β luminosity corrected for extinction and W(H_β) the equivalent width. HII galaxies cover the absolute magnitude range from -10 to -24 with a strong peak close to $M_B = -18$. The large blue luminosities are mostly a reflection of the large SFRs of HII galaxies and do not necessarily imply that these are *giant* irregular galaxies. In fact, the evolutionary corrections for these objects are 3-4 magnitudes (Terlevich and Melnick, 1981). However, with their present luminosities, HII galaxies are likely to be included in large numbers in deep photometric surveys and therefore I would like to suggest that an important fraction of the faint blue galaxies detected in these surveys may, in fact, be low to moderate redshift HII galaxies.

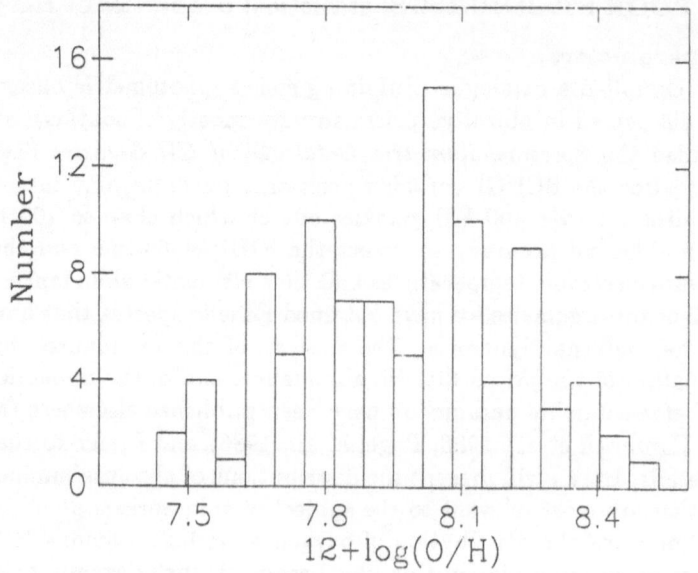

Fig. 2. Distribution of oxygen abundances among HII galaxies.

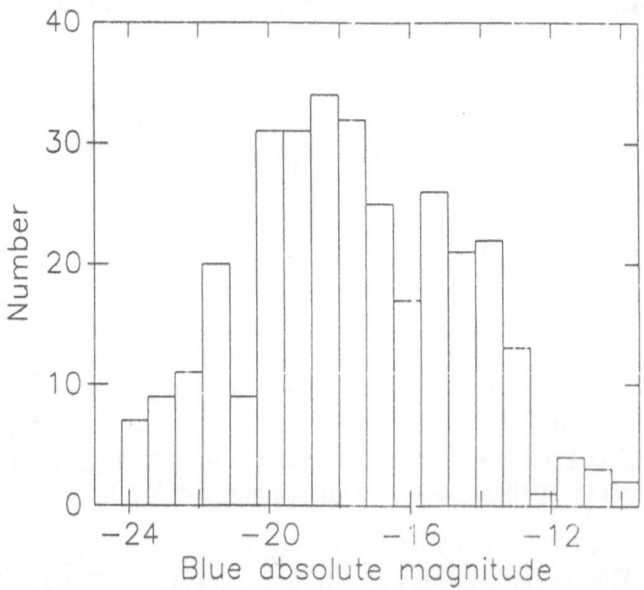

Fig. 3. Blue luminosity function of HII galaxies.

3.2 Optical morphology of HII galaxies

On the basis of a deep CCD survey of a small subsample of galaxies from the SCHG we find that about 20% of the objects appear to be parts of interacting systems while 30% show multiple morphology. Close to 50% of the HII galaxies surveyed are starlike and show no traces of companions or other components (Melnick, 1987a). I think that these compact HII galaxies are probably the best candidates to look for *young* galaxies.

Surprisingly, however, the abundances of these objects tend to be close to the mean of HII galaxies. Figure 4 shows the blue part of the spectra of 4 representative objects of the compact class. The abundances of these 4 galaxies range from $12+\log(O/H)=7.76$ for 1543+091 to 8.00 for 1053+064. The equivalent widths of the [OIII] lines reach values close to 2000 Å and, in fact, the optical continua of these objects show no traces that could reveal the presence of an underlying population of old stars. Even in the near IR there is no evidence for the presence of an old stellar component in these objects (Melnick, Moles and Terlevich 1985 and references therein). Their relatively high abundances, however, are indicative of substantial evolution. Since, given their low masses, a significant fraction of the metal enriched material may have been expelled in the form of galactic winds, it is quite possible that these compact HII galaxies have undergone many cycles of star formation in the past. However, there is no indication in the data that star formation has been going on in these objects for more than, say, 10^8 yrs, and therefore, if "young" galaxies are galaxies of ages much lower than the Hubble time, then I would say that star-like HII galaxies are indeed very young systems.

Interactions are often quoted to be the mechanism that triggers violent star formation activity. Our observations of HII galaxies show, however, that there are many cases, perhaps as much as 50% of the total, where massive violent star formation activity does not seem to be related to interactions. But IIZw40, one of the prototypes of the HII galaxy class, seems to have formed as a consequence of the interaction of two disk-like neutral gas clouds (Brinks and Klein, 1987). Thus, although the optical morphology of IIZw40 clearly shows indication of interaction (Baldwin, Spinrad and Terlevich, 1982), it seems quite possible that some, perhaps most, *compact* HII galaxies are formed by collisions of neutral gas clouds. It is therefore very important to map these objects in 21cm to gain further insight into the physics of the *starburst* phenomenon.

3.3 The clustering of HII galaxies

HII galaxies are gas-rich, low mass, irregular galaxies and their clustering properties are very interesting for a number of cosmological applications. In particular, independently of dark matter, HII galaxies should be better tracers of the overall mass distribution in the Universe than large galaxies because, *a fortiori*, bigger galaxies must form in bigger density fluctuations.

122

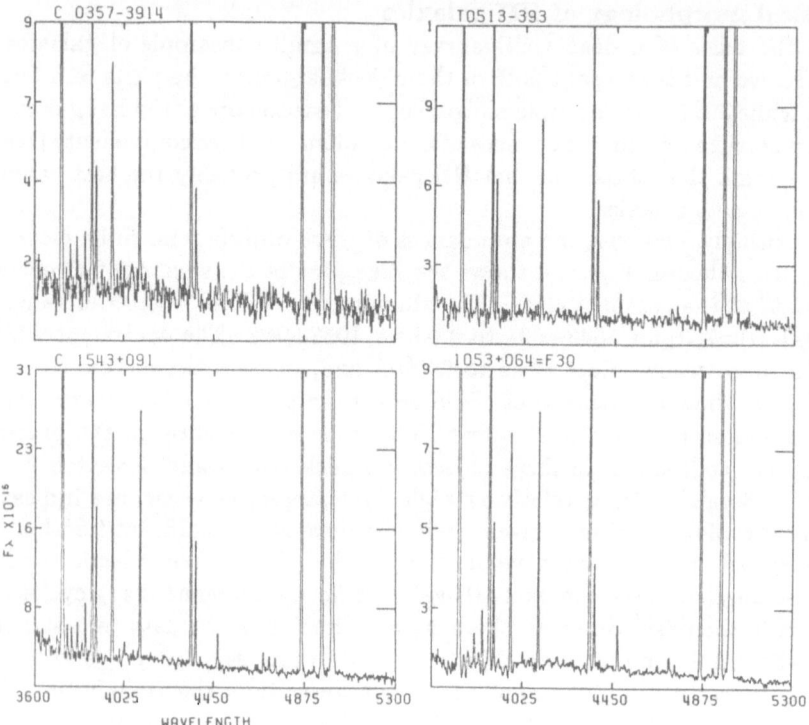

Fig. 4. Blue portion of Las Campanas *Reticon Scanner* spectra of 4 representative star-like HII galaxies.

Since HII galaxies are gas-rich, one does not expect to find them near the cores
of rich clusters and therefore, *a priori*, HII galaxies are expected to be less clustered
than normal galaxies and in particular less clustered than elliptical galaxies. In
collaboration with Peter Shaver and Angela Iovino at ESO we are studying the
clustering properties of HII galaxies. This has turned out to be a tricky problem
because the objects in the SCHG concentrate in relatively narrow cones in the sky.
Figure 5 presents an attempt to study the clustering of HII galaxies based on the
method described by Shaver (1984). In order to mimimize the "cone" effect the
redshifts have been limited to the range $0.01 < z < 0.1$, which reduces the sample
to 258 objects. HII galaxies are compared in the figure to the AAT galaxies from
the survey of Shanks et al. (1983) that covers approximately the same redshift
range.

The comparison indicates that HII galaxies are less clustered than normal
galaxies. A more detailed analysis done with Shaver and Iovino using different
techniques to define a comparison sample shows that HII galaxies are roughly two
times less clustered than AAT galaxies. The results are still preliminary and we
have not yet analysed the possible cosmological implications. The lower clustering

of HII galaxies, however, points to a possible method to discriminate them from normal galaxies in deep surveys.

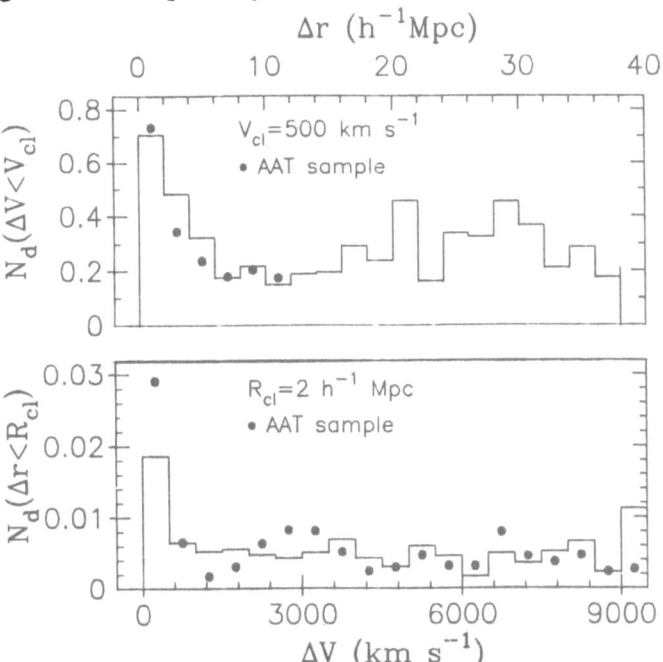

Fig. 5. Comparison between the clustering of HII galaxies with the AAT sample of normal galaxies of Shanks et al. (1983).

3.4 The IMF of violent star formation regions

VSFRs are probably the only places where the upper mass range of the initial mass function (IMF) can be sampled independently of stochastic fluctuations (Melnick, 1987b).

On the basis of a comparison of the observed global properties of HII galaxies with *starburst* models, Terlevich and I (1983) have found that the IMF of VSFRs changes systematically with chemical composition. The key obervations supporting this result are reproduced in Figure 6 where the electronic temperature of the ionised gas (T_e) is plotted against oxygen abundance O/H. The correlation between T_e and O/H is steeper than what is expected from the dependence of cooling on O/H. This implies that the "hardness" of the ionising spectrum must be inversely correlated with abundance. A systematic increase of the upper end of the IMF, from $M_U \sim 20 M_\odot$ for $O/H \sim 0.3(O/H)_\odot$ to $M_U \sim 120 M_\odot$ for $O/H \sim 0.03(O/H)_\odot$, is required to explain the observations. This result is to a large extent independent

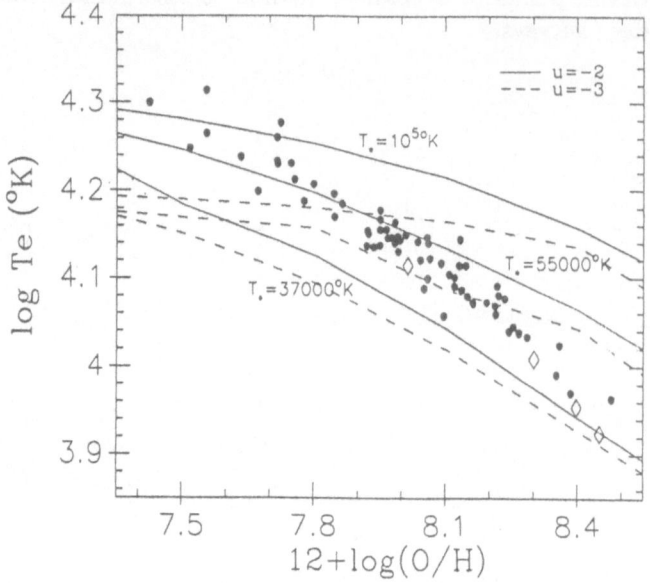

Fig. 6. Plot of electronic temperature of HII galaxies as a function of oxygen abundance. The lines represent models for three different stellar temperatures (T_*) and two different ionization parameters (u).

of the atmospheric models used because the emergent continuum of massive
stars in the 1-3 Ryd range is not too sensitive to wind effects (Melnick and Terlevich,
1987). Televich (1985) has analysed the systematics of the H_β equivalent widths of
giant HII regions and HII galaxies and has concluded that a systematic change of the
slope of the IMF is required to simultaneously explain the correlations of $W(H_\beta)$
and T_e with oxygen abundance. I think that the evidence for the variability of the
IMF is very *strong* and consequently that galaxy evolution models invoking a fixed
IMF are not realistic.

3.5 The $L\alpha$ problem

Only the most metal poor HII galaxies show L_α emission lines in their ultraviolet spectra and this is generally interpreted as evidence for the presence of dust
mixed with the nebular gas. (Meier and Terlevich 1981; Kunth and Weedman, 1987
and references therein). Only a small amount of dust suffices to destroy L_α and
indeed the Balmer decrements indicate 1-2 mag of internal visual absorption in
these systems. Thus, the detection of L_α in high redshift galaxies (Djorgovsky, this
volume) may be an indication of very low, maybe primordial, abundances. However, L_α is readily detected in the ultraviolet spectra of Seyfert nuclei which are
very metal rich. Thus, in principle, the presence of L_α in emission line regions is
only an indication of low dust content but not *necessarily* of low abundances.

4. METAL RICH STARBURSTS: narrow line Seyfert galaxies

As mentioned in the introduction, galactic nuclei and "E+A" galaxies are possibly the only places where super metal rich *starbursts* are observed. Terlevich and I (1985) have shown that Seyfert 2 nuclei and LINERS are stages in the evolution of nuclear VSFRs that occur when the most massive stars reach the WARMER phase.

This is illustrated in figure 7 that reproduces a typical diagnostic diagram where evolutionary tracks for *starburst* photoionised nebulae have been superposed. Models for two different total *starburst* masses different by a factor of 10 are presented. Low mass nuclear *starbursts* evolve from HII regions to LINERS while massive ones go through a Seyfert 2 phase before becoming LINERS. We have qualitatively followed the evolution of nuclear *starbursts* into the supernova phase which we identify with the transition to, and further development of, the Seyfert 1 class (Terlevich and Melnick, 1985; Terlevich, Melnick and Moles, 1987).

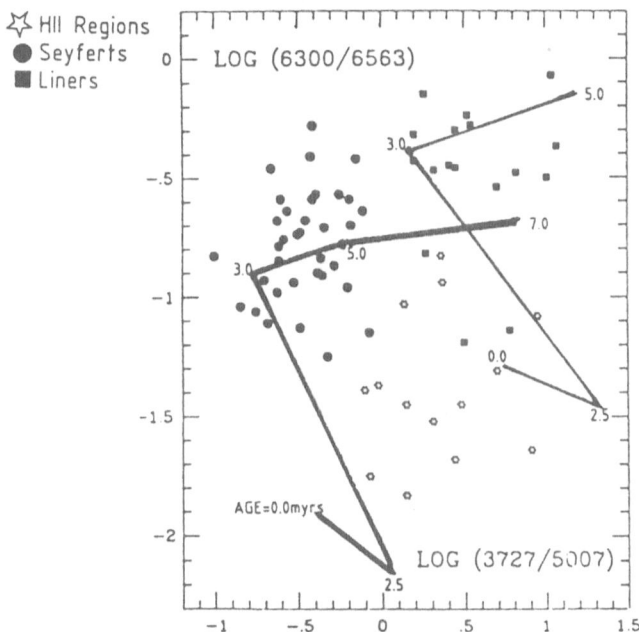

Fig. 7. The evolution of nuclear *starbursts* of two different masses in a typical BPT (Baldwin, Phillips and Terlevich, 1981) diagnostic diagram.

If the "E+A" phenomenon is the consequence of violent star formation covering a significant fraction of the total mass of large (metal rich) galaxies (Gunn, this volume) then the progenitors of "E+A" galaxies must have spectral characteristics very similar to those of luminous Seyfert galaxies. I would like to suggest, therefore, that an important fraction of the cluster population of active galaxies and QSOs may, in fact, be associated with "young" E+A galaxies.

REFERENCES

Baldwin, J.A., Phillips, M.M., Terlevich, R., 1981, *Pub. A.S.P.*, **93**, 5 (BPT).

Baldwin, J.A., Spinrad, H., Terlevich, R., 1982, *M.N.R.A.S.*, **198**, 535.

Brinks, E., Klein, U., 1987, *preprint*.

Campbell, A., Terlevich, R., Melnick, J., 1986, *M.N.R.A.S.*, **223**, 811.

Kunth, D., Weedman, D.W., 1987, 1987 in, *Exploring the Universe with the IUE Satellite*, eds., Y. Kondo et al., Reidel, p. 350.

Maeder, A., 1981, *Astr. Ap.*, **102**, 401.

Meier, D., Terlevich, R., 1981, *Ap. J.*, **246**, L109.

Melnick, J., 1987a in, **Xth Rècontres de Moriond,** *Workshop on Starbursts and galaxy evolution*, eds. Th. Montmerle, T.X. Thuan and and J. Tran Thanh Van, Frontières, in press.

Melnick, J., 1987b in, ESO workshop on *Stellar evolution and dynamics of the outer halo of the Galaxy*, eds., M. Azzopardi and F. Matteucci, in press.

Melnick, J., Moles, M., Terlevich, R., 1985, *Rev. Mex. Atron. Astrophys.* **11**, 91.

Melnick, J., Terlevich, R., 1987 in, proceedings of the *Tenth European Regional Meeting IAU/EPS*, Prague, August 1987, in press.

Melnick, J. *et al.*, 1987a, *M.N.R.A.S.*, **226**, 849.

Melnick, J., Terlevich, R., Moles, M., 1987b, *M.N.R.A.S.*, **submitted,**

Pagel, B.E.J., Terlevich, R., Melnick, J., 1986, *Pub. A.S.P.*, **98**, 1005.

Shanks, T., et al., 1983, *M.N.R.A.S.*, **274**, 529.

Shaver, P.A., 1984, *Astr. Ap.*, **136**, L10.

Terlevich, R., 1985 in, *Star Forming Dwarf Galaxies and related objects*, eds. D. Kunth, T.X.Thuan and J. Tran Thanh Van, Frontières, p. 395.

Terlevich, R., Melnick, J., 1981, *M.N.R.A.S.*, **195**, 839.

Terlevich, R., Melnick, J., 1985, *M.N.R.A.S.*, **213**, 841.

Terlevich, R., Melnick, J., 1983, ESO preprint **264**.

Terlevich, R., Melnick, J., Moles, M., 1987 in, IAU symposium **121**, *Observational Evidence of Activity in Galaxies*, eds. E.Ye. Khachikian, K.J. Fricke and J. Melnick, p. 499.

Terlevich, R., Melnick, J, Masegosa, J., Moles, M., 1987, *Spectrophotometric Catalogue of HII Galaxies*, in preparation.

Weedman, D.W., 1987 in, *Star formation in galaxies* , eds., C.J. Londsdale Persson, NASA CP 2466, p. 459.

RING GALAXIES AS PROBES OF STARBURST EVOLUTION

S. R. Majewski [1,2]
Yerkes Observatory
P.O. Box 258
Williams Bay, Wisconsin 53191
U.S.A.

ABSTRACT. A method is described whereby the unique properties of ring galaxies are utilized to calibrate the evolutionary properties of starbursts. The method is demonstrated using data from a newly discovered, ultraviolet-bright ring galaxy at a redshift of 0.24.

1. INTRODUCTION

The study of the evolution of the spectral energy distribution of galaxies identified as experiencing a "starburst" is important for understanding galaxies at high redshift. It is of great interest to know how the appearance of a galaxy changes with time and, in particular, to know what young galaxies look like. Features of galactic spectral energy distributions, e.g. broad-band colors, need to be calibrated with age so that the internal and cosmological timescales can be compared with the purpose of deriving cosmological parameters. Sensitivity to cosmological problems is increased with redshift, and therefore it is essential to understand galaxies in the very earliest stages of evolution.

Our task then is to determine what galaxies looked like during the first phases of large-scale star formation. The direct approach is to find galaxies undergoing the initial phases of star formation, whether they be "primeval" galaxies at high redshift or relatively low redshift galaxies that are latecomers in the sense that they have only recently begun grand-scale star formation. In either case the number of candidates at present is small, and therefore the construction of a picture of the evolution of young systems from this sample must necessarily be spotty.

---

[1] Visiting Astronomer, Kitt Peak National Observatory, National Optical Astronomy Observatories, operated by the Association of Universities for Research in Astronomy, Inc., under contract with the National Science Foundation.
[2] Visiting Astronomer at the Infrared Telescope Facility, which is operated by the University of Hawaii under contract from the National Aeronautics and Space Administration.

R. G. Kron and A. Renzini (eds.), Towards Understanding Galaxies at Large Redshift, 127–132.
© 1988 by Kluwer Academic Publishers.

A second approach is to look for analogues to young evolving galaxies in other systems possessing young stellar populations. Some examples are studies of the Magellanic Cloud (MC) clusters in the optical and the near infrared by Searle, Wilkinson, and Bagnuolo (1980) and Persson *et al.* (1983, hereafter PACFM), and work on blue compact dwarf galaxies by Thuan (1983; 1985). Comparisons between, for instance, the color evolution of these systems and a grand picture for galaxies may be valid, but one must be sure that the analogue does truly represent an evolutionary sequence. Another difficulty in relating observables to an age sequence in this way is the scatter inherent in a collection of objects that necessarily span a range of metallicities and reddenings. The initial mass function (IMF) and star formation rate (SFR) may also vary within the sample.

2. RING GALAXIES

The ideal galactic evolution experiment would be to watch one galaxy at a time evolve completely from initial starburst to some final, stable state, and describe for each set of initial conditions the time evolution of observables. (In this discussion I will be mainly concerned with broad-band colors, but certainly many other features of interest could be monitored.) The vector space of observables could then be applied to large data sets to infer cosmological information. Given the impossibility of the ideal experiment we must look for a viable substitute.

Ring galaxies are unique laboratories for the study of starburst color evolution for many reasons. The starburst activity in ring galaxies is generally very strong. In their study of A0035, Fosbury and Hawarden (1977) reported the existence of very luminous HII regions and the presence of $\sim 3 \times 10^6$ OB stars. Appleton and Struck-Marcell (1987, hereafter AST) noted the high far infrared luminosities of many of the ring galaxies (some are among the brightest galaxies known), indicating vigorous star formation, most likely concentrated in the ring itself. These galaxies typically have high surface brightness because of their strong starburst activity and are therefore easy to observe.

In addition, unlike most active galaxies, the cause of the starburst in ring galaxies is known. Ring galaxies are formed by the penetration of a compact galaxy at nearly normal incidence through the disk of a larger late-type galaxy, resulting in an outwardly propagating density wave and associated bursts of star formation (Lynds and Toomre 1976; Theys and Spiegal 1976). The starburst activity is very extended, being coherent over typical diameters of ~ 10-20 $(100/H_0)$ kpc (AST). Ring galaxies are attractive because of the simplicity and openness of their structure. The single outwardly propagating density wave allows far easier interpretation of the resultant stellar population than the jumble of populations produced by rotating spiral arms or the heavily obscured and complicated details found in nuclear starbursts. Also, unless the ring has propagated past the HI radius of the galaxy, one can be assured that star formation is in fact still occurring near the front of the density wave.

Because of the expanding density wave, each ring galaxy portrays a time lapse sequence of a starburst as a function of radius. This is perhaps the most useful feature of ring galaxies to the understanding of starbursts. A study of the characteristics of ring galaxies from the outer shock front at the ring perimeter inwards is, in effect, a study of the evolution of large-scale, density-wave-induced star formation from initial burst to an age equal to the time since collision. This age can be deduced by comparing the distance of the penetrator to the disk with their known relative velocities; in the cases studied so far, the age ranges cover some of the most interesting phases of starburst evolution, up to ages of around several 10^9 years (Fosbury and Hawarden 1977; Few, Madore, and Arp 1982; Joy and Harvey 1987).

Since each ring galaxy has its own particular metallicity and may have different IMFs and SFRs, we should try to calibrate the radial dependence of observables for as many ring galaxies as possible in order to understand the influence of these properties. However, we may need to proceed with some caution, as the metallicity, IMF, and SFR may also vary within each ring galaxy. In the best situation of a well-mixed precursor galaxy, these functions will be constant; at worst they should be monotonic with radius. A promising note is that AST found no correlation of far infrared flux to ring diameter, indicating perhaps that the SFR is dependent on the precursor rather than on the age of the ring.

3. AGES FROM OPTICAL-NEAR INFRARED COLORS

It has become evident that optical colors alone are inadequate for discriminating age effects in stellar populations. In particular, models of evolving stellar populations yield UBV color-color diagrams which give little hope of separating age effects from those of the star formation history since the differences are small and often comparable to observational errors (Searle, Sargent, and Bagnuolo 1973; Huchra 1977; Larson and Tinsley 1978, hereafter LT). As was pointed out by LT, there are the added complications of disentangling metallicity and reddening effects, since in both cases the correction vectors lay parallel to the evolutionary tracks.

Models by Struck-Marcell and Tinsley (1978; hereafter SMT) demonstrate that the combination of optical and near infrared colors give a much better age discriminator than optical colors alone. A comparison of the (U-B, B-V) models shown in LT with the (U-B, V-K) models in SMT shows that the V-K baseline is about three times longer than B-V, allowing a much better separation of the tracks representing the evolution of models with different star formation histories, and a better separation of ages along each track. The traditional view has been that the optical bands are sensitive to young stellar populations while the near infrared bands are sensitive to a cooler, evolved population of K and M giants (cf. Aaronson 1978; Balzano 1983; Thuan 1983; Telesco, Decher and Gatley 1985; Hunter and Gallagher 1985). The combination of optical and near infrared colors is then a measure of the relative importance of these two populations and is therefore sensitive to the age of the last burst of star formation.

However, recent observational data as well as new theoretical population synthesis models suggest that perhaps the traditional view may be too simplistic. For instance, work by Campbell and Terlevich (1984) on "violent star formation regions", and by PACFM on young MC clusters, shows quite dramatically the influence that red supergiants can play in very young starbursts. These groups find that red supergiants can contribute a very large fraction of the K flux (in some cases more than 90%) and more than half of the bolometric luminosity. Hence the V-K colors for the young MC clusters were found to be redder than the theoretical predictions of SMT by up to 2 magnitudes. (SMT warned that the lack of red supergiants in their models might result in colors that were too blue.) Another example is asymptotic giant branch stars, particularly carbon stars, which Renzini and Buzzoni (1985) predict should make a significant contribution to intermediate-aged stellar populations. This is confirmed in the intermediate-aged MC clusters where there exists a discontinuity in V-K due to the presence of carbon stars which dominate the flux in the near infrared (PACFM). Rather than limit the usefulness of combined optical-infrared colors to the determination of galactic ages, these results show how this data can be even more sensitive to age effects.

4. AN EXAMPLE

As an example, I present data on a previously unpublished ring galaxy candidate in SA57. A more complete study of this object will be presented elsewhere (Majewski *et al.* 1988). Discovered as radio source 52W036 by Windhorst, van Heerde, and Katgert (1984), the optical counterpart has been known from Mayall 4m plates for some time (Koo and Kron, personal communication). A spectrum by Koo, Kron and J. Munn shows narrow OII, OIII, and Hβ emission indicative of a hot, young stellar population, and gives a redshift of z = 0.24, which would make it the most distant ring galaxy known. This redshift puts the radio power at 3.0×10^{23} W Hz^{-1} at 1.4 GHz (all calculations in this section will assume $q_0 = 0.1$, $H_0 = 50$ km s^{-1} Mpc^{-1}). Heckman (1983), in a survey of radio emission in merging galaxies defines radio loud as > 10^{22} W Hz^{-1} at 1.4 GHz; the radio power of 52W036 is equivalent to the brightest object in his sample.

Deep optical UBVRI images were obtained with Koo and Kron on the Mayall 4-m + RCA prime focus CCD in June, 1986. A near infrared K band image was obtained with M. Hereld on the NASA IRTF 3-m with the new University of Chicago 2 micron imaging camera in March, 1987. These images (Figure 1) show a dramatic change in morphology as a function of wavelength. The object is clearly horseshoe-shaped in the blue bands, while in the red and near infrared the object is a double peaked source with the brighter peak contained inside the blue horseshoe. The western half of the galaxy clearly shows a shifting outward in the peak luminosity as the wavelength decreases, probably due to an outwardly propagating density wave of star formation. The same may be occurring in the eastern half of the galaxy (making it truly a *ring* candidate), however the structure on this side is complicated by what may be the intruder galaxy (the second red peak) seen in projection. The visible extent of the well-defined ultraviolet ring corresponds to ~75 kpc in diameter, however, there seems to be visible K band structure over ~120 kpc, perhaps the emission from the precursor.

Preliminary analysis of the integrated broad-band photometry gives V=17.67 ± .03, (U-B) = -0.88 ± .04, (B-V) = 0.91 ± .04, (V-K) = 3.11 ± 0.1; using K-corrections (without evolution) for Sm/Im galaxies provided by Koo (personal communication), these become in the rest frame (U-B) = -0.79, (B-V) = 0.44, (V-K) = 3.24. No corrections for reddening have been made; they will primarily make the optical colors bluer. Obviously, the UV flux of 52W036 is very high, giving M_U = -23.7, while the visual magnitude of M_V = -23.4 makes it comparable in brightness to first-ranked cluster galaxies.

In Figure 2a, the pixel-to-pixel colors of the entire galaxy are plotted in a (U-B, V-K) diagram (compare to SMT figure 2). The registration of the CCD images is estimated to be accurate to a 0".1, while the seeing is at worst 1".3 in any image. Note the amount of structure in Figure 2a, and the overall similarity of parts of the scatterplot to the evolutionary paths in SMT.

In order to understand better the origin of this structure, the same diagram is plotted in Figure 2b for only the western subsection of 52W036 that shows the outwardly propagating wave (see vector in Figure 1). Assuming that age is directly correlated with radius, we see that as we progress outward toward younger stellar populations, both the (U-B) and (V-K) colors get bluer. This is as predicted by SMT. However, unlike SMT, the very youngest stellar populations, at the edge of the ring where the starburst activity is most recent, show a signifgicant excursion towards bluer (U-B) and *redder* (V-K). *I tentatively identify this excursion with the emission from an evolving, mixed population of red and blue supergiants.* This excursionary behavior in the (U-B, V-K) diagram is similar to what is seen by PACFM in young MC clusters (compare their Figure 10), a phenomenon they attribute to the presence of red supergiants. Clearly spectroscopy or CO (2.29 μ) imaging will help to confirm this conclusion. In any case, the amount of structural information present in the color-color diagrams suggests that this technique is useful for understanding starburst evolution, and may provide a way to calibrate color-age relationships for galaxies with recent star formation activity.

This presentation would not have been possible without the gargantuan effort by Mark Hereld to produce a working 2 micron array for our use at IRTF. I would also like to thank Mark Hereld and Richard Kron for helpful criticism of the draft of this report, Richard Dreiser for help with the figures, and Eric Persson for useful discussions. Jeff Munn did the analysis of the spectral data collected by Dave Koo and Richard Kron.

References

Aaronson, M. 1978, Ap. J. 221, L103.
Appleton, P. N., and Struck-Marcell, C. 1987, Ap. J. 312, 566.
Balzano, V. A. 1983, Ap. J. 268, 602.
Campbell, A. W., and Terlevich, R. 1984, MNRAS 211, 15.
Few, J. M. A., Madore, B. F., and Arp, H. C., MNRAS 199, 633.
Fosbury, R. A. E., and Hawarden, T. G. 1977, MNRAS 178, 473.

132

Heckman, T. M. 1983, Ap. J. **268**, 628.
Huchra, J. P. 1977, Ap. J. **217**, 928.
Hunter, D. A., and Gallagher, J. S. 1985, A. J. **90**, 1457.
Joy, M., and Harvey, P. M. 1987, Ap. J. **315**, 480.
Larson, R. B., and Tinsley, B. M. 1978, Ap. J. **219**, 46 (LT).
Lynds, R., and Toomre, A. 1976, Ap. J. **209**, 382.
Majewski, S. R., Hereld, M., Munn, J. A., Koo, D., and Kron, R. 1988, in preparation.
Persson, S. E., Aaronson, M., Cohen, J. G., Frogel, J. A., and Matthews, K. 1983, Ap. J. **266**, 105 (PACFM).
Renzini, A., and Buzzoni, A. 1985, in "The Spectral Evolution of Galaxies", ed. C. Chiosi and A. Renzini (Dordrecht: Reidel).
Searle, L., Sargent, W. L. W., and Bagnuolo, W. G. 1973, Ap. J. **179**, 427.
Searle, L., Wilkinson, A., and Bagnuolo, W. G. 1980, Ap. J. **239**, 803.
Struck-Marcell, C., and Tinsley, B. M. 1978, Ap. J. **221**, 562 (SMT).
Telesco, C. M., Decher, R., and Gatley, I. 1985, Ap. J. **299**, 896.
Theys, J. C., and Spiegel, E. A. 1976, Ap. J. **208**, 650.
Thuan, T. X. 1983, Ap. J. **268**, 667.
Thuan, T. X. 1985, Ap. J. **299**, 881.
Windhorst, R. A., van Heerde, G. M., and Katgert, P. 1984, A&A Suppl. **58**, 1.

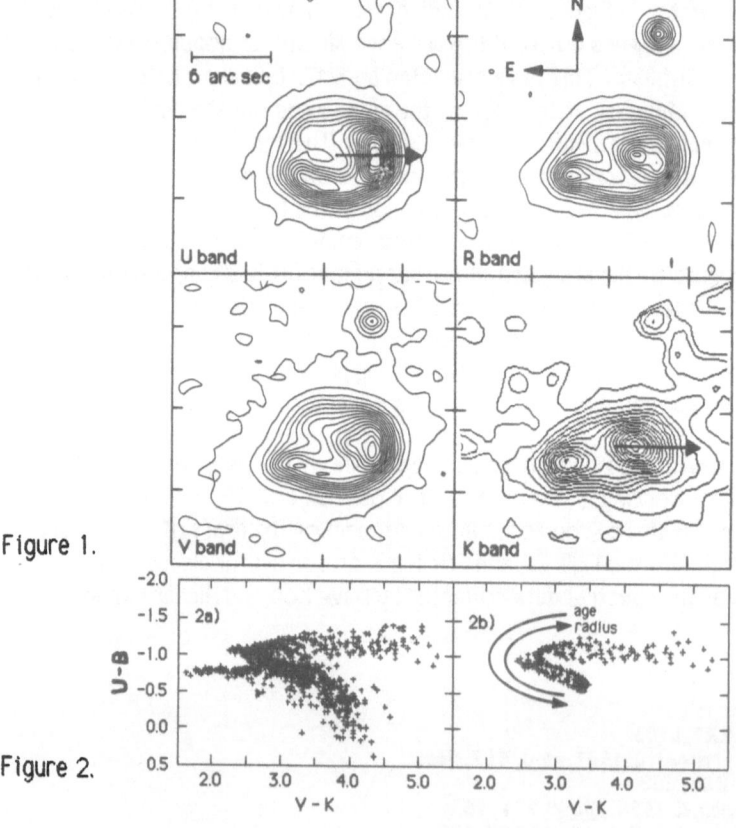

Figure 1.

Figure 2.

LARGE-SCALE VARIATIONS IN THE LUMINOSITY FUNCTION OF GALAXIES

Riccardo Giovanelli[1] and Martha P. Haynes[2]
NAIC[3], Arecibo Observatory and Astronomy Dept., Cornell University
USA

ABSTRACT. We report observational evidence that the blue luminosity function depends on the local density of galaxies. While high luminosity galaxies are segregated to the high density regions, low luminosity ones are more diffusely distributed into regions of lower density contrast. This segregation, also seen in other properties such as surface brightness, supports the predictions of theories of biased galaxy formation and argues against the validity of a universal luminosity function (LF) for galaxies.

1. INTRODUCTION

Some of the cosmological scenarios that have become popular in recent years demand that the universe be flat. Except in models that invoke a large value of the cosmological constant, the requirement that the universe be flat imposes the need for a dominant form of matter that is invisible to current observations. Furthermore, this unseen component must be distributed relatively smoothly throughout the universe. In contrast, the distribution of luminous matter, as traced by galaxies and X-ray emitting cluster gas, is highly inhomogeneus; superclusters and voids yield large-scale deviations from the mean density contained in luminous matter by factors of five to ten.

The contrast between the clumpy distribution of visible matter and the smooth one implied for the unseen matter has prompted speculation on mechanisms that may have made the process of galaxy formation strongly sensitive to small amplitude, large scale fluctuations in the primordial density perturbation spectrum. Within such schemes, massive galaxy formation would have occurred efficiently in the higher density fluctuations, and would have initially been inhibited in the surrounding, lower density regions. As a result of this bias, small mass density contrasts would result in a largely amplified luminosity density contrast (see Dekel and Rees 1987 for recent review). Among the predictions of this theoretical scheme is one that suggests that the galaxies that collapsed slowly, i.e. low luminosity, low surface brightness objects should be segregated in space from the bright, massive galaxies. An extreme corollary of this proposition is that low luminosity galaxies may fill the observed voids which themselves are defined only by the brightest objects.

In a previous paper (Giovanelli, Haynes and Chincarini 1986: GHC), evidence was presented that strong morphological segregation exists on supercluster scales. Here, we propose that the distribution of luminosity and surface brightness is also segregated. We first define how the density scale is obtained in Section 2, and then discuss the dependence of galaxian parameters in section

---

[1] Arecibo Observatory, P.O. Box 995, Arecibo P.R. 00613, USA
[2] Astronomy Dept., Space Sciences Building, Cornell University, Ithaca, NY 14853, USA
[3] The National Astronomy and Ionosphere Center is operated by Cornell University under cooperative agreement with the National Science Foundation.

R. G. Kron and A. Renzini (eds.), Towards Understanding Galaxies at Large Redshift, 133–138.
© *1988 by Kluwer Academic Publishers.*

3. Throughout this report we usually express distances in terms of a radial velocity in km s^{-1}, and when necessary, assume a Hubble constant of $H_0 = 75$ km s^{-1} Mpc^{-1}.

2. DEFINITION OF SPACE DENSITIES

The data base used consists of 4700 galaxy redshifts in the region of the Pisces-Perseus supercluster, in large part the result of our survey with the Arecibo 305 m telescope (Giovanelli and Haynes 1985; Giovanelli et al. 1986 and further results in preparation). The goals of the survey are completeness to both $m = 15.7$ and angular size greater than 1'. At this time, for bins of 0.1 mag, completeness is higher than 80 % to $m = 15.4$, drops to about 55 % at $m = 15.6$ and below that figure at $m > 15.6$. Completion statistics is somewhat better for spirals than for earlier types. Limiting to $m \leq 15.6$, one can obtain nearly complete volume limited subsamples, such as one to absolute magnitude $M = -19.0$ out to a redshift of 6250 km s^{-1}, or to absolute magnitude -18.4 out to a redshift of 4750 km s^{-1}. Within the limits of RA = 22^h to 04^h, Dec = 0^0 to 45^0, and with the exclusion of regions of extinction higher than 0.5 mag, we obtain a subsample of 1027 galaxies brighter than $M = -19.0$. The volume sampled is then the wedge bound by the above mentioned solid angle and by the sphere of radius 6250 km s^{-1}. We can then subdivide that volume with a cartesian grid of origin at Earth, x-axis toward (RA,Dec) = $(22^h,0^0)$, y-axis towards (RA,Dec) = $(04^h,0^0)$, z-axis toward Dec = 90^0. For all purposes in this paper, the grid cells are 375 km s^{-1} to the side. The number of galaxies in the volume limited subsample is counted within each cell. The resulting average density is $< n >= 4.7 \times 10^{-3}(H_0/75)^3$ galaxies Mpc^{-3}. Note that the number density n thus derived, unweighted by luminosity or mass, refers to the number of bright galaxies alone, $i.e.$ those with M < -19.0. By comparison, the space density that would be obtained using Felten's (1985) favorite LF (one with Schechter parameters $\Phi^* = 0.011$ Mpc^{-3}, $\alpha = -1.35$ and $M^* = -20.05$ at $H_0 = 100$) is $7.8 \times 10^{-3}(H_0/75)^3$. The difference reflects in part the incompleteness of our sample and perhaps real large scale variations in the mean density of galaxies. The volume occupied by the subsample includes the Pisces-Perseus filament and a large void in its foreground. On the other hand, the average space density of galaxies obtained from a subsample limited by absolute magnitude -18.4, out to 4750 km s^{-1} (a volume not fully including the Pisces-Perseus filament), is 5.9×10^{-3} galaxies per Mpc^3, which is a factor 2 smaller than the value predicted by Felten's LF.

It should be pointed out that before the calculation of $n(x, y, z)$, radial velocities were corrected for local large-scale deviations from Hubble flow, conservatively adopting the mean solution of Hart and Davies (1982), and for virial distortions within cluster cores. The latter correction was applied to galaxies within 1 Abell radius (or an equivalent quantity for clusters not listed in the Abell catalog) with velocities $|V_{cl} - 3\sigma| < V < |V_{cl} + 3\sigma|$, where V_{cl} and σ are respectively the systemic velocity and the line-of-sight velocity dispersion of the cluster; the correction consists of reducing the difference $(V - V_{cl})$ by the factor $150/\sigma$, where the value of 150 km s^{-1} is assumed to roughly represent a "cosmic" velocity dispersion.

We then estimate the dependence of galaxian parameters on the value of the local density n. The value of n associated with each galaxy is that obtained as described above for the cell within which each galaxy is located. It should be noted that the densities obtained from subsamples limited by different absolute magnitudes are not directly commensurable, as they result from counting to different luminosity depths.

3. SEGREGATION

3.1. Luminosity

The determination of a luminosity dependence on any parameter that may be related to distance is threatened by a Malmquist bias. Because the linear scale of the density inhomogeneities is comparable with that of the volume sampled (see GHC for a graphic realization), high and low density regions are not "well mixed", *i.e.* the median distance of cells of low n and that of those of high n are not the same. As a result, a plot of luminosity versus density that included all the galaxies in the sample would exhibit a Malmquist bias, in the sense that higher luminosities would be obtained for the densities with the higher median distance. In our sampled volume, the higher density regions, which tend to be found in the Pisces-Perseus supercluster chain, tend to be somewhat farther than the low density ones. In order to avoid the bias introduced by this difference, we analyzed separately different redshift intervals, such that within each interval the high and low density cells have similar median redshifts. Only two intervals will be considered here, the one within 4000 km s^{-1} and the one between 4000 and 6200 km s^{-1}. In order to optimize statistics, in each we use different density scales: in the first (n_1) that defined by the subsample of galaxies brighter than -18.4, in the second (n_2) that defined by galaxies brighter than -19.0. This nuisance is justified in an arena where problems of undersampling tend to be the rule.

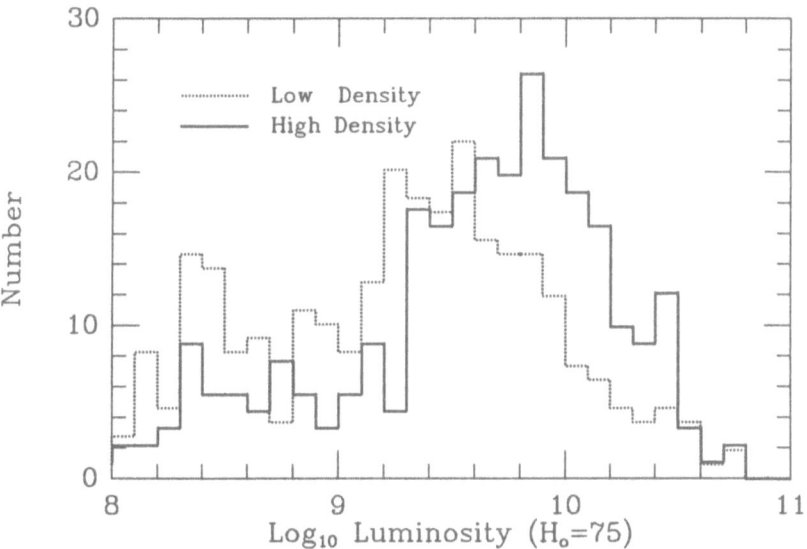

Figure 1. Histograms of photographic luminosity, expressed in solar units, in the velocity interval 0-4000 km s^{-1}, separately for galaxies located in high and low density cells. Space densities were determined from a subsample nearly complete to absolute magnitude -18.4, which yields a mean density $< log_{10}n_1 >= -2.2$, with n_1 expressed in Mpc^{-3}. The "high density" LF was obtained from galaxies located in cells with $log_{10}n_1 \geq -2.0$, the "low density" LF from those in cells with $log_{10}n_1 < -2.1$.

In Fig. 1, we display LF's obtained in the velocity interval 0-4000 km s^{-1}, for galaxies in the highest and lowest local density regimes. Approximately 270 galaxies were counted in each regime. The low density regime includes all galaxies in cells with $n_1 \leq 10^{-2.1}$ Mpc^{-3}. This region contains few high density cells, and the "high" density regime includes galaxies in cells with $n_1 \geq 10^{-2.0}$); only a handful are located in truly high ($n_1 \geq 10^{-1.5}$) density regions. Notice the marked deficiency of bright galaxies in the low density regions. Reasonable completeness in this subsample corresponds to a luminosity of about $10^9{}^7L_\odot$. Venturing cautiously in the muddy waters of incomplete data, we

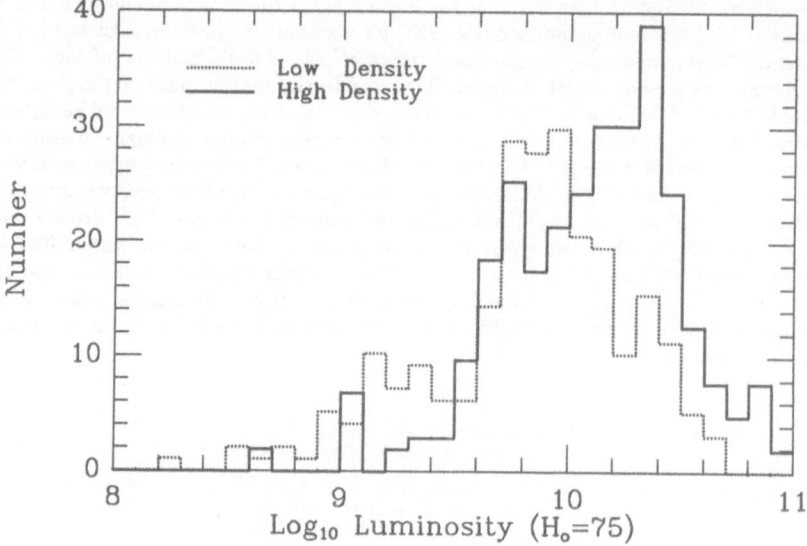

Figure 2. Histograms of photographic luminosity of galaxies in the velocity interval 4000-6200 km s^{-1}. Space densities were determined from a subsample nearly complete to absolute magnitude -19.0, which yields a mean density $< log_{10}n_2 >= -2.3$, with n_2 expressed in galaxies, brighter than -19.0, per Mpc^3. The high density LF was obtained from galaxies located in cells with $log_{10}n_2 > -1.5$, while the low density one from galaxies in cells with $log_{10}n_2 < -2.1$. Note that if the same LF were valid in all regimes, this density scale would correspond to values of $log_{10}n_1 \simeq log_{10}n_2 + 0.3$, because of the different depths counted by n_1 and n_2. Reasonable completeness is available here for galaxies brighter than $10^{10}L_\odot$.

can nonetheless note that faint galaxies form a relatively larger fraction in the low density regimes. Luminosities are corrected for galactic and internal extinction as described in Haynes and Giovanelli (1984). Figure 2 is the analog of Fig. 1, for the velocity interval 4000 to 6200 km s^{-1}. This region has many galaxies in high density cells; as a result, the "high density" LF is derived from 300 or so galaxies that reside in much higher density regions than the corresponding "high density" galaxies in Fig. 1. The "high density" galaxies in Fig. 1 do in fact match better the density distribution of the "low density" ones in Fig. 2. Because of a Malmquist bias, comparison of LF's in Fig.1 with those in Fig. 2 is a delicate affair. Although fewer low luminosity galaxies are counted in the more distant redshift interval, the shift in the LF between high and low n regimes noted in Fig. 1 is still present in this completely independent data set.

3.2. Surface Magnitude

In Figure 3 the dependence of surface magnitude on local density is displayed. Within each local density window ($log_{10}n_2 < -2.1$ and $log_{10}n_2 > -1.5$ respectively for low and high windows), all galaxies within 6200 km s^{-1} are included. Surface magnitude is defined as the magnitude of the galaxy, corrected for extinction and galactic absorption, divided by the square of the major angular diameter of the blue image, as measured on the PSS, in units of *mag arcsec*$^{-2}$. The effects of

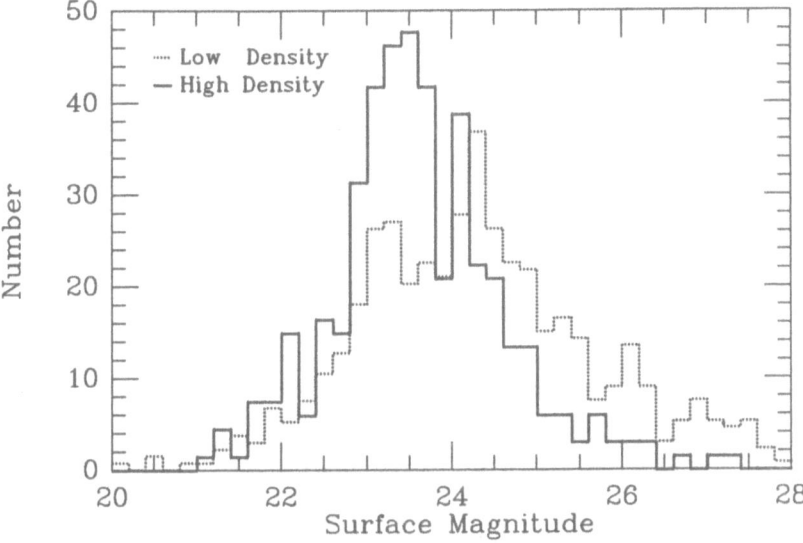

Figure 3. Histograms of surface magnitude, in *mag arcsec*$^{-2}$. High and low density regions are separated according to the same criteria as those applied for Fig. 2.

Malmquist bias, while not negligible, are less marked than for luminosity. Break down by velocity range reveals the same features visible in Fig. 3: galaxies in high density regions, which tend to be brighter, also have higher surface magnitude.

3.3. Discussion

Bothun *et al.*(1986) and Oemler (1987) have approached before the question of luminosity and surface brightness segregation between high and low density regions, with different conclusions than those reached in this report. They concluded that the outline of voids, in terms of the radial velocity histograms of galaxies within the solid angles subtended by the voids themselves, does appear substantially the same whether high or low luminosity (or surface magnitude) galaxies are plotted. We maintain that the density contrast between high and low density regions decreases as the luminosity - or the surface magnitude - of the galaxies decreases. Whether a luminosity - or surface magnitude - level is reached at which the galaxian distribution appears more or less homogeneous requires studies that reach levels of satisfactory completeness at least a few magnitudes fainter than the present one. Why do our results disagree with those mentioned above? We can only advance some hypotheses. For individual regions, previous studies are typically characterized by small number statistics, with associated larger uncertainties. In the case of the study of Bothun *et al.*(1986), higher number statistics is obtained by averaging over large regions of sky. The lack of evidence for a difference in clustering properties between high and low surface brightness galaxies in their study is in disagreement not only with the results of this report but also with those of Davis and Djorgovski (1985) and GHC. Because the subdivision of Bothun *et al.* between high and low surface brightness galaxies is largely one of morphological class, segregation should be present in their data base. One possible explanation of the lack thereof is that the averaging over large solid angles mixes high and low density regions within a given velocity bin, and partly obliterates the noisy signature of segregation. We produced similar velocity histograms to those of Bothun *et al.* from our data base and find barely perceptible differences in the clustering properties of high and low surface magnitude galaxies. Our conclusion is that velocity histograms of wide areas are not sensitive tools to study the differential clustering properties of galaxies of different types.

The question of whether low luminosity galaxies fill the voids (*i.e.* the density contrast is reduced to random, small scale fluctuations) still remains unanswered, although the failure by numerous groups to detect a conspicuous population of objects with high HI mass-to-light ratio suggests that such objects may not provide the filling stuff. A definitive solution will require redshifts of complete samples to fainter limits than the present one.

Whether faint galaxies do or do not "fill" the voids, it appears that efficient biasing mechanisms at the time of galaxy formation need be invoked to explain the observed marked morphological and luminosity segregation. Binggeli (1987) has suggested that the concept of a "universal" luminosity function should be revised, in order to allow for large-scale morphological segregation and the differences in the luminosity functions of galaxies of different morphological types, observed in the Virgo cluster region (Sandage, Binggeli and Tammann 1985). The general luminosity function would then be expressed as the sum of individual luminosity functions applicable to each morphological type, each multiplied by a weight that depends on the local density. We propose to go further and suggest that the general luminosity function should be expressed in the form

$$\Phi(L,n)dL = \Sigma_t \Phi_t(L,n)dL$$

where the summation is over all morphological types t, and $\Phi_t(L,n)$ is an explicit function of n.

4. CONCLUSIONS

We have presented observational evidence for luminosity and surface magnitude segregation, over a large range of local galaxian densities. The density contrast between high and low density regions decreases with decreasing galaxian luminosity and surface magnitude, thus infringing the concept of universality of the luminosity function. The effect supports the proposals that advocate biased galaxy formation mechanisms.

5. REFERENCES

Binggeli, B. 1987,*preprint*
Bothun, G.D., Beers, T.C., Mould, J.R. and Huchra, J.P. 1986, *Ap. J.* **308**,510
Davis, M. and Djorgovski 1985, *Ap. J.* **299**,15
Dekel, A. and Rees, M.J. 1987, *Nature,* in press
Felten, J.E. 1985, *Comments Astrophys.* **11**,53
Giovanelli, R. and Haynes, M.P. 1985, *A. J.* **90**,2445
Giovanelli, R., Haynes,M.P., Myers, S.T. and Roth, J. 1986, *A. J.* **92**, 250
Giovanelli, R., Haynes, M.P. and Chincarini, G.L. 1986, *Ap. J.* **300**,77 (GHC)
Hart, L. and Davies, R.D. 1982,*Nature* **297**,191
Haynes, M.P. and Giovanelli, R. 1986, *Ap. J. (Letters)* **396**,L55
Haynes, M.P. and Giovanelli, R. 1984, *A. J.* **89**,758
Oemler, A. 1987, *preprint*
Sandage, A., Binggeli, B. and Tammann, G.A. 1985, *A. J.* **90**, 1759

SOME REMARKS ABOUT MODELLING THE CHEMICAL EVOLUTION OF GALAXIES

M. Tosi
Osservatorio Astronomico
C.P. 596
I-40100 Bologna
Italy

1. INTRODUCTION

Understanding galaxies at large redshift requires somehow to understand their evolution, and models of galactic chemical evolution are among the best tools in the process of such understanding. In fact, they provide useful information on some of the most significant quantities regulating the evolution of galaxies of the various morphological types: initial mass function (IMF), star formation rate (SFR), stellar nucleosynthesis, etc. However, there are so many parameters involved in the chemical evolution models that a special care is necessary not only in computing the models but also in interpreting their results.

Here, we point out a couple of issues that must be taken into account to avoid misleading conclusions when dealing with the models: i) the importance of recognizing on which parameters mostly depend the various quantities resulting from the models, and ii) given the present lack of a unique solution to the problem of galactic chemical evolution, the need of minimizing the actual range of possibilities. Some examples and the more general conclusions related to these two issues are discussed in sections 2 and 3, respectively.

For a more schematic analysis of both arguments, the various parameters involved in the model computations can be divided into two groups:

<u>Stellar Parameters</u>, resulting from the theories of stellar evolution and nucleosynthesis. The most uncertain of these quantities are related to the treatment of convection in stellar interiors, and are the mixing length parameter α in stellar convective envelopes (e.g. Renzini and Voli 1981) and the overshooting parameter λ in stellar convective cores (e.g. Bertelli et al. 1985). Among the other uncertain parameters of this group are the rate of stellar mass loss and the $^{12}C(\alpha, \gamma)^{16}O$ reaction rate. None of these is actually a free parameter since the comparison of the stellar theories and observations has strongly restricted their possible range of variability. For instance, α formally ranges between 0 and 2, but is reasonably $1.2 \leqslant \alpha \leqslant 1.7$ (see e.g. Greggio and Tosi 1986), and the cross section of the carbon-oxygen reaction ranges between its standard value (Fowler, Caughlan and Zimmerman 1975) and three times that value (Kettner et al. 1982).

R. G. Kron and A. Renzini (eds.), Towards Understanding Galaxies at Large Redshift, 139–146.
© 1988 by Kluwer Academic Publishers.

Galactic Parameters, related to more global properties of the galaxies, like the SFR, the IMF, gas flows, etc. We know very little, if anything, of these processes, and therefore usually treat them as free parameters.

Since the only galaxy for which the number of observational constraints is larger than the number of parameters is our own, in the following we present and discuss only results relative to the Milky Way. However, most of the arguments and conclusions can be generalized at least to the other spirals. The details of the models and of the observational constraints are described by Tosi (1987a, hereinafter TO87).

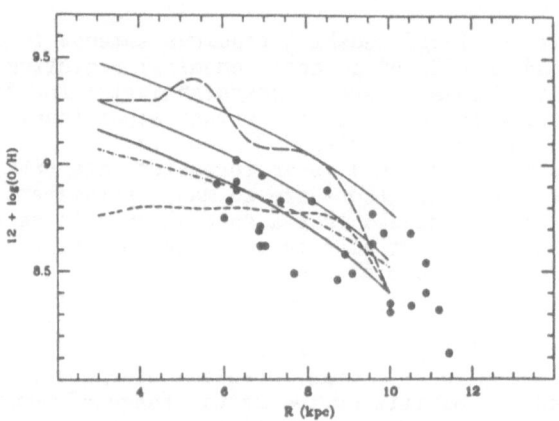

Fig.1. Radial distribution of the present oxygen abundance in the galactic disk. The average error on the observational data is ±0.2 dex. See text for details.

2. DEPENDENCE OF THE MODEL RESULTS ON THE PARAMETERS

Fig.1 shows the radial distribution in the galactic disk of the present oxygen abundance by number, as resulting from model predictions and observational data. The dots represent the abundances derived by Shaver et al. (1983) from radio and optical observations of H II regions. The lower solid line corresponds to a model with exponentially decreasing but almost constant SFR (e-folding time 15 Gyr), Tinsley's (1980) IMF, and an infall of gas on the disk with uniform density $F=0.004$ $M_\odot kpc^{-2} yr^{-1}$ and constant after the time of disk formation. Hereinafter we will mention this one as the reference model. The dash-dotted line corresponds to a model with more rapidly decreasing SFR (e-folding time 5 Gyr), and infall uniform and slowly decreasing with time (f= F exp[-t/ϑ], with F=0.002 and ϑ=100 Gyr). Both models fit quite well the observed distribution.

The short-dashed line shows the effect of keeping unchanged all the parameters of the reference model except the infall initial density F, which instead of being uniform is now assumed to increase steeply toward the galactic center. Since the infalling gas is assumed to be metal poor (see Tosi 1987b), in the inner regions its larger density produces a larger dilution of the metallicity of the interstellar medium and a lower oxygen abundance. This causes the flatness of the curve of Fig.1. For the long-dashed line, instead, the infall density is uniform again (F=0.0015, ϑ=50 Gyr), the SFR is proportional to the gas density, and the IMF is taken from Güsten and Mezger (1982), who suggest that on spiral arms only stars more massive than 3 $M_\odot$ can form. The bump shown in Fig.1 at R$\simeq$5$\div$6 kpc reflects in fact the oxygen overproduction due to the corresponding larger fraction of massive stars formed on the Sagittarius spiral arm. Neither of these two latter models is consistent with the observed distribution.

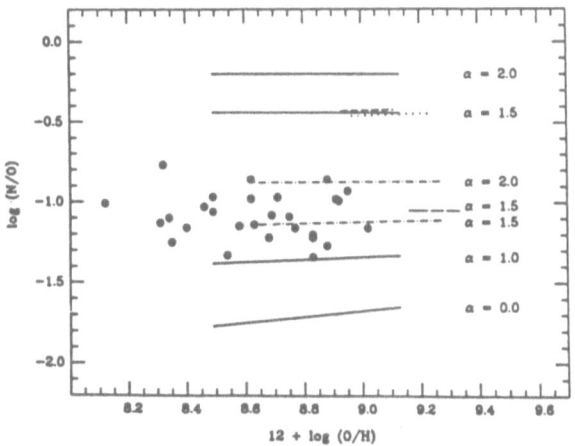

Fig.2. Present distribution in the galactic disk of the nitrogen/oxygen abundance ratio vs. oxygen abundance. The average error on the observational data is ±0.2 dex. See text for details.

The different predictions of these four models point out the sensitivity of both the slope of the radial gradient and the absolute abundances on the adopted galactic parameters.

Vice versa, if we consider the reference model, keep fixed all the parameters, but take overshooting from stellar cores into account (i.e. λ=1 instead of λ=0), the predicted oxygen abundance is increased by the larger oxygen production in the larger convective core of massive stars, and is represented by the middle solid line. If, instead of including overshooting, we assume that mass loss from stars never takes place (which is obviously not the case), the oxygen overproduction is even more conspicuous (upper solid line of Fig.1) and inconsistent with the data. In any case, the slope of the gradient is totally unaffected by these changes in the stellar parameters.

Fig.2 shows the present distribution in the galactic disk of the nitrogen/oxygen abundance ratio by number with respect to the oxygen abundance. The dots correspond to the same data as in Fig.1 and the solid lines to the reference model of Fig.1 for the indicated choices of α. The amount of mixing in stellar envelopes does not affect oxygen, which is mainly synthesized in the cores of massive stars, but deeply influences nitrogen, whose primary component is exclusively produced during the envelope burning of intermediate mass stars (Renzini and Voli 1981). For this reason, the choice of α has very important effects on the N/O ratios predicted by the models. If instead of $\lambda = 0$ we assume $\lambda = 1$, not only oxygen is enhanced, as mentioned before, but nitrogen is strongly depleted, and the two top solid lines of Fig.2 are replaced by the two dash-dotted lines.

The long-dashed line of Fig.2 is the result of substituting in our reference model Tinsley's IMF with a flatter one with Salpeter's (1955) exponent for all stellar masses. The larger fraction of massive stars resulting from this IMF produces an oxygen enhancement, and the corresponding smaller fraction of intermediate mass stars implies a smaller production of nitrogen, with a net result of strongly reducing the N/O ratio.

The dotted line and the short-dashed line show, instead, that the N/O ratio is unaffected by variations in the galactic parameters. The dotted line, in fact, corresponds to the reference model with no infall of gas after the disk formation, and the lack of diluting material is reflected by the extreme oxygen overabundance. The short-dashed line represents a model where not only infall has been removed but also the e-folding time of the SFR has been changed from 15 to 5 Gyr. Despite the strong effect on the absolute abundance, in neither case the N/O ratio is different from that predicted by the reference model.

All the above arguments can be generalized and summarized as follows:

a) The predicted abundance gradients depend only on the galactic parameters (SFR, IMF, infall and radial flows). Therefore, the comparison of these model predictions with the corresponding observed data is a test of the validity of the chemical evolution models.

b) The predicted abundance ratios depend only on the stellar parameters (α, λ, reaction rates) and on the IMF. Therefore, the comparison of these model predictions with the corresponding observed data is a test of the validity of the stellar nucleosynthesis models and of the IMF.

c) The predicted absolute abundances depend both on stellar and on galactic parameters. Therefore, they do not represent by themselves a good test of any theory. However, any self-consistent model should also reproduce all the observed element abundances.

3. THE PROBLEM OF UNIQUENESS

The reason why people worry about the lack of a unique solution to the problem of modelling the chemical evolution of galaxies is that even if different models might happen to give similar results for a given epoch (e.g. the present time) they must diverge at different epochs. It would then be dangerous to use these models to predict future or past galaxy conditions, like photometric properties or stellar populations.

Since we cannot reduce the number of parameters, in the present scenario of theories, observations, and relative uncertainties, the only critical approach to this problem is to compare the model results with all the available observational constraints, and reject any model whose results disagree with some observed data. In such a way, we can minimize the current range of possibilities and proceed toward the finding of a unique class of models consistent with all the constraints.

This approach has been followed (TO87) by comparing the predictions of some of the most popular models of the disk of the Galaxy with the corresponding observed data. If we derive the χ^2 corresponding to each comparison, only models with significantly good χ^2 for all the predictions can be considered representative enough of the actual evolution of the Galaxy.

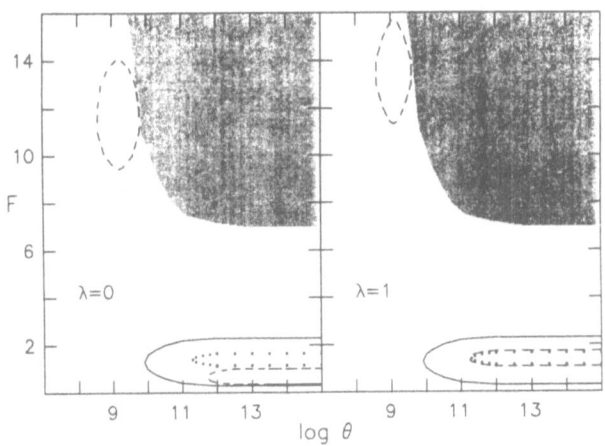

Fig.3. Contours of 90% confidence level for a model with SFR proportional to the gas density and uniform infall. The contours correspond to the age-metallicity relation (solid lines), and to the radial distribution of the global metallicity (dotted lines) and of the oxygen abundance (dashed lines). F is in units of 10^{-3} $M_{\odot}kpc^{-2}yr^{-1}$ and ϑ in Gyr.

Figs 3 and 4 provide some examples of the results of this method. For simplicity, we refer to the comparison of only three sets of model predictions with the corresponding observed data: the radial distribution of the present oxygen abundance (dashed line) as derived by Shaver et al. (1983), the radial distribution of the present total metallicity (dotted line) as derived by Panagia and Tosi (1981) and by Luck (1982), and the age-metallicity relation in the solar neighbourhood (solid line) as derived by Twarog (1980). Both figures show the projection of the surfaces of minimum χ^2+4 on the plane of the two free parameters F and ϑ of the infall. These contours correspond to the 95% confidence level on each axis and to the 90% confidence level on their combination.

The left panel of Fig.3 corresponds to a model with SFR proportional to the gas density, infall uniform and exponentially decreasing with time, inward radial flows of half the mass of infalled gas, $\alpha=1.5$ and $\lambda=0$. The shaded region in the upper part of the diagram is forbidden because the corresponding combination of infall parameters implies negative masses of the Galaxy at some epoch (see T087). It is apparent that the predictions in better agreement with the observations are obtained with a low infall density constant all over the disk lifetime. Only the observed oxygen distribution (dashed lines) can be reproduced also with an infall rate more rapidly decreasing with time (but ϑ longer than the e-folding time of the SFR) and large initial density F.

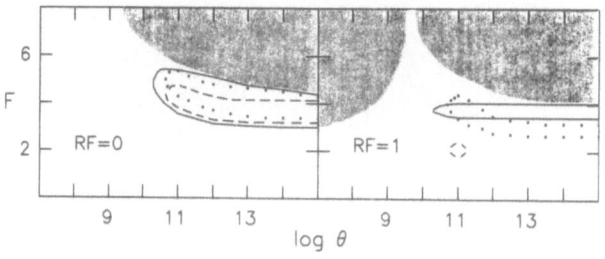

Fig.4. Contours of 90% confidence level for the reference model. RF indicates the amount of radial flows normalized to the infalling gas. Symbols are as in Fig.3.

The right panel of Fig.3 shows the effect on the same model of including overshooting ($\lambda=1$ instead of $\lambda=0$). Overshooting affects only the individual abundances, so that all the lines are as in the left panel, except the oxygen contours which are now shifted upwards. This is because the larger oxygen production due to overshooting requires a larger density of diluting infall gas to be compensated. This shift upwards happens to be just of the right amount to allow the intersection of all the three regions of minimum χ^2 in the right bottom corner of the diagram. Had we compared the predictions of this model only with the observed oxygen abundances, we would have reached the wrong conclusion that an infall actually decreasing with time ($\vartheta=1$ Gyr) is also consistent with the observational constraints. Instead, the good fit regions intersect only for $\vartheta>100$ Gyr and $\lambda=1$. In any case, there are other observed data (e.g. the radial distribution of the SFR described by Lacey and Fall 1985) which are not reproduced by this kind of model, that is, therefore, to be rejected (T087).

The left panel of Fig.4 shows the analogous diagram for the reference model of Fig.1. The top shaded region corresponds again to forbidden combinations of the infall parameters leading to unphysical conditions. Since the e-folding time of the SFR is longer than in the previous case, the amount of infalling gas required at any epoch to dilute properly the abundances of the interstellar medium is larger than in Fig.3. For the three observational constraints of the previous case, a good agreement is obtained for $F \simeq 0.004$ $M_\odot kpc^{-2}yr^{-1}$ and $\vartheta \gtrsim 50$ Gyr, i.e. longer than the e-folding time of the SFR.

If we keep unchanged all the parameters of the reference models, except for assuming that all the infalled gas moves toward the galactic center (RF=1), the contours of acceptable χ^2 become those of the right panel of Fig.4. To reproduce the oxygen distribution a lower infall density is now sufficient, but the age-metallicity relation still requires F=0.004, and the three contours are far from intersecting. Note that the forbidden region now extends also to short ϑ. It is apparent then that large amounts of radial gas flows are inconsistent with the available data.

In the present scenario of theories and observations, the results obtained in T087 with this approach can be summarized as follows:

a) The SFR is not simply proportional to the gas density. We are not able yet to discriminate among the other suggestions of various authors (see references in T087) that it is proportional only to the H_2 gas component, to the square gas density in special dynamic conditions, or any other law somehow reminiscent of an exponential function of both time and galactocentric distance.

b) The SFR has not decreased rapidly with time, i.e. the e-folding time must be longer than 5÷8 Gyr.

c) The IMF has not strongly varied in space and time. In particular, an IMF strongly dependent on metallicity (as suggested by Terlevich and Melnick 1984) predicts too few massive stars in metal rich regions and does not allow to reproduce the observed oxygen distribution. The same inconsistency, but for opposite reasons, is found for IMFs with too many massive stars, like the IMF with Salpeter's exponent for all masses, Güsten and Mezger's (1982) and Larson's (1986) IMFs.

d) The infall rate decreases in time more slowly than the SFR, and its present value ranges between 0.3 and 1.8 $M_\odot yr^{-1}$ for the whole disk. The infall density does not increase steeply toward the galactic center.

REFERENCES

Bertelli, G., Bressan, A.G., Chiosi, C. 1985, Astron.Astrophys. <u>150</u>, 33

Fowler, W.A., Caughlan, G.R. and Zimmerman, B.A. 1975, Ann.Rev. Astron.Astrophys. <u>13</u>, 69

Greggio, L. and Tosi, M. 1986, Astron.Astrophys. <u>156</u>, L1

Güsten, R. and Mezger, P.G. 1982, Vistas Astron. <u>26</u>, 159

Kettner, K.U., Becker, H.W., Buchmann, L., Gories, J., Krawinkel, H., Rolfs, C., Schmalbrock, P., Trautvetter, H.P., Vlieks, A. 1982, Zeit.Phys. A308, 73

Lacey, C.G. and Fall, S.M. 1985, Astrophys.J. <u>290</u>, 154

146

Larson, R.B. 1986, M.N.R.A.S. 218, 409
Luck, R.E. 1982, Astrophys.J. 256, 177
Panagia, N. and Tosi, M. 1981, Astron.Astrophys. 96, 306
Renzini, A. and Voli, M. 1981, Astron.Astrophys. 94, 175
Salpeter, E. E. 1955, Astrophys.J. 121, 161
Shaver, P. A., McGee, R. X., Newton, L. M., Danks, A. C., Pottasch,
      S. R. 1983, Monthly Notices Roy.Astron.Soc. 204, 53
Terlevich, R. and Melnick, J. 1984, preprint
Tinsley, B. M. 1980, Fund.Cosmic Phys. 5, 287
Tosi, M. 1987a, preprint (TO87)
Tosi, M. 1987b, Astron.Astrophys. in press
Twarog, B.A. 1980, Astrophys.J. 242, 242

DISTANT GALAXIES:
LIMITS ON COSMOLOGY AND EVOLUTION

Richard S Ellis
Physics Department
University of Durham
England

ABSTRACT: Detailed studies of intermediate redshift ($0.1 < z < 0.6$) galaxies are changing our ideas of galaxy evolution. In rich clusters, extended star formation histories are found in *some but not all* early-type objects. Together with variable blue fractions, the activity observed suggests local processes dictate the appearance of an individual galaxy at high redshift and thus the concept of a generalised *evolutionary correction* can no longer be meaningful. In field samples, spectroscopic surveys indicate that if luminosity evolution is responsible for the steep galaxy number count relation, it must be of a luminosity-dependent form. Brief spectacular enhancements in star formation activity in otherwise low luminosity galaxies may explain the excess faint galaxies at $B \sim 22$. The similarity between the cluster and field activities is striking. The absence of high luminosity primeval galaxies may now be understood via the extended star formation observed *in both samples*, but a physical mechanism for the bursts of star formation seen remains to be found.

1. INTRODUCTION

My definition of a distant galaxy will be one whose redshift is greater than 0.1. This may seem hardly beyond the Local Group to some participants of this conference, but it reflects a conviction presented here that studies of statistically complete samples of galaxies at intermediate redshifts ($z \sim 0.1$ - 0.6) are as important in changing our views of galaxy evolution as those at $z \sim 1$ - 2, where the signal/noise is often poor and the pedigree of the object under consideration is rarely well-understood. Exactly how we should use distant galaxies in cosmological and evolutionary studies depends on the assumptions made. Progressively more of these assumptions are being discarded with experience. It is illustrative to begin by noting some previous assumptions and why they were abandoned.

Initially it was believed that galaxies without present evidence of star formation, e.g. luminous cluster ellipticals, would remain unchanged over large look-back times in an old Universe. The apparent magnitude - redshift relation (Hubble di-

R. G. Kron and A. Renzini (eds.), Towards Understanding Galaxies at Large Redshift, 147–160.
© *1988 by Kluwer Academic Publishers.*

agram) for such objects would trace the cosmic decceleration q_o as defined in the standard Friedmann models. Subsequently, it became clear that stellar evolution could easily affect galaxy luminosities by a substantial amount over recent look-back times and this would bias the derived q_o. Tinsley (1968) convincingly demonstrated that such a *evolutionary correction E (z)*, might be as important in the Hubble diagram as the sought-after difference between the open and closed world-models. This realisation led to a significant change in direction for faint galaxy work.

The interplay of evolution and curvature was demonstrated clearly by Lilly and Longair (1984) in their infrared (K) magnitude-redshift relation for powerful radio galaxies. Assuming such objects have a well-defined time-independent luminosity, they derived a formal value of $q_o = 3.8$. For reasonable Hubble constants, such a high decceleration would lead to an embarassingly short cosmic age ($\tau \sim 7$ Gyr for $H_o = 50$ km s^{-1} Mpc^{-1}), but the result can be readily reconciled with more satisfactory world models if all radio galaxies were substantially more luminous in the past. The evolutionary correction required however, ($1^m.5$ at $z \sim 1$) is much larger than the difference in the K-z relation for the open and closed models. Since luminosity evolution in the near-infrared passband should be *less* than that in, say, the optical and ultraviolet bands, this implies *all* Hubble diagrams will be affected by this problem.

In principle, a precise determination of the evolutionary correction, $E (z)$, for a certain class of galaxy would follow from a knowledge of its time-dependent star formation rate, $\Psi (t)$. For early types, where on-going star formation is virtually absent today, the simplest assumption would be that all the gas was converted into stars at the formation redshift, z_f and thus that $\Psi (t)$ is a delta function; these models are termed *initial burst* models. When more elaborate models were examined (e.g. Bruzual 1983), it was found, however, that whilst the evolutionary correction could be readily predicted in the initial burst case (for a given world model), a wide variety of other star formation histories could also explain the present-day colours of early type galaxies - including some with extended star formation histories (the Bruzual μ models).

Observations at a variety of look-back times (e.g. photometric observations of similar classes of objects at different redshifts) are thus needed to solve for the evolutionary correction; model fits of present day galaxies being unable to yield a unique answer. Note however, that in combining observations at different epochs we do assume that *all* galaxies of this restricted class share the same form for $\Psi (t)$ driven by some *cosmological clock*.

Quite apart from the difficulty of defining taxonomically similar objects at different epochs, a dilemma has arisen in such arguments based upon the the assumption that evolution is driven solely by some cosmological clock. If $\Psi (t)$ was sharply peaked at the formation redshift, calculations show such primaeval galaxies would be readily visible as spectacular luminous emission line objects. Null results in searches for such objects (Baron & White 1987, Koo 1986b) and the growing evidence for on-going star formation in cluster and field surveys (see below) suggests that the initial burst model cannot be a good representation. More likely, star formation is more extended in duration, even for early-type galaxies.

Now, the identification, by Koo (1981) at $z = 0.54$, and later Hamilton (1985) to $z \sim 0.8$, of a population of distant objects with rest-frame colours equivalent to present-day ellipticals, presents a further constraint on the form of past star formation $\Psi(t)$, and one that is difficult to reconcile with the null primaeval galaxy searches. At more than half the Hubble time, there appear to be at least *some* luminous objects which have completed the bulk of their star formation. Exactly what proportion remains unclear but if the assumption is maintained that *all* early-type galaxies share the same star formation history, then effectively all star formation ceased in these galaxies at or before $z \sim 1$. Unless $z_f > 5$, the star formation rate at $z \sim 2$ - 4 would most likely be high enough in a large proportion of normal galaxies for a positive detection of line emission - even when dust eventually shrouds the forming galaxy (Cowie 1986).

The theme of this article is based on the fact that the dilemma posed above could be readily resolved if *some* early type galaxies complete their star formation more rapidly than others. This might be expected if, for example, extended star formation histories are induced from nearby processes, but external to the *closed system* assumed in the galaxy evolution models. A physical illustration of such a picture of star formation has been given by Scalo (this conference). The implication is clear: the evolutionary correction is likely to be a quantity particular to each individual galaxy. Combining data for seemingly homogeneous objects observed at different epochs is fraught with uncertainties until the origin and size of such local effects are understood.

The article is structured as follows. In § 2 I discuss how we might determine what proportion of objects at high z will have completed most of their star formation by z = 1. In §3 and § 4 I present recent evidence, respectively in clusters and the field, for the more complex asynchronous picture of galaxy evolution discussed above. Finally in § 5 I outline a possible path to q_o using supernovae of Type Ia seen in clusters at high redshift.

2. DISTANT CLUSTER SAMPLES

A normal galaxy, with luminosity L^*, can only be studied spectroscopically with a 4 metre telescope to a redshift of about 0.5. To probe further we must devise some short cut using good fortune, realising that such objects are exceptional not only in luminosity but probably in some other way as well. Hamilton's red objects are a good example; we know they exist but we have little idea how representative they are of early type galaxies at that redshift and thus whether evolutionary effects are uniform from object to object in populations that are relatively homogeneous today.

A natural way to proceed is to examine the colour and spectroscopic properties of luminous early type galaxies in remote clusters, some of which can be seen to z ~ 0.8. However, just as reliable samples of galaxies cannot readily be constructed beyond z ~ 0.5, so it is important to understand the processes that lead to the identification of a remote cluster.

The largest distant cluster samples (Gunn *et al* 1986, Ellis *et al* 1988) have been found by number density enhancements on panoramic photographic and CCD survey material. Figure 1 shows the optical contrast of a cluster comparable to

Coma, against a faint background of unclustered field galaxies, as a function of redshift (ignoring the effects of luminosity evolution). The background field assumed is that limited at $B_J \sim 25$ and $R_F \sim 23$; the calculation uses parameters appropriate to the source material used for the AAO cluster survey (details in Couch *et al* 1984).

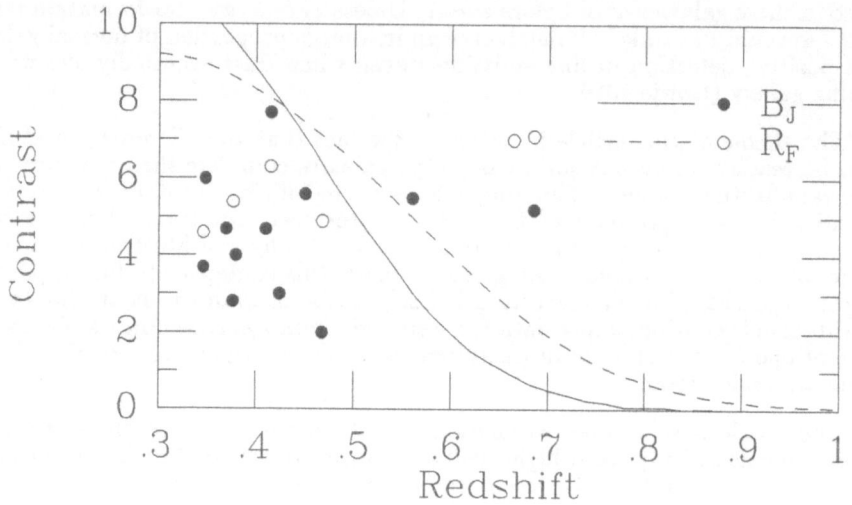

Figure 1: *Selection effects in probing clusters with $z > 0.5$. The visibility of a cluster comparable to Coma ignoring luminosity evolution, via the contrast (in σ) against a deep background of field galaxies, for two passbands is plotted against redshift. Some data from the AAO distant cluster survey are shown.*

The detectability of a cluster depends on the passband used for searching, the presence or otherwise of any luminosity evolution and any structure in the foreground field component. In the simplest case of no evolution, implied by Hamilton's observations to $z \sim 0.8$, it is surprising that *any* clusters beyond $z \sim 0.7$ are found at all. Coma would be a 2 σ perturbation in a R = 23 survey and virtually undetectable to B = 25.

Note particularly that if *some* clusters contain galaxies that have largely completed their star formation, they will dominate those found in red and infrared searches, whereas those found at B or V would only be ones showing strong luminosity and colour evolution. Evidently, if there is a range of evolutionary timescales, extreme caution needs to be exercised in analysing distant cluster statistics.

A good example of the range in properties of clusters of similar richness at the same redshift is provided by the diverse blue fractions claimed for 0016+16 (Koo

1981, Ellis *et al* 1985) at z = 0.54 (whose blue fraction is as low as that found in present-day systems), and 0054-28 (Couch *et al* 1985) at z = 0.58 (which has one of the highest blue fractions known). As might be expected, 0016+16 was found on a IIIa-F Kitt Peak plate, 0054-28 on a AAT IIIa-J plate.

The combination of spectroscopy (to weed out numerous foreground field objects) and CCD photometry could unravel how typical Hamilton's objects might be. The most straightforward analysis would be to examine the colour-luminosity relation at various wavelengths and its scatter. If Hamilton's objects are untypical, there will be a weak *red envelope* defined by present-day colours (appropriately redshifted) and a substantial scatter to the blue for the evolving objects.

Gunn and colleagues (this conference) have begun a related approach using the 4000 Å break index, D_{4000}, as a measure of the instantaneous star formation rate. In analysing the distribution of D_{4000}, however, it is important to ensure that the colour-luminosity relation is adequately sampled in the multislit samples. Any claim for a weakening of the D_{4000} break with redshift must first determine what fraction of red objects were actually spectroscopically studied. Systematic magnitude-limited samples represent the safest approach in such circumstances.

Progress will be slow: Absorption line spectroscopy beyond R = 22 is simply too difficult for 4 metre telescopes yet some distance indicator is essential to clarify whether a bluer object is an evolving early-type member or a foreground galaxy. In a few of the AAT distant clusters, substantial foreground groups have been found, and typically 4 redshifts are needed to identify the cluster redshift proper. Once this task is done, an alternative approach would be to image the cluster in many CCD passbands and form a magnitude limited sample substantially deeper than that which can be studied spectroscopically (c.f. Couch *et al* 1983, Ellis *et al* 1985). Assuming enhanced star formation affects primarily the shorter wavelengths, the early-type members can be found by examining colour luminosity relations in one or more colours at wavelengths *longward* of the 4000 Å break. The scatter at wavelengths *shortward* of the break then indicates what fraction have undergone recent star formation activity.

Such multi-band analysis of distant clusters is very successful at identifying red members, particularly when combined with limited spectroscopy for identifying the cluster redshift and probable contaminating groups. In a recent comparison between 86 spectroscopic and photometric redshifts for diverse galaxies with $0 < z < 0.6$ to R = 22 (MacLaren *et al* 1987), *no* significant errors were found in the photometric redshift estimates for early-type cluster members. Surprisingly, the method works *better* at high redshift, the difference between the spectral energy distribution of a cluster member and a typical foreground L^* field galaxy is then enlarged. The method will become more powerful yet with the addition of J,H,K measurements from the newly-commissioned infrared array detectors, since infrared colours are expected to be reliable redshift indicators regardless of their optical SEDs (c.f. Ellis and Allen 1983).

3. CLUSTER GALAXIES AT INTERMEDIATE REDSHIFT

If star formation is extended in duration, even in some proportion of early type galaxies, this might produce observable effects at intermediate redshifts (z ~ 0.3 - 0.5) where samples are less affected by selection problems and where spectra and photometry of high quality can be scrutinised.

Hints that many ellipticals had a secondary burst of activity as recently as 3-5 Gyr ago were first proposed by O'Connell (1980) and more recently in a detailed study by Rose (1985). The optical colours of E and S0 galaxies show remarkable homogeneity for a fixed metallicity because colours are only sensitive to *on-going* star formation. Spectral features present in massive stars (including Balmer lines which weaken the 4000 Å break) are valuable indicators of recent activity but only so long as the predominant stellar contributors remain on the main sequence (< ~ 1 Gyr). Ultraviolet colours have a good sensitivity to both on-going and past star formation (< ~ 2 Gyr) but suffer added complications because of the unknown contribution from hot evolved stars and the very small apertures used with IUE in such studies (c.f. Bertola *et al* 1985). Rose's analysis of gravity-sensitive line indices placed a tighter constraint on the absence of star formation in Virgo ellipticals than that using the IUE data. He also claims that at least *some* ellipticals suffered star formation episodes at epochs within ready reach of large telescopes.

One difficulty, of course, is that a secondary burst of star formation super-imposed on an otherwise uninteresting elliptical would transform its appearance. Without morphological information it might be difficult to distinguish from a normal spiral. A combination of spectral and/or colour indices is required that can yield the ratio of current to average star formation rate.

Fortunately, progress in instrumentation means that many of the techniques useful in studies at z ~ 0 can now be applied at intermediate redshift. Specifically, the useful ultraviolet window (λ_{rest} = 2500 Å) comes within range of optical CCD detectors for z > 0.4 and integrated photometry of ellipticals at this epoch is therefore feasible. Additionally, multiobject spectroscopy with fibre optics (Ellis and Parry 1987) now routinely provides spectra of 4 Å resolution for up to 70 galaxies simultaneously at limits of B ~ 21.5 - 22. Figure 2 shows colour-luminosity relations for a complete sample of over 100 galaxies in the distant cluster Abell 370 (z = 0.37) from the analysis of MacLaren *et al* (1987). As described in §2, those galaxies whose long wavelength colours are consistent with an early type component can be selected and it is gratifying to note they define a remarkably tight colour-luminosity relation similar to that expected from the studies of Sandage and Visvanathan (1978). Note the substantial ultraviolet enhancements in *some but not all* of this restricted class of galaxy in the observed U - 685 nm colour (U corresponds to 2500 Å in the cluster restframe).

Ultraviolet excesses (UVX) are not found at this rest-wavelength in present day E/S0s. However, a similar UVX phenomenon was also identified in the distant cluster 0016+16, a particularly significant result as its blue fraction is much lower than that of Abell 370. Limited spectroscopy of the UVX galaxies by Dressler and Gunn (in 0016+16, Gunn, this conference) and by Mellier *et al* (in Abell 370, this conference) suggests post-starburst features like Hδ may be more common than in the

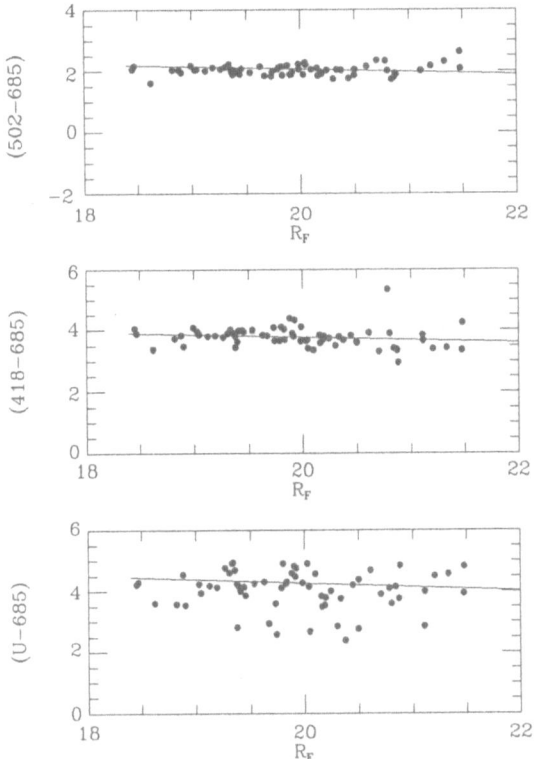

Figure 2: Colour-luminosity relations for objects classed as early-type members in Abell 370 (z = 0.37) from the multi-passband analysis of MacLaren et al (1987).

remaining red galaxies; thus it seems reasonable to attribute the UVX phenomenon to additional young stars.

Further support for the contention that *some* early-types enjoyed recent star formation is also provided by the extensive magnitude-limited spectroscopic study in 3 z = 0.31 clusters by Couch and Sharples (1987). Using the combination of Hδ and broad-band colour, again they identified some red galaxies with anomalously strong Hδ and a remainder indistinguishable from present day cluster E/S0s. Analysis of their composite spectrum for the latter class provides an upper limit for current star-formation activity at z = 0.31 almost as tight as that of Rose's today, and therefore consistent with Hamilton's interpretation at higher redshift.

Whilst much early attention in distant clusters focussed on the *blue* members, it is becoming clear that the recent star formation in the *red* galaxies is not unrelated to the Butcher-Oemler (B-O) effect (Butcher and Oemler 1978). MacLaren *et al* present a sequence whereby some cluster members suffer a burst of star formation for some reason, become optically active (blue B-O galaxies) and decay via the post-starburst phase (prominent Balmer lines) to become optically red galaxies with ultraviolet excesses. About 3-5 Gyr after the burst, there is little evidence of any of the earlier activity. Observations in Abell 370 and 0016+16 show galaxies in various stages of this sequence simultaneously in the same cluster, strongly suggesting that the bursts are locally induced. The fact that 0016+16 at the higher redshift has more galaxies in the advanced stage of this sequence (red or UVX) c.f. Abell 370, clearly demonstrates factors other than the cluster redshift are important in the evolution of galaxies.

Couch and Sharples suggest, from dynamical and other considerations, that the star-forming galaxies in their $z = 0.31$ clusters are destined to become today's S0 galaxies, following arguments of gas-stripping also discussed by Dressler and Gunn (1983). In the absence of any other mechanism for inducing the observed activity, one would therefore expect to detect a much stronger intermediate age population in present day cluster S0s than in ellipticals. However, preliminary studies by Rose, Ellis and Sharples (in preparation) using a large database at 4 Å resolution in the nearby cluster Abell 2670 ($z = 0.07$, Sharples, Ellis and Gray 1987) failed to find any significant difference between the two populations.

If there is a sequence of activity which becomes difficult to identify after a few Gyr, the blue fraction observed at any time ($\sim 0 - 30 \%$) is then of course a lower limit to the number have undergone this sequence during the cluster lifetime, unless we somehow preferentially select clusters displaying the effect (c.f. Koo, this conference). More likely, therefore, we are seeing a phenomenon that affects most early-type cluster galaxies, rather than simply a path for producing present day S0s.

It is now important to determine whether gas stripping e.g. due to infalling spirals colliding with the intercluster medium, is consistent with the dynamical and spatial distribution of the active objects. The curious circular distribution of UVX objects some 300 kpc from the centre of Abell 370 is suggestive of some dynamical process, but the same distribution is not seen in the UVX objects in 0016+16. More statistics are clearly needed via further combinations of spectroscopy and deep wide-band imaging. Another crucial observation would be to morphologically resolve galaxies at various stages in the activity sequence. In the case of Abell 370, where 1 arcsec is equivalent to ~ 3 kpc, no evidence for a difference in the UV and red image profiles can yet be found. Adequate imaging for morphological information cannot yet be done convincingly, though important progress has been demonstrated by Thompson (1987).

4. SURVEYS OF FIELD GALAXIES

For 10 years it has been realised that photographic galaxy number-magnitude counts detect more faint galaxies than the best no-evolution models suggest. Within the photographic region, the comparison project of Koo and Ellis (see Ellis 1987) whereby all number-counters were sent the same photographic data set for analysis, revealed good agreement in detection and photometry within $21 < B_J < 23.5$. At Durham, Jones et al (1987) have now analysed a homogeneous set of 6 prime focus AAT B_J plates each calibrated with precision CCD sequences. The mean slope of these counts, in the same magnitude window, is $d \log N / dm = 0.435 \pm 0.026$, considerably steeper than the best estimate no-evolution slope of 0.34 (see Ellis 1987, King and Ellis 1985).

CCD counts have become available through direct field surveys (Hall and Mackay 1984, Tyson 1987, Metcalfe et al 1987) and Majewski, this conference) and via indirect programmes associated with distant clusters where control fields are required (Yee 1987, Oemler 1987). These studies have their own peculiar problems and a similar comparison of counts in the same field using different detectors/image processing techniques would be well-worth doing. Although Tyson discusses the limiting magnitude in detail, counts fainter than $B_J \sim 26$ are affected by crowding problems for which model-dependent corrections are required. Until fainter counts can be independently checked, it seems safe to adopt $B_J = 26$ as a reasonable deep CCD limit.

This gain in magnitude limit over photographic studies leads to a $\sim$ 10-fold increase in surface density but this is offset considerably in the analyses by a collecting area some 250 times smaller. The referenced CCD surveys contain typically 20 frames, so the slope estimates to, say, $B_J \sim 26.0$ remain rather noisy at present, though the indication is of no significant change; Tyson quotes a mean slope of 0.45 consistent with the photographic data.

The mean $(B_J - R)$ colours of faint galaxies becomes monotonically bluer with faintness and this trend also continues with the CCD data. Frequently this has been claimed as evidence for large numbers of high redshift galaxies undergoing their initial burst of star formation, but both the absence of features in the counts and the detection of a large proportion of faint galaxies in Majewski's (this conference) deep U frames gives a strong indication of more extended star formation.

In deriving quantitative estimates of luminosity evolution from models reproducing the steep count slopes, workers have either made the assumption that all galaxies share the same evolutionary correction (Shanks et al 1984), or at least that the evolutionary correction is fixed for all galaxies of a certain type and $regardless$ of $luminosity$ (Peterson et al 1979, Koo 1986a). On the other hand the no-evolution models might be missing some population of intrinsically faint blue galaxies which would make a major contribution to the counts only at faint magnitudes.

Faint galaxy spectroscopic surveys have begun to address these important questions (Broadhurst et al 1987, Koo and Kron, this conference). CCD colours are also being used as $photometric$ redshift indicators (Loh and Spillar 1986) and will eventually be a powerful extension of genuine spectroscopy. However, careful comparisons of photometric and spectroscopic redshifts are needed, especially for blue field galaxies whose spectral energy distributions have weak/absent 4000 Å breaks (c.f. examples in Figure 4 of Ellis 1987). The major drawback, at present, is that estimating redshifts and properties from colours can only be done for a $priori$ known classes of objects using present-day or model SEDs; in short, the technique could not discover any new kind of object such as any postulated to fit the steep count slope.

The spectroscopic surveys currently achieve completeness at $B_J \sim 21.5$ where the count slope is already steep. Using the AAT fibre optic coupler, Broadhurst *et al* 1987 have completed a spectroscopic survey of over 200 galaxies with $20 < B_J < 21.5$ in 5 fields (see Ellis 1987 for a preliminary discussion). Surprisingly, the mean redshift for each of the 5 fields agrees closely with the no evolution prediction, and the steep count slope appears to be associated with a subset of strong emission line galaxies covering the same redshift range (Figure 3). Restricting the sample to galaxies with rest-frame equivalent widths, W_λ, for [O II] 3727, > 20 Å, we find a slope of 0.6 ± 0.2 over the interval $20 < B_J < 21.5$ c.f. 0.2 ± 0.1 for the remainder. Whilst only a marginal 2σ result, there is at least some suggestion that the emission line population may be responsible for the faint excess.

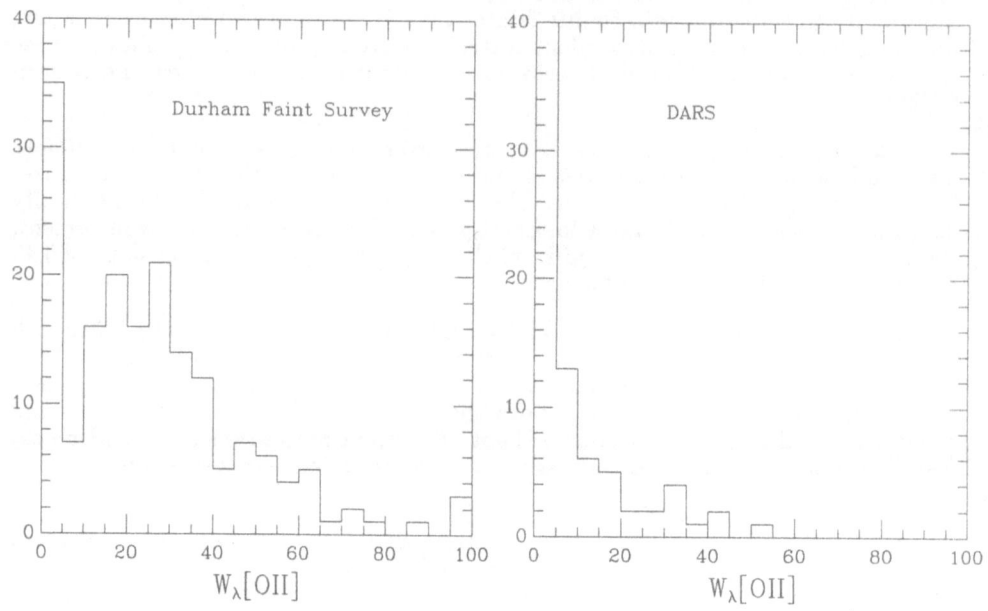

Figure 3: Rest-frame equivalent width distribution for the emission line [O II] 3727 Å in (a) a nearby sample of field galaxies complete to $B_J = 17$ (Peterson et al 1986), and (b) in the AAT faint sample to $B_J = 21.5$ (Broadhurst et al 1987).

The mean spectrum of a subset of the faint emission line galaxies is shown in Figure 4, together with a modelled fit using a passively evolving galaxy upon which a 5 % burst by mass has been added in 10^8 yrs. Such models show a good match in colour and absorption line strengths, *only if the galaxies are being observed during a short-lived burst*. A burst of a few percent by mass in such a short period implies a high star formation rate and a significant temporary increase in the galaxy luminosity of about 2-3 magnitudes. Such galaxies would not otherwise be counted in magnitude limited surveys; they temporarily become visible and represent the excess count at faint magnitudes. Presumably this has to be an evolutionary trend in the sense that the *proportion* of galaxies lighting up increases with z. When more redshifts are available, it will be interesting to examine the spatial distribution and luminosities of the faint emission line galaxies with respect to the remainder. What induces this activity remains unclear but its similarity to the seemingly random behaviour observed in the rich clusters is striking.

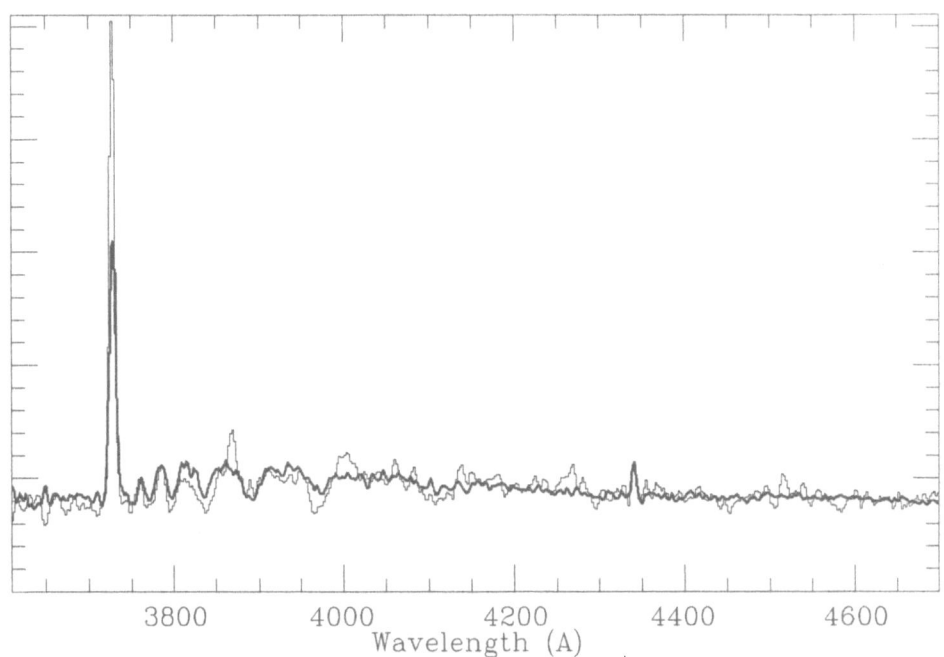

Figure 4: Mean restframe spectrum for faint galaxies in the AAT survey selected with equivalent widths, W_λ ([O II]3727) > 40 Å. The solid line is a model prediction for a 5 % burst of duration 0.1 Gyr superimposed on a passively evolving old galaxy.

5. DISCUSSION AND ANOTHER PATH TO q_0

Evidence is accumulating, from studies in both distant clusters and faint field samples, that extended star formation occurs in restricted classes of objects. The similarity between many aspects observed in the two samples has been noted by Koo (this conference) and used to suggest a novel explanation for the B-O effect

158

as primarily a field evolutionary effect. However, so long as one connects the blue B-O phenomenon to the various kinds of post-starburst activity (Balmer spectra, UVX etc), the evidence for recent star formation in the cores of rich clusters at high redshift seems fairly secure (c.f. Couch and Sharples 1987). Nonetheless, the key to a better understanding of the field and cluster phenomena is better spectra of samples at current magnitude limits, rather than low dispersion spectra of yet fainter objects.

One important distinction in the field samples already apparent, and discordant with traditional ideas invoked to explain the steep galaxy count slope, is that any luminosity evolution might have to be itself luminosity-dependent. This dependence might have a natural explanation if the gas associated with each burst has a characteristic mass. However, ignoring the dependence would yield an incorrect value for q_o from the Loh and Spillar curvature test, since in that method a luminosity density, ϕ^*, is determined at high z on the assumption of an unchanging *shape* for the galaxy luminosity function.

This point brings us back to the original motivation for studying faint galaxies - namely q_o. The results presented suggests no individual galaxy can be assigned a meaningful evolutionary correction; star formation is neither synchronous nor, apparently, well-behaved - even in the most homogeneous early-type galaxies. Abandoning galaxies leaves us with only quasars and supernovae (SNe) as sufficiently luminous to map the curvature by traditional methods. If QSOs could be identified in a systematic way irrespective of their luminosity and/or if some feature in their two-point function could be monitored with redshift, a promising approach might be found. However, the question of biased samples applies just as in the work with galaxies, particularly as it is known their luminosities are strongly evolving (Shanks 1988). On the other hand, SNe are *events*, rather than objects, and so their study offers a refreshing independence from the selection difficulties that have plagued the galaxy work. The idea is not new (c.f. Wagoner 1977) and of course, SNe may themselves evolve (via chemical or other changes), but in principle spectra of the SNe concerned could delineate these effects.

Using a sample of about 70 clusters with $0.2 < z < 0.7$, we have begun a new search for Type Ia SNe via a frequent CCD monitoring programme at the Danish 1.5m telescope on La Silla. At z = 0.3, a SN Ia has a peak luminosity of V = 21.6; including time dilation effects it decays to V = 22.6 within 20 days. Our Danish collaborators have demonstrated the feasibility of a detection to this limit via realistic simulations (Norgaard-Nielsen *et al* 1987). Provided the peak luminosity can be determined from follow-up photometry to within the cosmic scatter ($0.^m25$), in the absence of systematics 15 SNe at z = 0.3 (or 5 at z = 0.5) would determine q_o to ±0.2.

Using van den Bergh *et al*'s (1987) revised SN Ia rate, ignoring galaxy evolution and integrating across the CCD search field of our average distant cluster, we expect to find 0.024 new SNe per comparison. Our "success" rate (*zero SNe*) after a year's painful scrutiny of 65 comparisons is not yet embarrassing but does seem to indicate a lifetime's commitment to the project!

Acknowledgements: Thanks are due to my collaborators Warrick Couch, Ray Sharples and Iain MacLaren on various aspects of the distant cluster work, and

Tom Broadhurst and Tom Shanks on the field redshift survey. I acknowledge useful discussions with Jim Gunn, David Koo, Gus Oemler and Jim Rose.

REFERENCES

Baron, M and White, S D M 1987 *Preprint.*
van den Bergh, S, McClure, R D and Evans, R 1987, preprint.
Bertola, F, Capaccioli, M and Oke, J B 1982 *Astrophys. J.,* **254**, 494.
Butcher, H and Oemler, A 1978 *Astrophys. J.,* **219**, 18.
Broadhurst, T J, Ellis, R S and Shanks, T 1987 *in preparation.*
Bruzual, G 1983 *Astrophys. J.,* **273**, 105.
Couch, W J and Sharples, R M 1987 *Mon. Not. R. astr. Soc.,* in press.
Couch, W J, Ellis, R S, Carter, D and Godwin, J 1983 *Mon. Not. R. astr. Soc.* **205**, 1287.
Couch, W J, Ellis, R S, Kibblewhite, E J, Malin, D F and Godwin, J 1984 *Mon. Not. R. astr. Soc.* **209**, 307.
Couch, W J, Shanks, T and Pence, W D 1985 (*Mon. Not. R. astr. Soc.* **213**, 215.
Cowie, L 1986, preprint.
Dressler, A and Gunn, J E 1983 *Astrophys. J.* **263**, 533.
Ellis, R S 1987, *Observational Cosmology, IAU Symposium 124,* eds. Hewitt et al, D. Reidel Publ. Co., p367.
Ellis, R S and Allen, D A 1983 *Mon. Not. R. astr. Soc.,***203**, 685.
Ellis, R S and Parry, I R 1987, *Instrumentation in Astronomy,* Santa Cruz Summer Workshop, in press.
Ellis, R S, Couch, W J, MacLaren, I and Koo, D 1985 *Mon. Not. R. astr. Soc.* **217**, 239.
Ellis, R S, Couch, W J, MacLaren, I and Malin, D F 1987, in preparation.
Gunn, J E, Hoessel, J and Oke, J B 1986 *Astrophys. J.* **306**, 30.
Hall, P and Mackay C D 1985 *Mon. Not. R. astr. Soc.* **210**, 979.
Hamilton, D 1985 *Astrophys. J.* **297**, 31.
Jones, L R, Shanks, T, Ellis, R S, Fong, R and Peterson, B A 1987 in preparation.
Koo, D C 1981 *Astrophys. J.* **251**, L75.
Koo, D C 1986a *Astrophys. J.* **311**, 651.
Koo, D C 1986b in *Spectral Evolution of Galaxies,* eds Chiosi, C and Renzini, A. (Reidel), p419.
King, C R and Ellis, R S 1985 *Astrophys. J.* **288**, 456.
Lilly, S J and Longair, M S 1985 *Mon. Not. R. astr. Soc.,* **211**, 833.
Loh, E D and Spillar, E 1986 *Astrophys. J.* **303**, 154.
Metcalfe, N, Shanks, T and Fong, R 1987, in *High Redshift and Primeval Galaxies, 3rd IAP Astrophysics Meeting,* eds. Bergeron *et al,* in press.
MacLaren, I, Ellis, R S and Couch, W J 1987 *Mon. Not. R. astr. Soc.,* in press.
Norgaard-Nielsen, H U, Hansen, L and Jorgensen, H E 1987 *ESO Messenger,* **47**, 46.
O'Connell, R 1980 *Astrophys. J.* **236**, 436.
Oemler, A 1987, in *High Redshift and Primeval Galaxies, 3rd IAP Astrophysics Meeting,* eds. Bergeron *et al,* in press.
Peterson, B A, Ellis, R S, Kibblewhite, E J, Bridgeland, M T, Hooley T, and Horne, D 1979, *Astrophys. J.* **233**, L109.
Peterson, B A, Ellis R S, Bean, A J, Efstathiou, G, Shanks, T, Fong R and Zou

Z-L 1986 *Mon. Not. R. astr. Soc.* **221**, 233.

Rose, J A 1985 *Astron J* **90**, 1927.

Sandage, A and Visvanathan, N 1978 *Astrophys. J*, **223**, 707.

Shanks, T 1988 in *IAU Symposium 130, Evolution of Large Scale Structures in the Universe*, eds. Audouze, J and Szalay, A S. D Reidel, in press.

Shanks, T, Stevenson, P R F, Fong, R and MacGillivray, H T 1984 *Mon. Not. R. astr. Soc.*, **206**, 767.

Sharples, R M, Ellis, R S and Gray, P M 1987 *Mon. Not. R. astr. Soc.*, in press

Thompson, L 1987 *Astrophys. J.* in press

Tinsley, B M 1968 *Astrophys. J.*, **151**, 547.

Tyson, A 1987, preprint

Wagoner, R V 1977 *Astrophys. J.* **214**, L5.

Yee, H C 1987, in *High Redshift and Primeval Galaxies, 3rd IAP Astrophysics Meeting*, eds. Bergeron *et al*, in press.

GALAXY MORPHOLOGY AT LARGE REDSHIFT

Gustavo Bruzual
Centro de Investigaciones de Astronomia (CIDA)
Apartado Postal 264
Merida 5101-A
Venezuela

ABSTRACT. The importance of studying the morphology of galaxies at
large redshift is considered. Some speculative estimates concerning
the possibility of recognizing the distant counterparts of the nearby
morphological standards are included. Even though the high angular
resolution data that the HST will provide is not yet available, there
are indirect ways to infer distant galaxy morphology. I review three
of the most frequently used methods: (a) comparing spectral energy
distributions of nearby and distant galaxies, (b) studying the
populations of galaxies in clusters of galaxies at various z, and
(c) comparing faint galaxy counts and color distributions with the
predictions of detailed models.

1. INTRODUCTION

A large fraction of the papers presented in this Workshop on the Understanding of Galaxies at Large Redshifts has addressed the problem of the
evolution of galaxy properties. Specifically, the evolution in time of
the luminosity, the color, and other photometric properties of galaxies,
as well as the evolution of their chemical composition, have been examined
in detail. The interpretation of the available data and the conclusions
that can be derived in all these subjects are model dependent. They also
depend to some extent in the cosmological model.

Even though these models have provided us with a framework for
studying galaxy evolution, there are still many uncertainties in the
models and many questions that remain unanswered. None of these models
can reproduce all the details contained in the observational data, whose
quality has increased impressively in the recent past. The evolutionary
models are continuously being refined with the inclusion of newer and more
accurate data on evolutionary tracks and spectral energy distributions
(SED's) of stars of most known types, and new constraints on chemical
evolution scenarios. These data are being obtained in ground based
observatories. This is also true for the large body of data on distant
galaxies that is being accumulated and which is indispensable to choose
among the many different evolutionary schemes that have been proposed.

R. G. Kron and A. Renzini (eds.), Towards Understanding Galaxies at Large Redshift, 161–176.
© *1988 by Kluwer Academic Publishers.*

The ability to gather data from the ground will increase considerably with the advent of the new generation of large telescopes.

2. MORPHOLOGY AT LARGE REDSHIFT

2.1. Motivation

Contrary to the properties of galaxies mentioned above, the shape and the surface brightness of a galaxy can be determined without reference to cosmological or evolutionary models of any kind. Studying galaxy morphology at intermediate and large redshift we can look directly for systematic changes in the morphological properties of galaxies with redshift. This will be, no doubt, a very important problem for extragalactic astronomers in the near future.

To study morphology of distant galaxies we need very high (less than 1") angular resolution. Ocassionally this resolution is obtained at some priviliged observatories with sub-arc-second seeing, or can be accomplished with the help of adaptive optics systems. However, the study of morphology for large samples of galaxies is obviously a task for the Hubble Space Telescope (HST). We can obtain conservative estimates for the limit in redshift up to which morphology will be studied with high signal-to-noise ratio using the HST.

It is expectd that in the V band a 2000s exposure with the HST Wide Field Camera (WFC) will reach a signal-to-noise ratio close to 3 at a surface brightness of 23 mag/square arcsec. If one assumes that about 10 pixels are required to obtain meaningful structures, one should be able to measure the isophotal radius at 23 mag/square arcsec up to:

$z = 0.2$ (for exponential disk spiral, $r_c = 1$ kpc, $q_o = 0$),
$z = 0.6$ (for standard r^{V_4} elliptical, $r_c = 1$ kpc, $q_o = 0$).

Some of the problems that can be studied with complete samples of galaxies of known morphology are the following:

- Correlation between the morphological type and the local density of galaxies,
- Limiting z up to which the morphology-density relationships are valid,
- Systematic counts of galaxies by morphological type,
- Differences between field and cluster galaxy morphology,
- Search for preferred orientations,
- Evolution of morphological types into the local types,
- Dependence of color gradients on morphology as a function of z.

2.2 Sources of Uncertainty in Morphological Classification at Large z

This is still a speculative subject, but some possible problems that may arise when we try to classify distant galaxies can be anticipated. The morphological groups, as we know them today, have been established locally, based on B and V images of bright galaxies. At large z some or all of the following effects will conspire to make galaxy classification uncertain:

- Galaxies will be faint, due to the $(1+z)^4$ dimming. Features will not be easily recognizable in distant galaxies.

   - The K and evolutionary corrections are different from bulges and disks. In general, bulges will become fainter with respect to the surrounding disk as we observe galaxies at large z (King and Ellis 1985). The classification criteria based on the bulge to disk luminosity ratio will have to be revised. The prediction of the value of this ratio that is expected at large z depends on evolutionary models, as well as in the cosmological model. Figure 1 illustrates the situation for a typical Sab

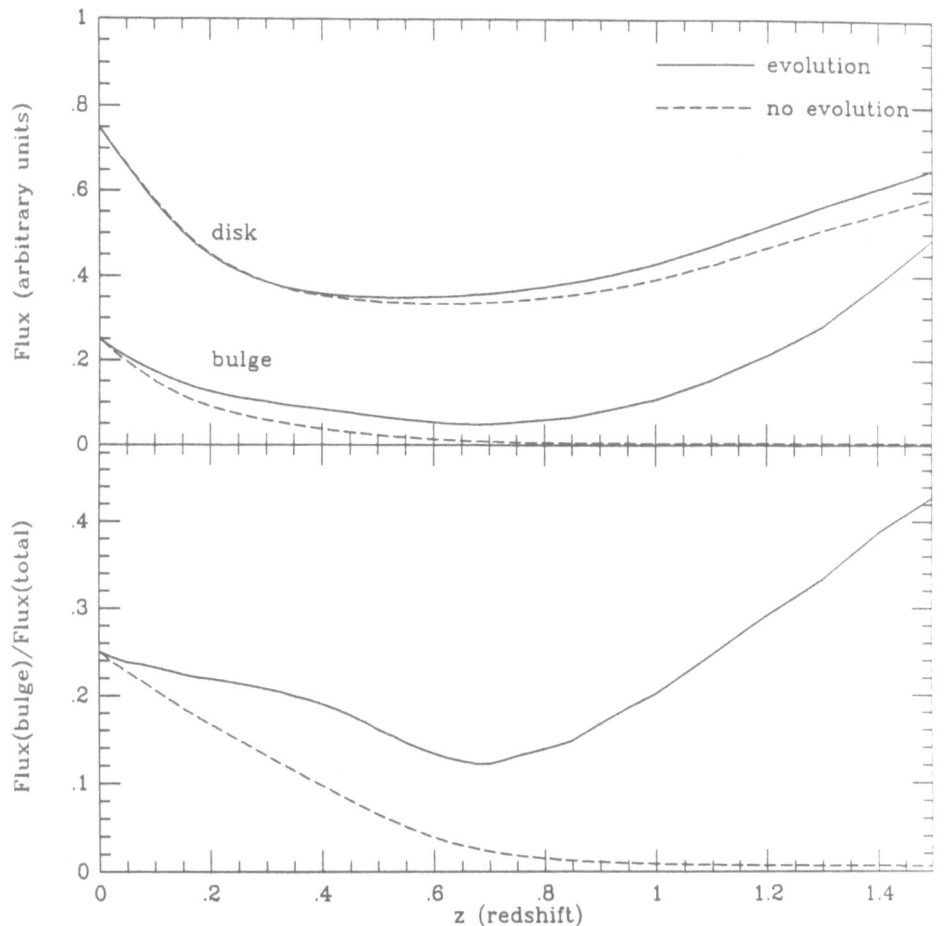

Figure 1. Lower frame: Bulge to disk luminosity ratio for a typical Sab galaxy as a function of z in the observer frame. At z = 0 this ratio is assumed to be 0.25. The solid line indicates the expected behavior if the bulge and the disk luminosities evolve according to the μ = 0.5 and μ = 0.01 models of Bruzual (1983), respectively, (H_0 = 50, q_0 = 0, Salpeter's IMF). The dashed lines include the effect of the K correction alone. Upper frame: Behavior of the bulge and disk luminosities for the same models.

galaxy. At $z = 0$ the bulge to disk luminosity ratio is 0.25. At $z = 0.6$ this ratio is 0.1, if we assume that the bulge and the disk luminosities evolve according to the $\mu = 0.5$ and $\mu = 0.01$ models of Bruzual (1983), respectively. I assumed $H_o = 50$, $q_o = 0$, and I used Salpeter's IMF in both models. This value of the bulge to disk luminosity ratio is close to the value observed in nearby Scd galaxies. If evolution is neglected, at $z = 0.6$ the ratio is close to 0.05 (dashed line). This low value implies that the bulge will be practically imperceptible compared to the disk (7 times brighter according to the evolving line in the upper frame of Fig. 1; at $z = 0$ the disk is only 3 times brighter than the bulge). There are elliptical galaxies that seem to evolve slower than the $\mu = 0.5$ model. If this is also the case for spiral bulges, then the predicted ratio at large z will be even lower than indicated. We can speculate that such galaxies will look doughtnut shaped, a relatively bright disk surrounding a dark central region. Similar conclusions are reached by King and Ellis (1985).

   - Color gradients have been observed in nearby galaxies. These gradients reflect metallicity gradients, but in spirals they may be affected by star forming regions. Spirals seen at high z will, on the average, be undergoing star formation at a higher rate than nearby ones. The galaxies will be bluer where star formation is highest (neglecting large amounts of reddening). If distant enough galaxies are observed, the color gradients may be reversed with respect to the known gradients in nearby galaxies, i.e. distant spirals may be bluer towards the center.

   The classification criteria for galaxy morphology may need to be revised and adapted to the new kind of data that will be available with the HST. One should not expect that the morphological groups determined locally will be enough to classify all large z galaxies.

3. INDIRECT WAYS TO INFER DISTANT GALAXY MORPHOLOGY

Nearby galaxies show a good correlation between the morphological type of the galaxy and its integrated SED (Pence 1976). If we assume (neglecting spectral evolution) that these correlations hold approximately for large z, we can use the observed SED's of distant galaxies to infer their morphological class. In practice, detailed SED's that allow us to perform this comparison are known for very few distant galaxies.

   Fluxes, measured through narrow filters, and photometric colors are related to the SED's and to morphology. These quantities can be measured for larger samples of galaxies, but assigning morphological types to the individual galaxies is not straightforward. Spectral evolution always introduces additional uncertainties in the results.

   I will discuss briefly the techniques that have been mentioned in some detail in this meeting by other participants (see papers by Ellis, Gunn, Koo and others in this volume). I will expand on the problem of the interpretation of faint galaxy number counts and color distributions in terms of simple models for the distribution of galaxies in the universe that include the evolution of the SED's of galaxies of different color classes (related to morphological type).

3.1. Comparison of Galaxy SED's

In this method, the morphology of the local galaxy whose SED matches the distant galaxy SED best, is assigned to the distant galaxy. Ellis et al. (1985) and Ellis (1988) used intermediate band filters to derive both morphological type and redshift for field and cluster galaxies. Loh and Spillar (1986) and Loh (1988) matched observed colors of faint galaxies to colors of a set of fiducial objects, deriving z (and type) for a large sample of galaxies. The following comments are pertinent:
    - This technique can be used to divide distant galaxies into different "spectral classes", e.g. E/S0, Sab, Sbc, Scd, Sdm. This is not a morphological classification properly, but provides us with an indication of the relative numbers of red, intermediate, and blue galaxies.
    - There is increasing evidence from nearby galaxies (Burstein 1986, 1988) that galaxies with identical morphology (E's, S0's) may have very different SED's in the UV region (sampled at high z). The comparison of SED's is useful for deriving the z of a galaxy, but it may be premature to assign a morphological class to a distant galaxy based on this comparison.
    - For distant galaxies one has to allow for some degree of spectral evolution. Models for spectral evolution of galaxies predict that as a galaxy ages its SED travels through the "sequence" Sdm-Scd-Sbc-Sab-ES0 just due to stellar evolution (cf. Figure 4 of Bruzual (1986) with Pence (1976) galaxy SED's). A parallel change in galaxy morphology is not expected. Thus, comparing to local SED's we may misclassify distant galaxies morphologically. The models for spectral evolution are still uncertain and do not provide more accurate morphological types.
    - Galaxies can be classified only according to the types present in the local template. Any incompleteness in this template implies that some galaxies cannot be classified or that will be misclassified. As a matter of curiosity, there are no E+A galaxies in the Loh and Spillar (1986) reference libray (Gunn 1988; Loh 1988).

3.2 Cluster Galaxies

It has become common practice to study distant clusters of galaxies in search of the BO effect (Gunn 1988; Ellis 1988; Henry 1988; Couch and Sharples 1987). From the projected density of galaxies in the cluster, and using the morphological mix-local density relation of Dressler (1980), one can predict the average mix E:S0:Sp expected for the cluster. Using K and evolutionary corrections one predicts the color distribution at the z of the cluster expected for a local counterpart cluster with a similar mix.
    - If the cluster shows the BO effect there will be a large fraction of blue galaxies, some times vaguely called "spirals".
    - Spectrophotometry (Dressler 1986) has shown that many of these blue galaxies are of the E+A type.
    From the morphological point of view it is likely that the galaxies that show the E+A spectrum are S0 galaxies using up the last amount of remaining gas. They could also be Sa spirals. Later spirals are too hot and will not show an SED of the E+A kind.
    As concluded before, star formation changes the SED's of galaxies. Morphological types derived from the SED's may be misleading.

3.3. Multicolor Photometry of Field Galaxies

This section is a brief summary of an ongoing collaboration between the
author and D. C. Koo. More details can bee found in Bruzual and Koo (1988)
and Koo and Bruzual (1988). The objective of the project is to use the
multicolor photometry of Koo (1986) to learn something about the properties of the galaxies in these samples. In particular, we will obtain
information about the distribution of galaxies in different color classes,
which are related to morphological types. The results are necessarily
model dependent, as I will discuss below.

In what follows I will use the data corresponding to CAT 68 in Koo
(1986) to compare the model predictions with the observations. This
catalogue contains photographic magnitudes in the U (3600 A), J (4800 A),
F (6200 A), and N (8000 A) bands for galaxies in the range $19 < J < 23.5$
over a field of 769 square arc minutes. At the bright end I have used
the J vs J-F color distribution of field galaxies in the range $14 < J < 19$
published by Butcher and Oemler (1985). Thus in (J,J-F) the data cover
the range from $J = 14$ to $J = 23.5$

3.3a. Modeling Multicolor Photometry

At a given apparent magnitude m, the number of galaxies of class i,
is given by

$$N(m,i) = \int F(M,i)(dV/dz)dz,$$

where

$$m = M + dm(q_o,z) + K(z) + E(z).$$

$F(M,i)$ is the luminosity function for galaxies of class i, $dm(q_o,z)$ is
the luminosity distance, and $K(z)$ and $E(z)$ are the K and evolutionary
corrections, respectively. The luminosity function is likely to depend
on galaxy type (Binggeli 1986) but for the illustrative purposes of this
talk I will use the Schechter (1976) luminosity function shifted in M by
different amounts depending on the galaxy J-F color at $z = 0$. Using a
variable $F(M,i)$ will not change the main conclusions reached below.

For each member of a family of SED's (observed, evolving models) we
compute the number counts in the U, J, F, and N bands in the range from
16 to 26 magnitude. The U-J, J-F, F-N, and U+J-F-N versus J color distributions (Koo 1986) are computed in the range from $J = 19$ to $J = 26$. In the
case of J-F the distribution is extended up to $J = 14$. The number counts
and color distributions are computed after the incompleteness of the
sample and the random errors in the U, J, F, and N fluxes have been
modeled independently for each band according to the parameters given
by Koo (1986).

We also compute the number counts and color distributions for the
CAT 68 sample. Next we use a Non-Negative Least Squares (NNLS) algorithm
(Lawson and Hanson 1974) to find the (positive) densities $w(i)$ that must
be assigned to each member of the family of SED's to obtain the best fit
to the observed counts and color distributions.

The method outlined above provides an objective choice of the density $w(i)$. We avoid assigning relative weights or fractions to each color class in a given bright-magnitude range based on observations of local galaxies. We also avoid the need to normalize the number counts at the bright end (Bruzual and Kron 1980; Koo 1981; Shanks et al. 1984). The quantity $w(i)$ represents the number of galaxies of class i per unit co-moving volume of the universe that are present in the sample being studied (for the given choice of the galaxy luminosity function).

3.3b. Results

In this section I discuss the results of a model fit to Koo (1986) CAT 68

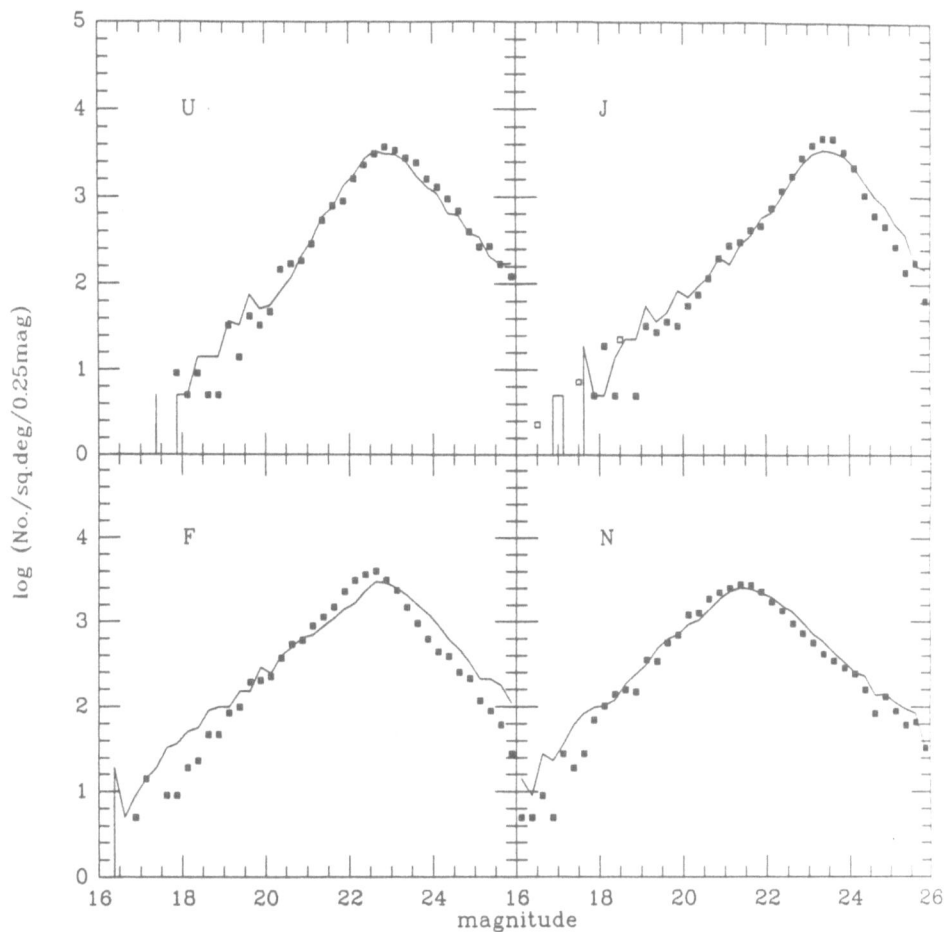

Figure 2. Differential counts of galaxies in the indicated bands plotted every 0.25 mag. The solid squares correspond to CAT 68 from Koo (1986). The open squares (J band) correspond to the Butcher and Oemler (1985) sample. The lines represent the predicted counts.

data. The model was built as indicated above starting from a set of 20 μ-models (Bruzual 1983), with μ in the range from 0.01 to 0.95, and using Salpeter's IMF. The age of galaxies was assumed to be 16 Gyr, and $H_0 = 50$, $q_0 = 0$. The NNLS algorithm assigned non-zero densities to 11 SED's. The predicted number of galaxies per square degree with $14.5 < J < 15.5$ is 1.45, which agrees well with the observed value of 1.5 used by Bruzual and Kron (1980). Even though it is not easy to assign a given morphological type to each model SED, one can group them according to their colors at $z = 0$. The following table gives the number of galaxies per cubic megaparsec required by the model. The agreement with the observed numbers quoted by King and Ellis (1985) is satisfactory.

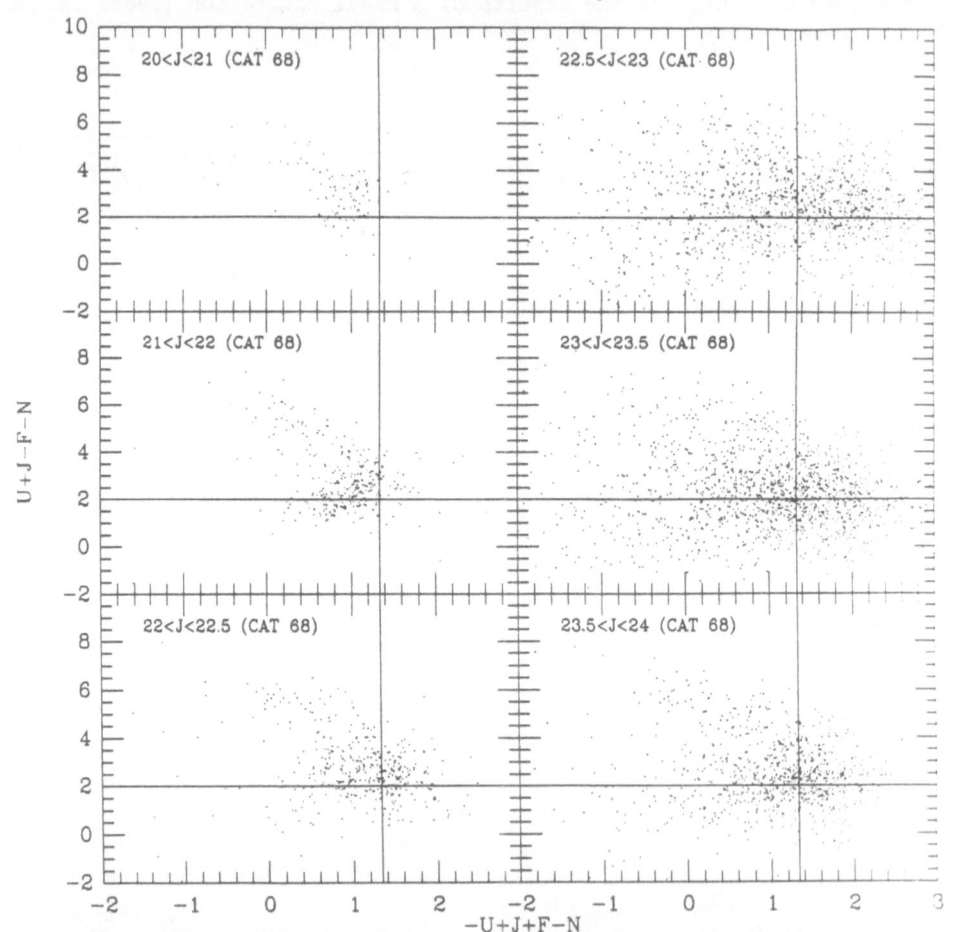

Figure 3. The U+J-F-N versus -U+J+F-N distribution of galaxies from CAT 68 in the indicated ranges of J magnitude. The lines which divide each plot into four rectangles intersect near the centroid of the observed distributions (Koo 1986).

Galaxy Type:	E/S0	Sab	Sbc	Scd	Sdm
King and Ellis (1985):	0.00179	0.00187	0.00187	0.00162	0.00162
This Model:	0.00100	0.00272	0.00244	0.00161	0.00119

Figure 2 shows the differential counts of galaxies per 0.25 magnitude interval in the U, J, F, and N bands for both CAT 68 and our model. The observed and predicted U+J-F-N versus -U+J+F-N distributions are displayed in Figures 3 and 4, respectively, for the J ranges indicated on each frame. Figure 5 shows the U-J, J-F, and F-N color histograms of galaxies for different ranges in J. The solid line corresponds to the model prediction and the dashed line to CAT 68 from Koo (1986). The vertical scale repre-

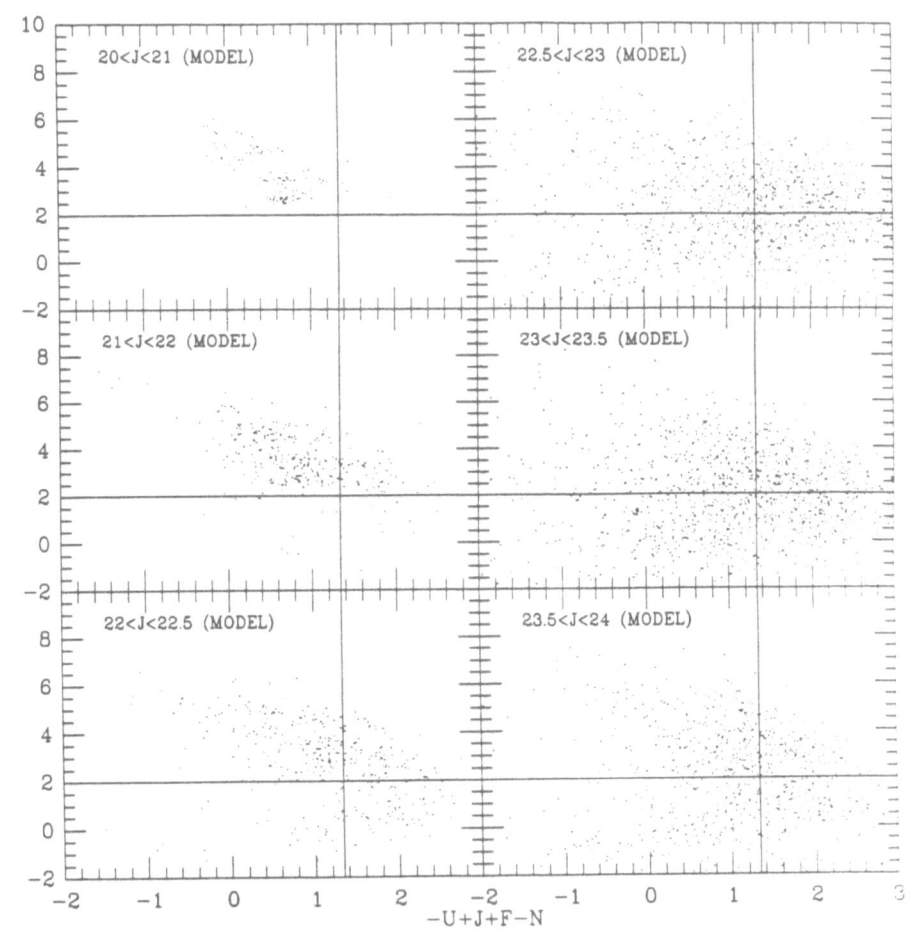

Figure 4. The U+J-F-N versus -U+J+F-N distribution predicted by the model in the indicated ranges of J magnitude. The lines which divide each plot into four rectangles intersect near the centroid of the observed distributions shown in Figure 3.

sents the fraction of the total number of galaxies in each 0.1 magnitude bin in color.

Despite the good agreement between the predicted and the observed number counts seen in Figure 2, we can see in Figure 5 that the observed distributions in U-J and J-F at the bright end (J < 22.5) are not well reproduced by the model. There are relatively bright galaxies in CAT 68 with 0 < J-F < 1 which are not included in the model. A careful comparison of Figures 3 and 4 also shows this lack of bright galaxies at the blue end of the predicted distributions.

In order to explore the nature of the galaxies that are missing in

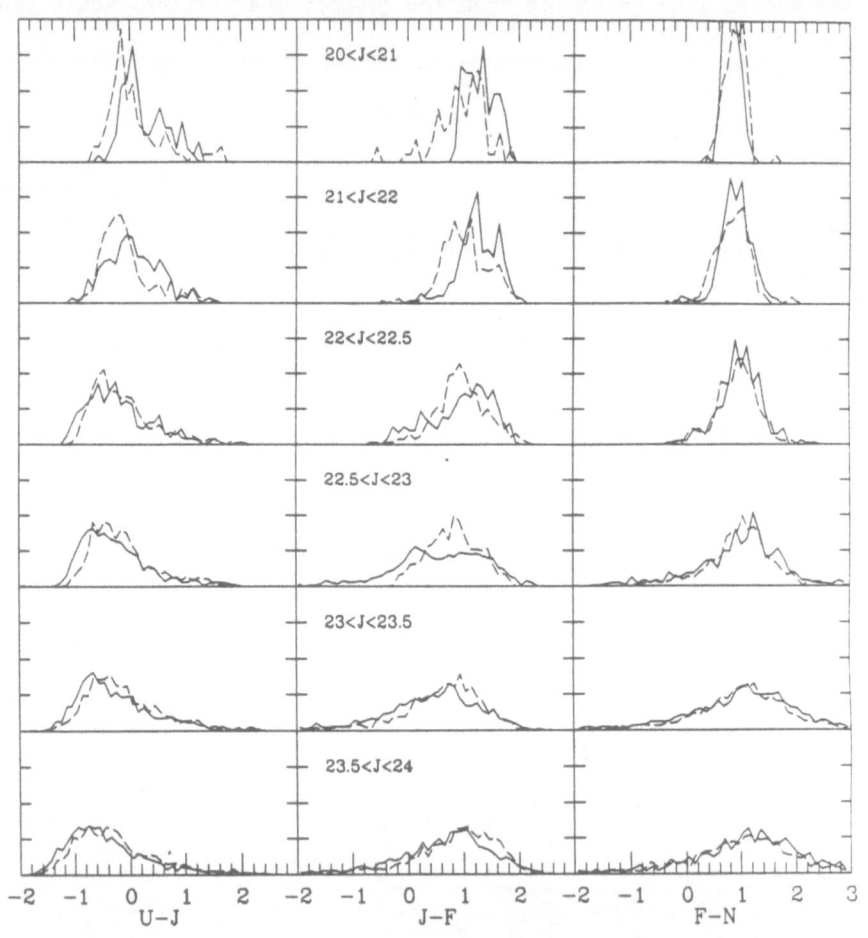

Figure 5. U-J, J-F, and F-N color histograms of galaxies for the indicated J ranges. The solid line corresponds to the predictions of the model and the dashed line to CAT 68. The vertical scale varies from 0 to 0.2 and represents the fraction of the total number of galaxies in each 0.1 magnitude bin in color.

the model it is instructive to plot the redshift distribution. Figure 6
shows the redshift histograms for model galaxies in the indicated ranges
of J. The high z tails (z > 0.5) seen in the predicted z distributions
for 21 < J <22.5 are not present in the redshift surveys that have been
carried out (Ellis 1988; Koo, private communication).

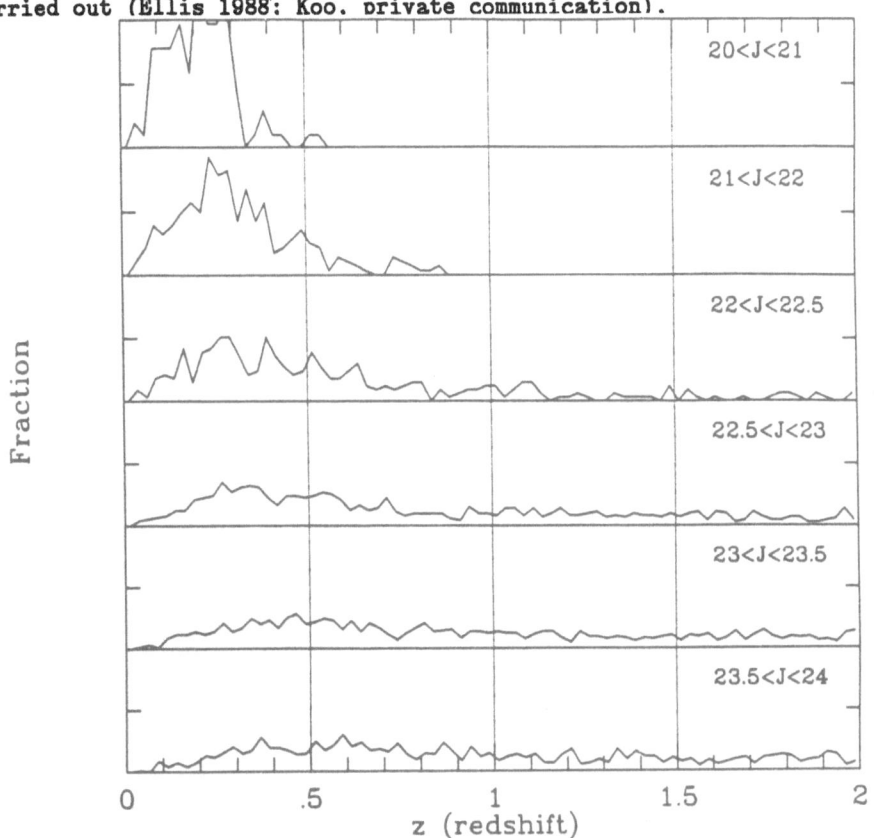

Figure 6. Redshift histograms in the indicated ranges of J for the
galaxies in the model.The vertical scale varies from 0 to 0.1 and
represents the fraction of the total number of galaxies in each 0.025
bin in redshift.

Figures 7 and 8 show the color versus redshift behavior for the SED's
used to build our model. Figure 7 shows the predicted color lines corresponding to each model SED. In Figure 8 each dot represents a galaxy. The
plotted colors were computed after the random errors were added to the U, J,
F, and N fluxes (Koo 1986). Only the galaxies that passed the incompleteness
test have been plotted. Most of the galaxies with 0 < J-F < 1 are at z > 0.5
These galaxies are fainter (on the average) than J = 22 and they do not sho
up in the color distributions in the first two rows of Figure 5. The NNLS
algorithm picks up the high z galaxies in order to get as close as possibl
to the observed color distributions. These galaxies will contribute to the

172

two upper rows of Figure 5 only if they are at the bright end of the
luminosity function. There are very few of these galaxies, which explains
the deficiency seen in Figure 5.

We have explored different shapes of the galaxy luminosity function,
other choices of the cosmological parameters H_o, and q_o, and different
galaxy ages and IMF's. In all cases the deficiency of blue galaxies at
$20 < J < 22.5$ remains. If we consider observed (non-evolving) galaxy SED's,
the deficiency increases. To avoid this problem we must include into the
galaxy count model galaxies that have a very different history of star
formation. They must be quite blue at low z and not unusually bright. It

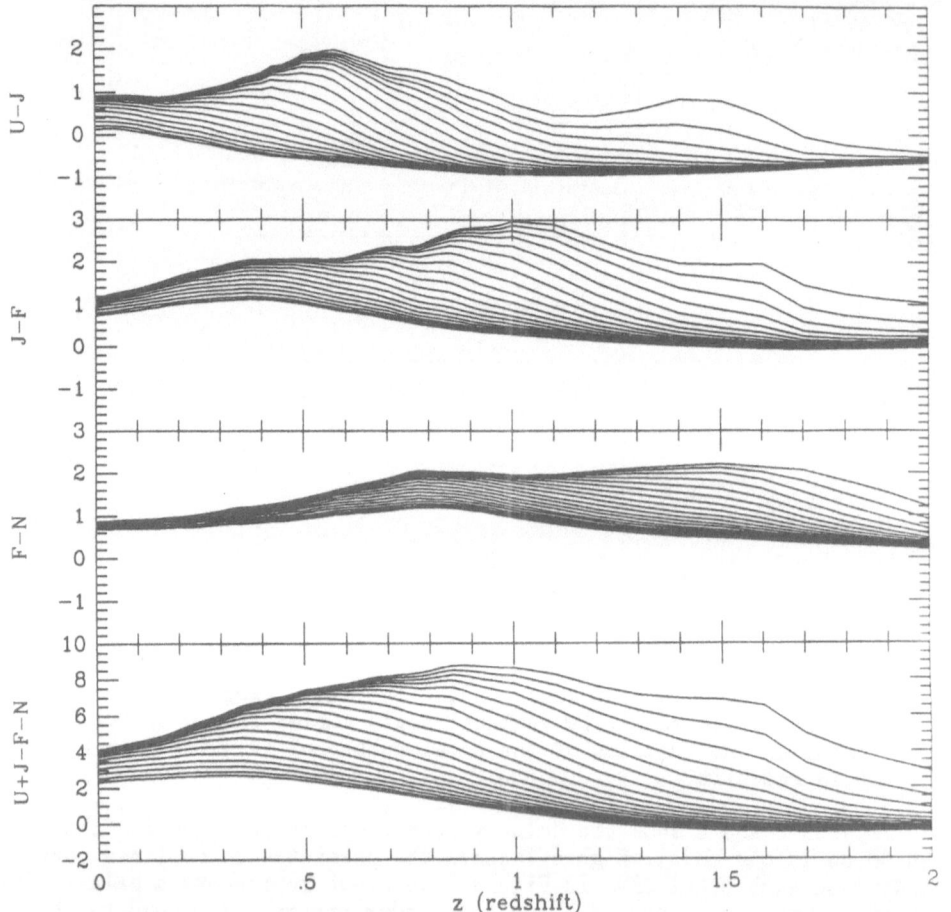

Figure 7. Behavior of the colors U-J, J-F, F-N, and U+J-F-N with
redshift for the SED's used in modeling the galaxy counts. The top
line on each frame corresponds to the μ = 0.95 model SED; μ decreases
by 0.05 for each successive line, except for the bottom line which
corresponds to the μ = 0.01 model. The galaxy age was assumed to be
16 Gyr, and H_o = 50, q_o = 0. The Salpeter's IMF was used.

is tempting to speculate that we need to include galaxies that are seen a short time (1Gyr) after a large burst of star formation took place in them. These galaxies could be the counterparts in the field of the E+A galaxies found in clusters (Dressler 1986). This problem is discussed in detail in Bruzual and Koo (1988) and Koo and Bruzual (1988).

For comparison with the results of Loh and Spillar (1986) Figure 9 shows the behavior of J with redshift for the galaxies shown in Figure 8. The envelope line shown by Loh and Spillar is consistent with the data in Figure 9 if we assume that J = their m(8000) + 1.5; which is reasonable. Casual inspection of both figures shows that at z < 0.1 there are fewer galaxies in our model than in Loh and Spillar sample. This fact is presumably related to the deficiency of blue galaxies discussed before.

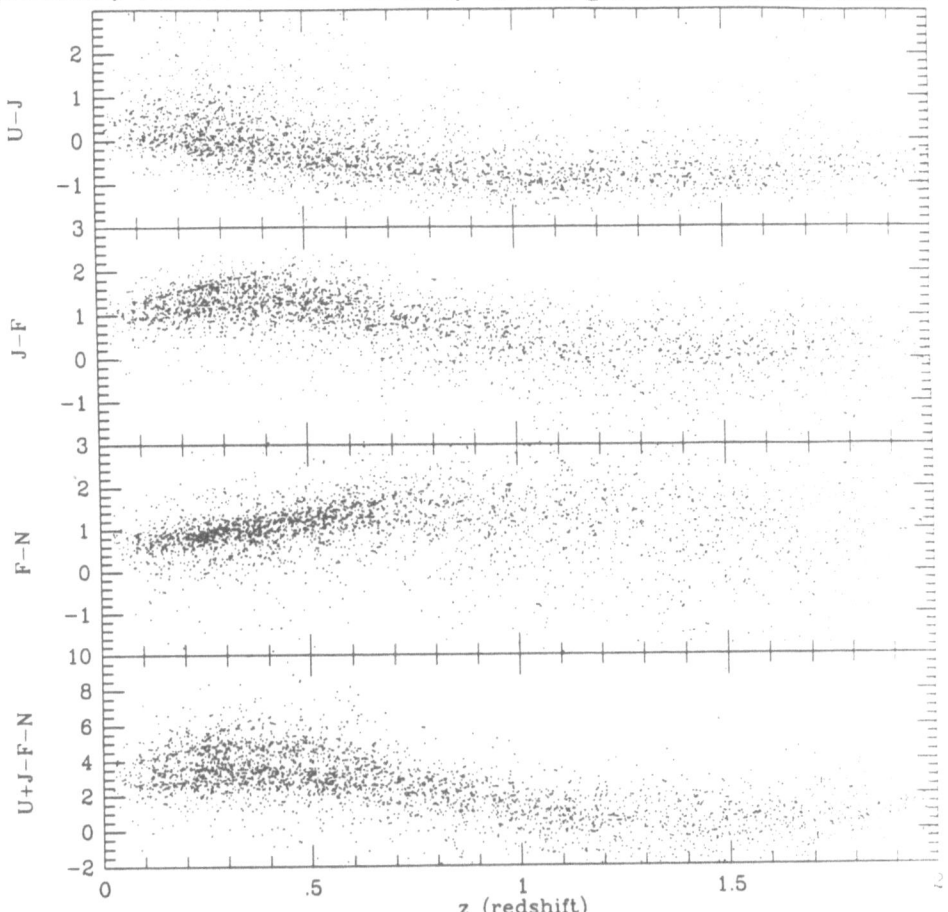

Figure 8. Behavior of the colors U-J, J-F, F-N, and U+J-F-N with redshift for the same SED's shown in Figure 7. Each dot represents a galaxy. The plotted colors were computed after the random errors were added to the U, J, F, and N fluxes (Koo 1986). Only the galaxies that passed the incompleteness test have been plotted.

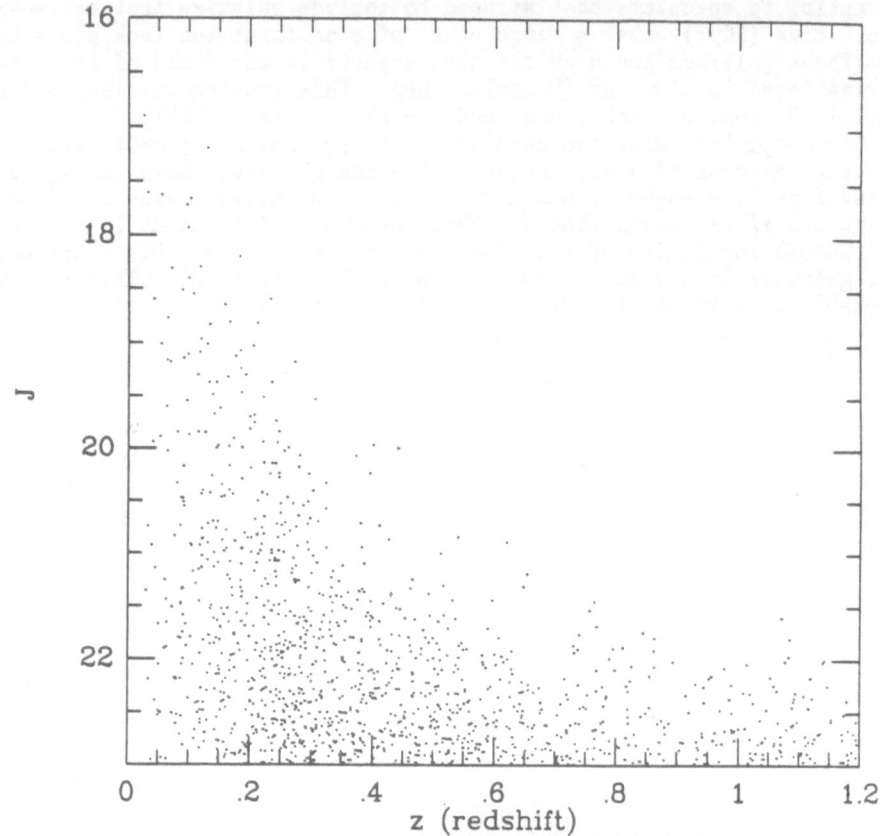

Figure 9. Behavior of J with redshift for the galaxies shown in
Figure 8. The envelope line shown by Loh and Spillar (1986) is
consistent with these data if we assume that $J = m(8000) + 1.5$.

4. CONCLUSIONS

At the present time the morphology of distant galaxies cannot be studied
in a systematic way using ground based instruments. However, by observing
the SED's of individual galaxies and performing multicolor photometry
of large samples of galaxies, we can, in principle, derive estimates of
the relative fractions of galaxies of each kind at different cosmological
epochs.

Comparing SED's of distant and nearby galaxies we obtain a rough
classification of galaxies into different spectral types. This method
has the drawback that the local template may not contain all the variety
of spectra seen in distant galaxies. Besides, spectral evolution is
necessarily neglected when we compare to a local galaxy SED sample.

Studying cluster galaxies people have derived indicators of the
relative number of blue and red galaxies as compared to local clusters.
Even though the BO effect is still controversial (Koo 1988), these

studies have helped identifying peculiar galaxies (E+A kind) not known locally. Little progress has been made concerning the morphology of these systems, but it seems likely that they are SO (or Sa spirals) undergoing strong bursts of star formation in the recent past. We do not know yet if these burst will change the morphology of the system.

Detailed modeling of multicolor photometry has shown that a coeval population of galaxies, evolving according to SED models with exponentially decaying star formation rates, standard luminosity functions and cosmologies, cannot reproduce the observed color distributions, despite the fact that they reproduce the observed galaxy number counts quite well.

The number counts alone do not provide enough constraints on the possible scenarios for galaxy evolution because the predicted counts are not very sensitive to the ingredients used in building the models. The color distributions provide extra information that allow us to build models that we expect to be closer to reality than the models based purely on galaxy counts. The observed color distributions indicate that there are galaxies in the universe that were very blue in the recent past. Possibly these are the field galaxy counterpart of the E+A galaxies seen in clusters.

Finally, reproducing the redshift distribution of the galaxies in the sample provides the most conclusive test for choosing among different models.

REFERENCES

Binggeli, B. 1986 in Nearly Normal Galaxies From the Planck Time to the Present, ed. S. M. Faber, (New York: Springer-Verlag), p. 195.
Bruzual A., G. 1983, Ap. J. 273, 105.
Bruzual A., G. 1986 in Nearly Normal Galaxies From the Planck Time to the Present, ed. S. M. Faber, (New York: Springer-Verlag), p. 265.
Bruzual A., G., and Koo, D. C. 1988 (in preparation).
Bruzual A., G., and Kron, R. G. 1980, Ap. J., 241, 25.
Butcher, H. R., and Oemler, A. 1985, Ap. J. Suppl., 57, 665.
Burstein, D. 1986 in Nearly Normal Galaxies From the Planck Time to the Present, ed. S. M. Faber, (New York: Springer-Verlag), p. 47.
Burstein, D. 1988 (this conference).
Couch W. J., and Sharples, R. M. 1987 (preprint).
Dressler, A. 1980, Ap. J., 236, 351.
Dressler, A. 1986 in Nearly Normal Galaxies From the Planck Time to the Present, ed. S. M. Faber, (New York: Springer-Verlag), p. 276.
Ellis, R. S. 1988 (this conference).
Ellis, R. S., Couch, W. J., MacLaren, I., and Koo, D. C. 1985, M.N.R.A.S., 217, 239.
Gunn, J. E. 1988 (this conference).
Henry, P. 1988 (this conference).
King, C. R., and Ellis, R. S. 1985, Ap. J., 288, 456.
Koo, D. C. 1981, Ph. D. thesis, University of California, Berkeley.
Koo, D. C. 1986, Ap. J., 311, 651.
Koo, D. C. 1988 (this conference).
Koo, D. C., and Bruzual A., G., 1988 (in preparation).

Lawson, C. L., and Hanson, R. J. 1974 in Solving Least Squares Problems,
    Prentice Hall.
Loh, E. D., and Spillar, E. J. 1986, Ap. J. (Letters), 307, L1.
Loh, E. D. 1988 (this conference).
Pence, W. 1976, Ap. J., 203, 39.
Schechter, P. 1976, Ap. J., 203, 297.
Shanks, T., Stevenson, P. R. F., Fong, R., MacGillvray, H, T. 1984,
    M.N.R.A.S.,206,767.

THE RED ENVELOPE AND THE AGE OF THE UNIVERSE

Robert W. O'Connell
Astronomy Department
University of Virginia
Charlottesville, VA 22903 USA

ABSTRACT. The photometric evolution of old stellar populations is slow in the optical/IR. Consequently, observational error is a strongly limiting factor in the accuracy possible for determining galaxy ages or estimating q_0 at high redshifts from indices such as the 4000 Å Break. Existing observations of the red envelope at $z \gtrsim 0.3$ are consistent with a wide range of evolutionary histories. The middle-UV spectral region, $\lambda_{rest} = 2000\text{-}3300$ Å, appears to promise less ambiguous results.

INTRODUCTION

The original motivation for observations of very distant galaxies was to study the large scale structure of the universe and in particular to estimate the deceleration parameter, q_0, which in combination with the Hubble constant, H_0, determines the age of the universe in the Friedmann models. The classical method (Sandage 1962) involves determining the dependence of apparent magnitude on redshift, z. Unfortunately, it is necessary to make a number of significant corrections (*e.g.* for redshift, sampling, aperture, and evolutionary effects) to the magnitude data on objects at $z \gtrsim 0.2$. These have rendered the cosmological problem less tractable than originally hoped (*e.g.* Gunn & Oke 1975, Tinsley 1977, Kristian *et al.* 1978, Spinrad & Djorgovski 1987), and progress toward an unambiguous measure of q_0 by these methods has been slow.

As our ability to obtain photometry and spectroscopy of distant systems has improved, an alternative approach has suggested itself—namely to use the *colors* or other spectral characteristics of luminous galaxies viewed at large lookback times to estimate their *ages* and thereby constrain H_0 and q_0. The first applications of this method have yielded large values for the age of the galaxies and hence low values of q_0 for a given H_0. Hamilton (1985) and Spinrad (1986) claim that the presently fashionable inflationary model, for which $\Omega = 1$ ($q_0 = 0.5$), would be excluded given the best current estimates of H_0, which are ≥ 40 km s^{-1} Mpc^{-1}.

While the spectrum/age-dating method is immune to some of the corrections

R. G. Kron and A. Renzini (eds.), Towards Understanding Galaxies at Large Redshift, 177–185.
© *1988 by Kluwer Academic Publishers.*

plaguing the $m - z$ method, it is, of course, not without its own difficulties. I want to consider some of those in this paper. The issues are identical to those involved in the question of how well we can judge the *evolutionary state* of galaxies at high redshifts and hence in the controversies over the interpretation of the Butcher-Oemler effect (*e.g.* Oemler 1986 and many contributions in this conference) and the ages of elliptical galaxies (O'Connell 1986a, Renzini 1986, Pickles 1987 & this conference). These are important problems in their own right, independent of cosmological questions.

AGE-DATING THE RED ENVELOPE

For cosmological applications of the spectum/age-dating method one clearly wants to choose systems which started forming stars at as early an epoch as possible. For technical reasons (*e.g.* O'Connell 1986a) it is also desirable that they *completed* star formation early or at least that star forming activity at more recent epochs was minimal. Now, it is well established that at any redshift up to $z \sim 0.7$ there is a well defined upper boundary or *red envelope* to the color distribution of galaxies. At $z = 0$ this occurs at $(B - V) \sim 1.0$ while at $z \sim 0.3$ it has shifted to $(B - V) \sim 1.7$ in the laboratory frame as a consequence of the K - correction (*e.g.* Kristian *et al.* 1978, Spinrad 1986). There are very few objects located redward of this envelope. At low redshifts, the envelope is occupied by the most luminous E/cD/S0 galaxies. At moderate redshifts, the most luminous objects also tend to fall near the envelope (*e.g.* Butcher *et al.* 1983, Lilly & Gunn 1985) and it is presumed that these are also E/cD/S0's, although morphological data is usually not available. For $z \gtrsim 0.4$ however, brightest cluster members exhibit a wider range of colors (Eisenhardt & Lebovsky 1987; Persson, this conference), and it is not clear whether all of these are E/cD/S0's or how they are related to the red envelope systems.

In the absence of anomalies such as nonthermal radiation, dust, or unusually high metal abundance, it is undoubtedly valid to assume that the objects consti-tuting the red envelope are the *least active* among the sample at a given redshift in terms of star formation over the preceding few Gyr. They are therefore the best systems with which to pursue the cosmological problem. However, to de-cide whether or not these objects completed star formation at an early epoch and have been completely quiescent since—*i.e.* whether they resemble globular clusters, which have been quiescent for ~ 15 Gyr—requires a careful analysis of their spec-tral properties. Proximity to the red envelope is not sufficient to guarantee this, as will be clear shortly.

To probe the evolutionary state of any galaxy through integrated light obser-vations one requires: *(i)* accurate data and *(ii)* a robust spectral synthesis technique. The reliability of spectral synthesis techniques is a bone of contention at present (*e.g.* Pickles, this conference), but for the purposes of this section I will *ignore* this and imagine that a perfectly robust technique is available (as it surely will be at some point in the future). I will focus instead on the first requirement and ask: what observational precision is necessary to make unambiguous inferences in this problem?

The properties of the integrated spectrum which are of interest are continuum slopes (or colors), continuum breaks, absorption line strengths, and so forth—which I will refer to collectively as *indices*. Numerous spectral synthesis models for such indices are available in the literature, and it turns out that for a quiescent stellar population which is aging in the absence of new star formation the models indicate their evolution can in most cases be well approximated by the simple relationship

$$Q(\mathrm{mag}) = a + b \log t_{\mathrm{Gyr}}$$

where Q is the index expressed in magnitudes, a and b are constants, and t is the age of the population. This approximation is best for $t \geq 1$ Gyr but is usually adequate for $t \geq 0.1$ Gyr.

It is evident from this expression that spectral indices evolve more rapidly at early times (0.1-1.0 Gyr) than at late times ($t \geq 3$ Gyr). Consequently, unless the coefficient b is large, the photometric evolution of old populations will be *slow*. In fact, b is not large for most optical/IR indices. Taking data from models by Larson & Tinsley (1978) and Rabin (1980), I find the following b's for $(U - V)$, $(B - V)$, and $(V - K)$, respectively: 0.80, 0.33, 0.47 (mags). Individual absorption line indices, *e.g.* Mg I, typically have shallower time dependences than the long-baseline continuum colors.

To illustrate the effects of slow photometric evolution on interpretation of high redshift data consider the case of the "4000 Å Break" index, which is often used as an evolution diagnostic (*e.g.* Bruzual 1983; Hamilton 1985; Spinrad 1986; Gunn, this conference). Values for the standard Break index, $D(4000)$, in old population models with solar abundance and e- folding times for star formation of 0.25 Gyr are given by Hamilton. I have converted these to magnitude form, $d(4000) \equiv 2.5 \log D(4000)$, and find that the b coefficient is 0.37 mag for $t > 3$ Gyr. Not surprisingly, the sensitivity of d to age is comparable to that of $(B - V)$.

This formulation of spectral index evolution may be combined with the look-back/redshift function in any cosmological model to derive the dependence of index on redshift. In Figure 1, I show the resulting $d(4000) - z$ diagram for $H_0 = 50$ km s^{-1} Mpc^{-1}, $\Lambda = 0$, and several assumed values of q_0 and z_{GF} (the formation redshift). (Note that the d index is measured in the rest frame of the galaxy and is not subject to a K-correction.)

The character of this diagram is governed mainly by the b coefficient. The fact that b is small implies, for example, that the four plotted curves are separated by only 0.11 mag at $z = 0$ despite an age difference from top to bottom of 7.8 Gyr. Changing the value of H_0 results in only a zero-point offset to the curves, which amounts to -0.07 mag if $H_0 = 75$ km s^{-1} Mpc^{-1} were adopted instead. In any attempt to estimate q_0 from such diagrams, one presumes that a good value of H_0 would be available from other observations.

The slopes of the curves depend on q_0 but only slightly. This implies that a determination of q_0 requires proper modeling of the absolute Q values or, equiva-

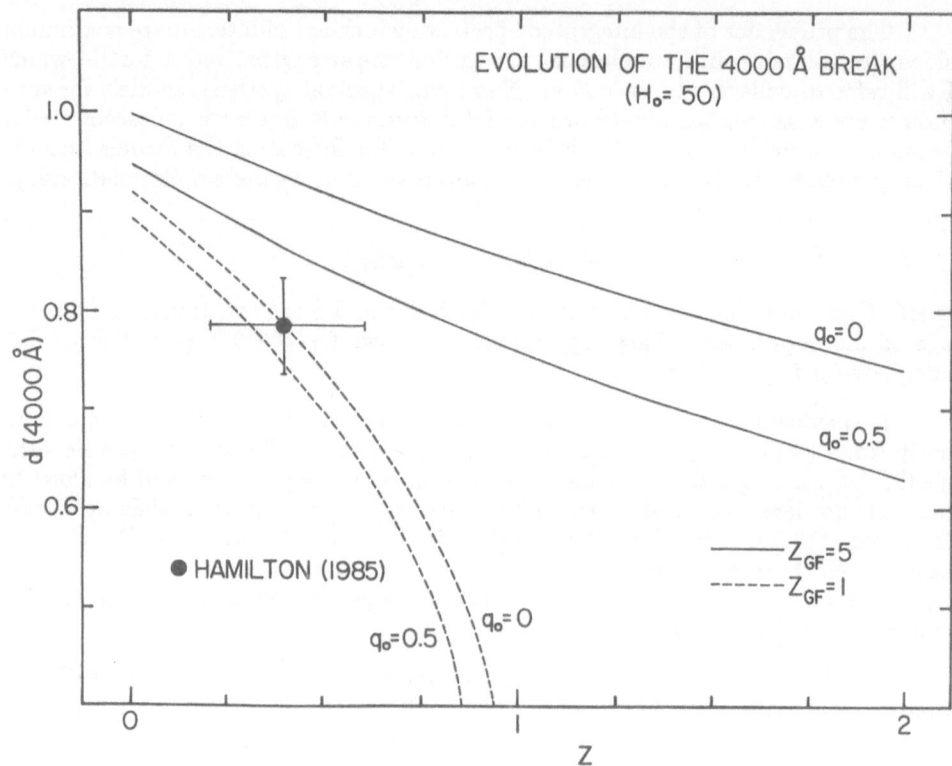

FIGURE 1: The evolution of the 4000 Å Break index (in magnitudes) for two values of q_o and the galaxy formation redshift, z_{GF}. Based on Hamilton's synthetic models, for which $a = 0.55$ and $b = 0.37$, as described in the text. The plotted point is the mean value for Hamilton's red envelope sample; the horizontal bar encompasses 80% of his objects.

Hamilton's data is plotted mainly to illustrate the presently achievable σ_{obs}; however its interpretation is worth a comment. The mean break index shown corresponds to an age of 4.3 Gyr in the context of his $\tau = 0.25$ Gyr model and falls on the $q_o = 0$, $z_{GF} = 1$ curve. Hamilton was unwilling to accept so recent a formation epoch and suggested that his model d indices were too large because of a mismatch in metal abundance. However, his $Z = Z_\odot$ models actually should represent integrated gE light well (Baum et al. 1986). Hamilton's data showed no strong trend with z, probably because of selection effects. The mean trend lines of the Spinrad (1986) and Gunn (this conference) samples, however, fall near the $z_{GF} = 1$ curve. No strong conclusions should be drawn, but the properties of the red envelope are not inconsistent with relatively recent formation of E galaxies.

lently, *absolute age determinations*. At $z \sim 1$, the $q_o = 0$ and $q_o = 0.5$ curves for $z_{GF} = 5$ are separated by only 0.09 mag, which implies that a 3σ determination of q_o to ± 0.1, for instance, requires a photometric precision of $\sigma_{obs} \sim 0.006$ mag.

This S/N requirement seems well beyond the current state of the art for optical/IR observations. The data point plotted in Figure 1 is from Hamilton's (1985) spectroscopy and represents the mean for 26 faint galaxies, a considerable investment of observing time. For this sample, $\sigma_{obs} = 0.05$ mag., far larger than required for an accurate q_o determination. To put this another way, this σ_{obs} corresponds to an age differential of $\pm 30\%$ or a $t(max)/t(min)$ ratio of 1.9. Since the dominant contributors to observational error on distant objects are not necessarily photon statistics but instead uncertainties in the sky background, flat-fielding, flux calibration, redshifts, and so forth, this situation is not readily ameliorated simply by using longer integration times.

The conclusion is that, quite apart from uncertainties in synthesis modeling, observational error is a *strongly limiting factor* in the accuracy with which one can age-date the red envelope or obtain q_o values at high redshifts from observations at (restframe) optical/IR wavelengths. In fact, since σ_{obs} at $z \sim 0$ is much smaller, Figure 1 suggests that this is better done on local objects than at high z. But at any redshift, one desires indices with larger b coefficients.

THE USEFULNESS OF THE MIDDLE-UV FOR AGE-DATING

Because of the strong dependence of UV flux on stellar temperature, spectral indices in the middle-UV region (2000-3300 Å) have larger b coefficients than the optical/IR indices discussed above. For instance, I find that the index $(m_{25} - V) \equiv -2.5 \log \left[F_\lambda(2500 \text{ Å}) / F_\lambda(5500 \text{ Å}) \right]$ has $b = 3.85$ mag, which is *ten times* larger than for $(B - V)$ or $d(4000)$. Such middle-UV indices therefore promise superior age-dating resolution. They are particularly worth considering for the high redshift problem because the restframe UV is shifted to wavelengths which are readily observable from ground-based observatories if $z \gtrsim 0.5$.

There are difficulties with the middle-UV, however, most of which have to do with our current lack of understanding of this region in composite light. I list some of these below more in the spirit of pointing out interesting astrophysical problems to be solved than as mere technical gremlins:

(*i*) Galaxies are fainter in the UV than in the visible. Observations by Smith & Cornett (1982) and Burstein *et al.*(1987) indicate that $(m_{25} - V) \sim 2$-4 mags for nearby E/S0 galaxies. Photometric S/N from source photon statistics will therefore be lower in the UV. Extinction from foreground dust will also be higher. However, the sky background for 3500-6000 Å will be fainter by 2-3 mags than in the near-IR, meaning that restframe UV data for distant objects is subject to less sky noise than restframe visible/IR data. One should compare the S/N resulting from these observational factors with the greatly improved segregation in the UV $Q - z$ diagram. If effects other than source photon statistics dominate the net S/N, then UV brightness may not be a serious limitation.

(ii) E/S0 galaxies contain a still-unidentified hot component which produces significant flux in the far-UV ($\lambda \leq 2000$ Å) region (reviewed in O'Connell 1986a). Its amplitude is apparently correlated with galaxy metal abundance (Burstein *et al.* 1987). I have computed expected $(m_{25}-V)$ colors for old stellar populations without this "UVX" component. When these are compared to the large aperture middle-UV data of Smith & Cornett (1982) or the nuclear *IUE* data of Burstein *et al.*, I find most galaxies exhibit a 1-2 mag excess in $(m_{25} - V)$ regardless of the age assigned to the old population. My models assumed $Z = Z_\odot$, and since metal abundances in E galaxies remain above solar to relatively large radii, this excess is unlikely to be a metallicity effect. It is reasonable to attribute it to the long-wavelength tail of the UVX component. The implication is that an understanding of the UVX component is important to interpreting middle-UV as well as far-UV data. If the UVX component is directly linked to the old population (*e.g.* post-AGB stars), which I think is likely, this problem will be less serious than if it is wholly independent (*e.g.* recently formed massive stars).

(iii) Smith & Cornett (1982) performed large aperture broad-band photometry on a number of Virgo E/S0 galaxies in the middle-UV and found for the S0's a strong color-luminosity dependence, $(m_{25} - V) \sim -0.70V + 11.4$, where V is the apparent V magnitude of the galaxy. The brightest E galaxies appeared *not* to follow this same relation. The photometry has relatively low S/N, but it does represent the largest homogeneous sample of integrated UV data on E/S0's, and it raises the spectre of significant luminosity and morphological effects on middle-UV colors.

(iv) Middle-UV spectra will be sensitive to metal abundance as a consequence of strong metallic line blanketing. Good estimates of blanketing effects are not available yet, but based on Kurucz's (1979) model atmosphere calculations for solar-type stars, I estimate $\delta(m_{25} - V) \sim 1.6 \log Z/Z_\odot$. This, again, is a relatively strong effect which must be properly calibrated in synthesis models. Good methods for determining abundances in high redshift systems are also evidently essential.

(v) Finally, most data on distant systems will be for their integrated light, and comparison to nearby galaxies therefore requires proper UV-color/aperture corrections. One expects significant gradients from the outward decline of metal abundance (*e.g.* Baum *et al.* 1986), but the available sample of UV photometry is too small and inhomogeneous to study aperture effects. Unfortunately, the *Hubble Space Telescope* is not well suited to obtaining off-nuclear UV surface photometry of E/S0 galaxies, leaving the three UV instruments of the *ASTRO* Spacelab missions as the only likely important contributors to our understanding of aperture effects for the next 5 years or beyond.

Does this long and sobering list of problem areas remove the middle-UV from consideration for studying the evolutionary state of distant galaxies? I think not, because most of these arise from the *increased sensitivity* of the UV to precisely those parameters we want to measure—namely turnoff temperatures and metal abundances—and the complicating factors must be considered in the context of the increased amplitude of all such photometric effects in the UV. Some of these issues—*e.g.* the nature of the UVX components—should be settled quickly by *HST* and *ASTRO*. Others are no more serious and, given the relative amplitude of the effects, may be less serious than comparable difficulties and ambiguities in the visible region (*e.g.* the problem of separating evolved from main sequence starlight).

To close on two positive notes, first Ellis (this conference) has already found evidence at $\lambda_{rest} \sim 2500$ Å of recent star formation in cluster E/S0's at moderate lookbacks which is not detectable at visible wavelengths. This will serve as an important probe of the "post-starburst" or "E+A" phenomenon. Second, absorption line features in the middle-UV are very sensitive to the age of the stellar population (see Figure 2), meaning that moderate resolution UV spectroscopy of nearby systems with *HST* and *ASTRO* should rapidly improve our understanding of this spectral region.

LATE EVOLUTION OF ELLIPTICAL GALAXIES

The nature of the red envelope is also a key element in the debate over the evolutionary history of E galaxies. There is a substantial body of evidence which indicates that significant star formation in nearby E/S0 galaxies continued until as recently as 5-8 Gyr ago. Since this has been covered at length in recent reviews (Burstein 1985; O'Connell 1986a, 1986b, 1987; Pickles 1987) I will not discuss it here. This is a controversial conclusion, however, and the existence of the red envelope at higher redshifts is often cited by critics as being inconsistent with this interpretation.

It should be clear from the preceding why the red envelope does not invalidate this late-evolution picture for E galaxies. First, the rapid evolution of color during the first 3 Gyr after star formation ceases implies that the ancestors of local E/S0's could readily have made the transition from the "blue population" at moderate redshifts to the $z = 0$ red envelope in the available time period. Such "quenched" evolutionary paths might be expected if mergers or other strong environmental interactions are common; examples were published some time ago by Tinsley (1980, Fig. 1). Second, the slow color evolution after 3 Gyr coupled with photometric errors and residual uncertainties in the evolutionary models implies that the red envelope at moderate redshifts is consistent with a *wide range* of evolutionary interpretations. Bruzual (1986, Fig. 6) and Wyse (1985) were able to fit the red envelope data for $z \lesssim 0.5$ with models having ages at $z = 0$ of 6-16 Gyr, depending on modeling assumptions and the metal abundance assigned. The envelope at $z \sim 0.5$ was fit by Bruzual's models with an age at that epoch of $\gtrsim 3$ Gyr. Overall, I think there is nothing in the available high redshift data which constitutes a serious challenge to the late-evolution interpretation of nearby E's, and there is a good deal which is consistent with it (see, for example, the discussion in the caption for Figure 1).

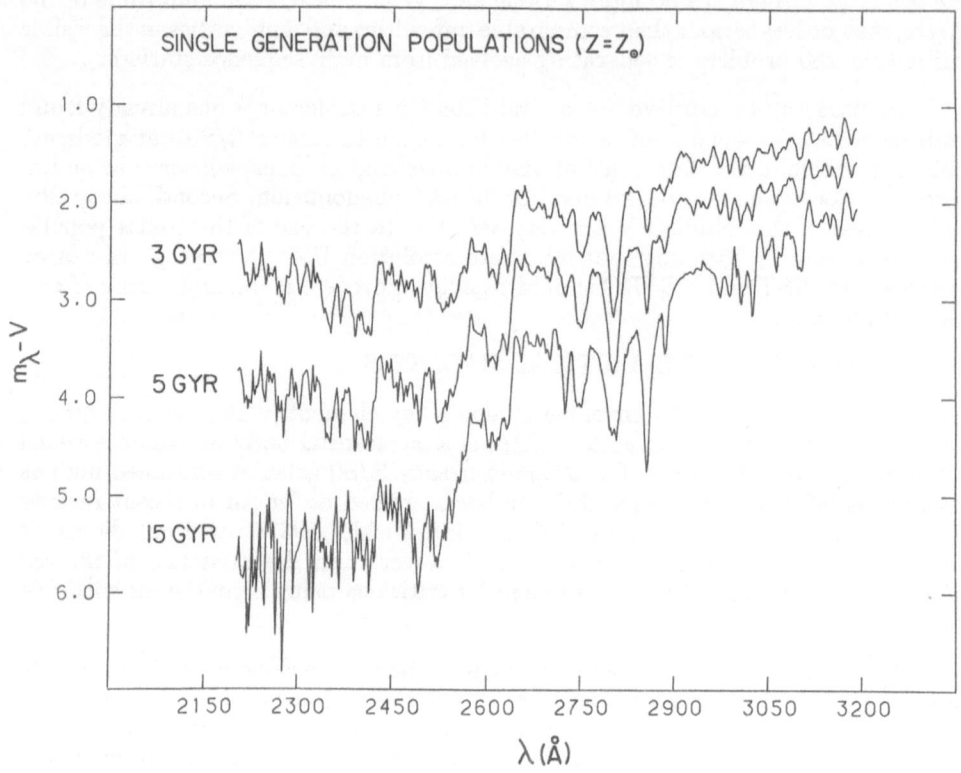

FIGURE 2: Synthetic middle-UV spectra for single generation, old populations with $Z = Z_\odot$, based on the *IUE UV Spectral Atlas*. All spectra are normalized at 5500 Å. Ages are indicated. Note the significant changes in both continuum level and line strengths with age. The optical/IR spectrum is much less sensitive to age.

CONCLUSION

The slow evolution of optical/IR photometric indices in old stellar populations places heavy demands on both observational precision and synthesis modeling techniques if the goal is to obtain a good estimate of q_o or the ages of high redshift galaxies. For indices such as the 4000 Å Break, one requires $\sigma_{\text{obs}} \lesssim 0.01$ mag, and this seems beyond the current state of the art. Consequently, the existing data on the red envelope is consistent with a wide range of evolutionary histories and, in particular, with the late-evolution interpretation of nearby E's. The restframe middle-UV promises much improved age-dating of old populations if a number of technical and astrophysical complications can be successfully addressed. None of these appears intractable. It remains to be seen whether such methods of estimating q_o, based on stellar evolution timescales, will be competitive with results based on nuclear chronometers (Fowler 1987).

Finally, I think it is important to have a better understanding of the physical sources of photometric dispersion in luminous red galaxies for $z \lesssim 0.5$ (e.g. Wilkinson & Oke 1978) and to establish the nature of the reddest outriders on the color distribution.

This work was partially supported by NASA grant NAG-700.

REFERENCES

Baum, W.A., Thomsen, B., and Morgan, B.L. 1986. *Ap. J.*, **301**, 83.
Bruzual, A.G. 1983. *Ap. J.*, **273**, 105.
Bruzual, A.G. 1986. In *Spectral Evolution of Galaxies*, eds. C. Chisoi and A. Renzini (Dordrecht: Reidel), p. 263.
Burstein, D. 1985. *P.A.S.P.*, **97**, 89.
Burstein, D., Bertola, F., Buson, L., Faber, S., and Lauer, T. 1987. In preparation.
Butcher, H., Oemler, A., and Wells, D.C. 1983. *Ap. J. Suppl.*, **52**, 183.
Eisenhardt, P.R.M., and Lebofsky, M.J. 1987. *Ap. J.*, **316**, 70.
Fowler, W.A. 1987. *Q.J.R.A.S.*, **28**, 87.
Gunn, J.E., and Oke, J.B. 1975. *Ap. J.* , **195**, 255.
Hamilton, D. 1985. *Ap. J.*, **297**, 371.
Kristian, J., Sandage, A., and Westphal, J.A. 1978. *Ap. J.*, **221**, 383.
Kurucz, R.L. 1979. *Ap. J. Supp.*, **40**, 1.
Larson, R.B., and Tinsley, B.M. 1978. *Ap. J.*, **219**, 46.
Lilly, S.J., and Gunn, J.E. 1985. *M.N.R.A.S.*, **217**, 551.
O'Connell, R.W. 1986a. In *Stellar Populations*, eds. C. Norman, A. Renzini, and M. Tosi (Cambridge: Cambridge Univ. Press), p. 167.
O'Connell, R.W. 1986b. In *Spectral Evolution of Galaxies* , eds. C. Chisoi and A. Renzini (Dordrecht: Reidel), p. 321.
O'Connell, R.W. 1987. In *Starbursts and Galaxy Evolution*, eds. T. Montmerle and T. Thuan (Paris: Editions Frontieres), in press.
Oemler, A. 1986. In *Stellar Populations*, eds. C. Norman, A. Renzini, and M. Tosi (Cambridge: Cambridge Univ. Press), p. 197.
Pickles, A.J. 1987. In *Structure and Dynamics of Elliptical Galaxies (IAU Symposium No. 127)* , ed. T. de Zeeuw (Dordrecht: Reidel).
Rabin, D.M. 1980. Ph. D. Thesis, California Institute of Technology.
Renzini, A. 1986. In *Stellar Populations*, eds. C. Norman, A. Renzini, and M. Tosi (Cambridge: Cambridge Univ. Press), p. 213.
Sandage, A.R. 1962. *Ap. J.*, **136**, 319.
Smith, A.M., and Cornett, R.H. 1982. *Ap. J.*, **261**, 1.
Spinrad, H. 1986. *P.A.S.P.*, **98**, 269.
Spinrad, H., and Djorgovski, S. 1987. In *Observational Cosmology (IAU Symposium No. 124)*, ed. G. Burbidge (Dordrecht: Reidel).
Tinsley, B.M. 1977. *Ap. J.*, **211**, 621.
Tinsley, B.M. 1980. *Ap. J.*, **241**, 41.
Wilkinson, A., and Oke, J.B. 1978. *Ap. J.*, **220**, 376.
Wyse, R.F.G. 1985. *Ap. J.*, **299**, 593.

EVIDENCE FOR LUMINOSITY AND COLOR EVOLUTION IN FAINT FIELD GALAXIES

J. A. Tyson
AT&T Bell Laboratories
Murray Hill, NJ 07974

ABSTRACT. This paper reviews the first results of a .36 to 1 micron CCD survey of faint galaxies, which reveals strong evidence for color and luminosity evolution. Galaxy counts at 25th B_J magnitude are a factor of 5-15 above no-evolution models. A strong bluing trend is seen for the fainter galaxies, suggesting that the UV excess from star bursts in these evolving galaxies has been redshifted into the blue band. Mild and continuous starburst activity in field galaxies seems to have occurred, and any global burst phase at $z < 6$ must have been shrouded in dust.

1. INTRODUCTION

It has long been argued that very young "primeval" galaxies would be luminous (Partridge and Peebles 1967, Meier 1976, Sunyaev et al. 1978, Shull and Silk 1979). Early stellar populations would form fast (Eggen et al. 1962), and very young galaxies would appear bright and UV-enhanced in their rest frame. In contrast to the number-redshift relation, number counts of galaxies, N(m), are relatively independent of the deceleration parameter in a Friedmann-Robertson-Walker universe. Any change in q_0 is equivalent to a slight change in timescale. For example, in Bruzual's (1983) $\mu = .6$ model, varying q_0 over the range 0.01 to 0.5 is equivalent to changing the galaxy timescale from 13 to 15 Gyr. Departures of galaxy counts from that expected in a non-expanding Euclidean universe (N ~ dex .6m) are dominated by the effects of expansion: the K-correction (Sandage 1961, 1975; Whitford 1975; Pence 1976).

With evolution, however, differential number-magnitude counts can be a more sensitive tool than the extragalactic background light. Schematic models of galaxy number counts, matching present colors and assuming universal (coeval) slow star formation starting at some universal formation redshift, predicted a relatively large enhancement in counts around 21-22 B mag, which is not seen (Tinsley 1977, 1978, 1980). Galaxy mergers (Toomre 1977, Djorgovski, et al. 1987a), dissipation and accretion probably play an important role (Gunn, 1982; Larson, et al. 1980; Tinsley, 1981). The size and luminosity of accreting galaxies are governed by uncertain gas infall rates, and realistic stellar population evolution models are limited by unknown luminosity functions and UV spectral history (Bruzual and Kron 1980; Bruzual 1983; Yoshii and Takahara 1987). Dust at early epochs may play an important role. A central

R. G. Kron and A. Renzini (eds.), Towards Understanding Galaxies at Large Redshift, 187–196.
© 1988 by Kluwer Academic Publishers.

issue is the timescale for disk formation. A weakness of all current models is their arbitrary assumptions regarding the morphology and timescale of star formation. Butcher and Oemler (1978) first suggested that there is an excess of blue galaxies in higher redshift clusters. Nearby "field" surveys (Kirshner, et al. 1979) indicated that, outside rich clusters, 70-80% of the galaxies are spirals. Blue galaxies will dominate faint galaxy counts, due to their small K-corrections: brighter parts of the luminosity function are sampled as one reaches fainter limiting magnitudes.

Little evidence for luminosity evolution has been found in spirals (Kron 1982, Gunn 1982), until recently. A small (0.3 mag) bluing trend in B-V colors of field galaxies in photographic surveys (Kron 1980, Tyson 1983) could have been due to luminosity function selection systematics rather than early phases of galaxy formation. Over the past 7 years, several studies have been made of selected areas of the sky, to the limit of photographic plates (Tyson and Jarvis 1979; Peterson, et al. 1979; Jarvis and Tyson 1981; Koo 1981, 1986; Shanks, et al. 1982; Infante, et al. 1986). Although the general agreement is fair to 23 B_J mag, it was unclear whether the scatter in number counts was due to galaxy number inhomogeneity or systematics in the photometry. Evidence for spiral evolution from the number counts brighter than 24th B_J mag is barely perceptible and is disputed. The observational difference between evolution and no-evolution models is largely in the proportion of galaxies detected with $z>1$. These photographic surveys, which reached 24th B_J magnitude, were useful however in showing that evolution was not synchronous or coeval (Tyson and Jarvis 1979, Peterson et al. 1979).

Here I review evidence for field galaxy starburst activity from deep CCD multicolor observations of a statistically complete sample of galaxies in 12 high-latitude fields. Galaxy counts fainter than 24 B_J mag exceed that expected if most stars were formed at redshifts greater than 6-10. Galaxy number counts from a photographic survey of 35 fields to 26 B_J mag arcsec^{-2}, done on the KPNO 4m telescope, are reviewed below, along with a deeper CCD survey. A complete report on these observations, including galaxy counts as a function of wavelength from .36 to 1 micron, will be published elsewhere (Tyson 1987).

2. DEEP CCD SURVEY

In 1983 Pat Seitzer and I began a program of 4-meter prime focus CCD observations at CTIO, with the aim to develop techniques for imaging and photometry to the theoretical limit (Tyson 1984) of that telescope and overall 30-60% efficiency: 27 B_J mag, 26 R mag, 25 I mag. The detector was an RCA 320x512 buried-channel, thinned, back-illuminated CCD, at 0.6 arcsec/pixel. The summed exposures total about 7500sec in each of the three bands: B_J=3600-5200Å, R=5800-7200Å, I=7800-11000Å. The definition of this B_JRI CCD photometric system in terms of calibration standard stars is given in Gullixson, et al. (1987). In order to derive a master sky frame, the telescope is moved randomly up to 20 arcsec between exposures, insuring that the resulting collection of images have sufficient information on the true background (sky + fringing) for each pixel to permit a unique determination of these systematic errors and subtraction. Details of the observing and finding charts for all the survey fields are given in Tyson and Seitzer (1987). These CCD images, typically 16 per band, are then automatically registered to fractional pixel accuracy and averaged by median filtering, producing the final images in each of the three bands B_J, R, and I. These image processing techniques are described in detail in Seitzer, Tyson, and Butcher (1987), and reviewed by Tyson (1986b).

These images are then input to the FOCAS v3.2 automated detector and classifier (Jarvis and Tyson 1981, Valdes 1982), creating a catalog of properties (isophotal, aperture, and total magnitudes, centroid positions, and several central moments) for each object. The surface brightness threshold is low enough that even the faintest galaxies are large enough to be distinguished from stars. Over 1500 objects are detected in every CCD frame, covering about 11 square arcminutes. In addition to the direct CCD images, artificial images are made by adding real star and galaxy images, dimmed many magnitudes, at nearly random locations on a real CCD sky background image. FOCAS is then re-run on this simulated data, in order to determine the probability of detection and photometric errors as a function of the magnitude of the coadded stars and galaxies. Many of these simulations are performed in order to reduce the errors.

We have had six runs on the CTIO 4-meter to date, and have high quality data on twelve high-latitude CCD fields. The fields were chosen by inspecting Schmidt survey prints, and selecting regions where there were no stars or galaxies brighter than about 20 B_J mag and 19 R mag. All 12 fields avoided significant IR cirrus. Average seeing FWHM ranged from .8 to 1.3 arcsec. Galaxies in the final position-matched catalog, which form our statistical sample, are conservatively selected to be real objects. Optimal filtering of 16 median-processed images, combined with a conservative threshold, assures that nearly all detected galaxies are statistically significant. Moreover, each galaxy in the final 3-band matched catalog must be reliably detected at the same location (within 2 arcsec) in all three bands. For detection, real objects are selected by the median filtering (reproducibility on most of sixteen 500 sec exposures) and are further constrained by the requirement that they have a minimum of 6 connected pixels above 3 sky σ, after convolution with an optimal filter. Under the hypothesis that protogalaxies should have been as bright in Lyman-α as most QSOs, they should be easily observable in our data. FOCAS automated photometry is performed on each final median image. Isophotal magnitudes are calculated out to the 3 σ detection isophote, which, for objects classified as galaxy and less than 5 arcsec in size, is about 29.0 B_J mag arcsec^{-2}. The corresponding limiting surface brightness for the R and I bands are: 27.9 R mag arcsec^{-2}, and 26.5 I mag arcsec^{-2}.

3. EVIDENCE FOR EVOLUTION

Figure 1 shows the raw differential number-magnitude plot, from 18 to 28 B_J mag. The results of both the photographic survey (18-24 mag) and our CCD survey (22-27 mag) are shown. Also shown are some previous photographic observations, and a range of recent no-evolution model calculations. Figure 2 shows the B_J-band CCD counts corrected by our simulations at the faint end, for the crowding effect. At the bright end, the CCD counts are deficient because we chose the fields on the basis of not seeing any objects on the Schmidt plates. Note that the counts follow the simple relation (Tyson and Jarvis 1979) log N = 4.25 - 0.45(J-24) from 18 to almost 26 B_J mag. Fainter than 26 B_J mag, there is evidence that the corrected counts flatten.

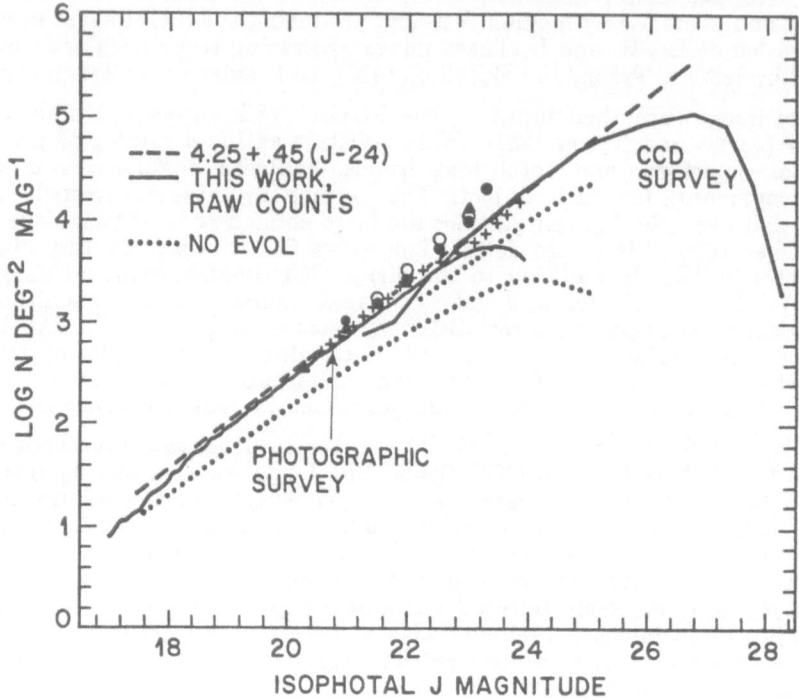

Figure 1. Differential number-magnitude plot for galaxies (log number per isophotal B$_J$ magnitude per square degree, to various limiting isophotes). The raw CCD deep survey counts (solid line at the faint end) in 12 fields are to 29 B$_J$ mag arcsec^{-2} surface brightness, and the 4-m photographic survey (solid line at the bright end) was complete to 26 B$_J$ mag arcsec^{-2}. Other photographic survey results are shown, along with a range of recent non- or passive-evolution predictions (dotted curves). The photographic survey was incomplete fainter than 23.5 B$_J$ mag, due to noise (Tyson 1983). The CCD deep survey encounters systematic undercounting due to overlapping galaxy images, two magnitudes brighter than the detection limit set by noise.

Are the many faint galaxies found at 26-27 B$_J$ mag at high redshifts, or are they an abundant population of relatively nearby dwarf galaxies? Counts of intrinsically faint and blue galaxies will be approximately Euclidean (N-m slope 0.6) out to a redshift of about 0.2. Thus, it is unlikely that the majority of these galaxies are dwarfs. Moreover, the colors of our faint galaxies appear different from morphologically-classified dwarf galaxies, either in surveys of nearby galaxies or within the deep CCD survey itself. These faint galaxies in the deep CCD survey may appear bluer due to redshifted UV from copious star formation in these young galaxies at high redshift.

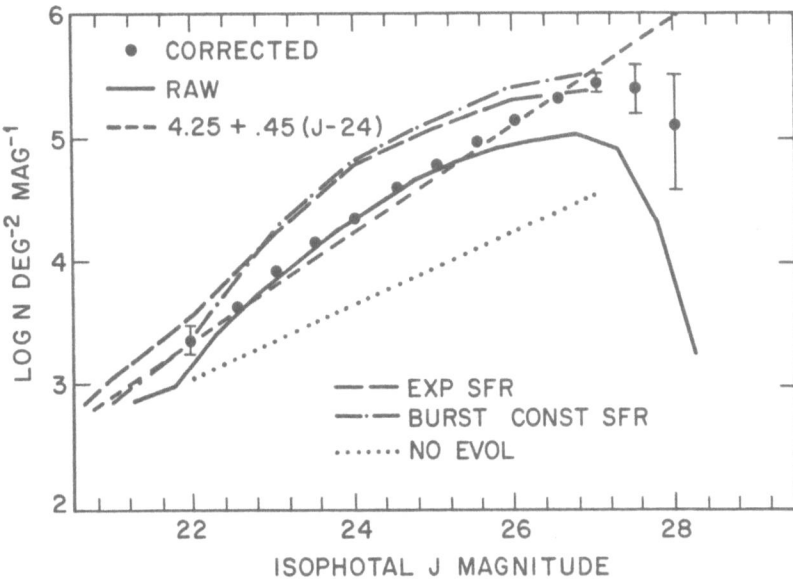

Figure 2. The B_J-band deep CCD survey differential number-magnitude counts (raw: solid line), corrected for undercount systematic by simulations (points). The error bars represent the 3 σ scatter among many simulations. The heavy dashed line is the continuation of the N(m) relation found at brighter magnitudes. The undercount at the bright end around 22 B_J mag is caused by observational selection: the CCD fields were chosen to be blank on the Schmidt survey plates. There is evidence for a flattening of the number counts for $B_J >$ 27 mag. Shown for comparison are three model calculations of Tinsley (1981): dash-dot line is initial burst followed by constant star formation rate; light dashed line is exponential SFR; dotted line is the no-evolution model. These corrected counts are fit well by a model by King and Ellis (1985).

Also shown in Figure 2 are the no-evolution and two evolution models of Tinsley (1980). The galaxy counts fainter than 24 B_J mag significantly exceed theoretical estimates for passive or no evolution. At 25 B_J mag, our B_J counts lie a factor of 5 - 15 above various theoretical passive evolution predictions (Ellis 1983, Gunn 1982, Koo 1981, Tinsley 1980), where the models are normalized to the bright galaxy counts at 16th B_J mag. This range of no- to passive evolution models is shown by dotted lines in Figure 1. Theoretical no-evolution slopes for N(m) at 23 B_J mag range from 0.3 to 0.36, and become flatter than 0.2 at 26th B_J mag. An increasing fraction of blue $(J-R < .5)$ galaxies at the faint end contributes to the steepening of the slope of the N(m) counts at shorter wavelengths. The observed excess in the B_J-band counts, over that predicted by no-evolution models, is fairly unambiguous evidence for galaxy luminosity evolution.

Is the high surface density of faint galaxies we have discovered due to a population of intrinsically faint dwarf galaxies, or are these galaxies at high redshift? The sub-Euclidean number counts make this very unlikely. We have checked to see if our sample of faint galaxies could be contaminated by faint dwarf galaxies of the type already known at lower redshift. This appears unlikely, since the mean color of 16 DDO and 40 UGC dwarf galaxies in the B_JRI system (Gullixson and Tyson 1987) is well separated from the track in the color-color plane followed by the faint galaxies in our CCD survey. These two samples of dwarf galaxies are indicated in the color-color plot in Figure 3. The sub-Euclidean number-magnitude counts, together with the non-dwarf colors, suggest that the faint blue galaxies found in the CCD survey for R$>$22 mag are high redshift galaxies with sufficient UV excess to overcome the color-redshift effect. It is likely, therefore, that we have detected field starburst galaxies at relatively high redshift. One conclusion is that the Schechter luminosity function, since it predicts significant contamination by dwarfs, may not be applicable to field galaxies. The field galaxy luminosity function may vary with epoch. Koo and Kron (1987) find correlation between color and redshift in a spot survey of galaxies with R$<$21 mag. This evidence suggests that galaxies in the region B_J-R$=$-.5 to .5 and R-I$=$.7 to 1.2 (corrected colors) are strong candidates for redshifts exceeding 1.

5. PROTOGALAXIES

Various models of primeval galaxies predict extended low surface brightness morphology. With the possible exception of one of our CCD survey fields, we find no consistent evidence for the expected numbers of these extended objects. However, their surface brightness may be below 30 B_J mag arcsec^{-2}. In the absence of significant dust, it is possible that many primevals would be mixed with the large numbers of faint blue galaxies. Ostriker and Heisler (1984) show that dust obscuration may become severe at redshifts 3-4. If galaxy formation began at z$>$4, it may be unreasonable to hope to detect either protogalactic Lyman-α clouds or very early primevals in the optical. The apparent optical counterpart of 3C326.1 (McCarthy, et al. 1986) is a primeval galaxy candidate. Most primevals are expected to be at least 2-3 magnitudes fainter than this, otherwise they would have been found in emission line surveys (Koo and Kron, 1980; Koo, 1986; Cowie, 1987). If, as our data suggest, star formation in early galaxies is continuous and spread out over at least 10^9 years, earliest galaxy formation redshifts can be high (say 5-10) but some first-generation star formation may be ongoing at a redshift of 2. Below I summarize our results for protogalaxies.

Brighter than 30 B_J mag arcsec^{-2}, there is no evidence of many 15-30 arcsec diffuse groups of unresolved dwarf primevals. There is also no evidence for an excess of red or near IR extended objects, as expected in low-dust theories where star formation first occurs soon after turn-around. Pregalactic Lyman-α emission clouds of size less than 100 kpc should be observable in our survey, if not blocked by dust associated with later epochs. For H_o=70, q_o=.15, a 100 kpc cloud would subtend an angle of 13 arcsec for 2$<$z$<$3. Only one object with this morphology was seen, in .037 deg^2 and 6000 galaxies surveyed.

4. COLOR EVOLUTION

Figure 3 shows the mean color-color plot for galaxies, averaged over 12 fields. The solid line is our stellar main and subdwarf sequence. Spiral galaxies occupy the region to the lower (bluer) part of the plane, while elliptical galaxies move up (redder in both colors) and away from the stellar sequence at increasing redshift, as shown by the dashed line. However, due mostly to K-correction effects, there are very few ellipticals in our faint galaxy sample. The numbers plotted in Figure 3 represent 3 sigma bounds for the mean color of galaxies in that R magnitude bin: $2 = 21 < R < 22$, $6 = 25 < R < 26$, etc. The trend in the color-color plane with R magnitude has been corrected for systematic errors, by the simulations. This correction is indicated by the vectors.

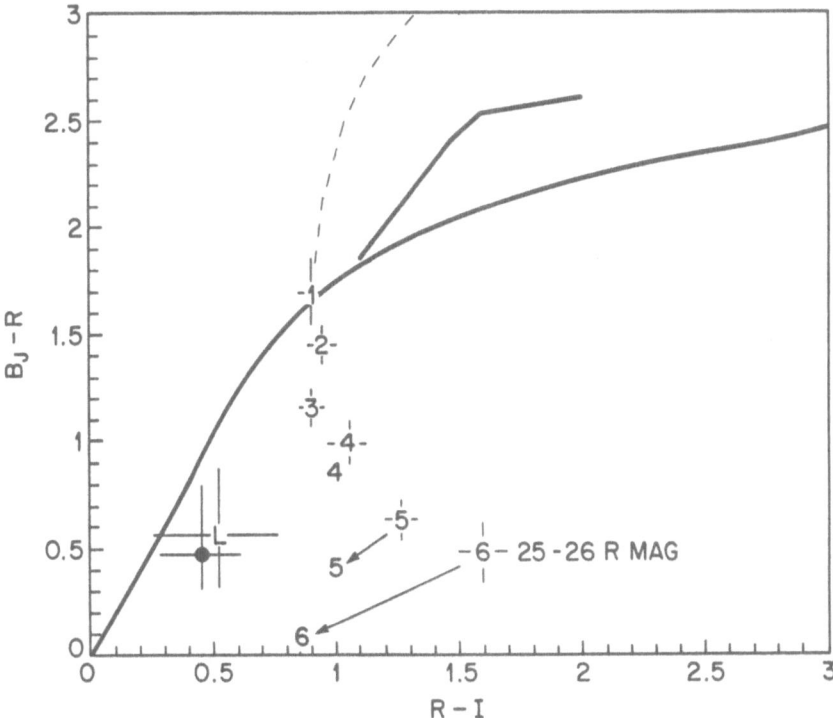

Figure 3. The B_J-R vs R-I color-color plot of our faint object CCD survey data. The color sequence followed by stars is shown as a solid line. The mean colors (error bars are 3 σ in the mean) of the galaxies in various R magnitude bins (21-26 R mag) are shown. Note the extreme bluing trend in B_J-R. These data have been corrected by simulations for a number of systematics. Also shown are the mean colors of low surface brightness galaxies (presumably dwarfs at $z=.1-.2$) in the deep CCD survey, and 40 dwarf galaxies (Gullixson and Tyson 1987) in the B_JRI system. Several clusters (mostly elliptical galaxies) of known redshift were studied on the B_JRI system, and their colors vs redshift trend is indicated by the dashed line.

6. DISCUSSION

Pure number evolution, is ruled out by the observed color-magnitude trend. The mean apparent colors cannot become as blue as B-R =0 in the case of pure number evolution. A mechanism for the color-color-redshift correlation could be redshifted UV continuum from rapid star formation entering the B_J filter, giving bluer B_J-R colors. Coeval models with synchronous galaxy evolution, and without a lot of dust (Tinsley 1977, 1978, 1980), predict a dramatic peak in galaxy counts which is not seen, now to $B_J < 27$ mag. This places the epoch of formation of most stars in such evolution models at roughly $z > 10$. However, this prediction is more an artifact of the universal formation redshift and resulting synchronous evolution. It is more likely that galaxy evolution is asynchronous, continues over a wide range in redshift, and may have begun at relatively high redshifts. What degree of asynchronicity is required for burst models? From Tinsley (1980), it appears that spreading initial bursts over a period of 3 Gyr could produce agreement with these data. This would force the earliest epoch of galaxy formation to $z=6$, or above, in such a low-dust model.

Silk and Norman (1981) point out that the absence of an early burst of star formation is also a consequence of dissipative slow disk formation. Thus, the required disk formation times of 10^9 - 10^{10} yrs in such models would place the formation redshift $z_f > 5$, if we are detecting disks at $z=2$, and if $\Omega > 0.3$. Silk and Wyse (1986) argue that dissipative scenarios imply gradual, perhaps hierarchical, star formation in order to form solar metallicity over 3-4 billion years. If solar metallicity were produced over a free-fall time of a massive galaxy, its luminosity would rival that of a quasar. The extreme bluing and the apparent saturation of the B_J counts for $B_J > 26$-27 mag suggest that one may be sampling the epoch of disk formation. This interpretation is consistent with the observed shallower slopes of the number-magnitude counts at longer wavelengths. If confirmed by redshift measurements, galaxy color-color data to even fainter magnitudes will provide new constraints on early galaxy evolution.

These trends are similar to those found for cluster galaxies. Butcher and Oemler (1978, 1984) detected more blue galaxies at high redshifts in clusters. Studies of selected radio galaxies appear to detect moderate to strong evolution at $z > 0.8$ (Windhorst 1984, Djorgovski, et al. 1985), depending on how the samples are selected. Statisical studies of faint galaxies clustered about QSOs of high redshift indicate strong luminosity evolution (Tyson 1986a). Djorgovski, et al. (1987b) may have detected a galaxy associated with PKS 1614+051 at a redshift of 3.2. Crude theoretical estimates of galaxy evolution have been made using stellar synthesis models. For example, one of Bruzual's evolution models ($\mu=0.5, q_o=0, H_o=50, z_{gf}=5$) shows R = 22.4 apparent magnitude for an L* field galaxy at a redshift of 1.5. Much of the flux from these evolving galaxies would appear in the blue part of the spectrum.

This kind of model, and an evolving galaxy spectrum rising to Lyman-α, is qualitatively consistent with our number count and color data, suggesting that many of these faint blue galaxies may have redshifts of 2-3. If some disks already exist at $z=3$, then most theories of galaxy evolution would put the epoch of first formation at redshifts $z > 5$. However, in the absence of direct imaging evidence for disks, the majority of the blue-excess galaxies found in this survey may be pre-disk systems. Detection of Lyman-α damped absorption line systems in the spectra of z=2-3 QSOs covering a significant fraction of the sky (Wolfe, et al. 1986) argues in favor of models with ongoing disk formation.

In summary, exponential SFR or burst plus constant SFR models with low dust must be spread out over several Gyr, and imply earliest epochs of formation z=6-7. Protogalaxies as luminous in Lyman-α as QSOs must have formed before z=6, or they were shrouded in dust. Detection of damped Lyman-α systems in QSOs of redshift 2-3 covering 20% of the sky imply the existence of disks. We must conclude that earliest formation redshifts were generally larger than 4, and that continuous non-coeval starburst activity is taking place.

It is doubtful that this project would have been done without Pat Seitzer's early collaboration.

References

Bruzual, G. (1983). Ap. J. Supp., **53**, 497; Ap. J., **273**, 105.
Bruzual, G. and Kron, R. G. (1980). Ap. J., **241**, 25.
Butcher, H. and Oemler, G. (1978). Ap. J., **219**, 18.
-------------------------- (1984). Ap. J., **285**, 426.
Cowie, L. L. (1987). private communication.
Djorgovski, S., Spinrad, H., and Marr, J. (1985). in "New Aspects of Galaxy Photometry" (J.-L. Nieto, ed., Springer Verlag, in press).
Djorgovski, S., Spinrad, H., Pedelty, J., Rudnick, L., and Stockton, A. (1987a). Astron. J. (in press).
Djorgovski, S., Strauss, M. A., Perley, R. A., Spinrad, H., and McCarthy, P. (1987b). Astron. J. (in press).
Eggen, O. J., Lynden-Bell, D., and Sandage, A. R. (1962). Ap. J., **136**, 748.
Ellis, R. (1983). in "The Origin and Evolution of Galaxies", B. J. T. Jones and J. E. Jones, eds. (Reidel), p 255.
Gullixson, C. A., and Tyson, J. A. (1987). Preprint
Gullixson, C. A., Boeshaar, P. C., Seitzer, P., and Tyson, J. A. (1987) Preprint
Gunn, J. E. (1982). in "Astrophysical Cosmology", eds. H. A. Bruk, G. V. Coyne, M. S. Longair (Pont. Acad. Sc., Vatican).
Infante, L., Pritchet, C. J. P., and Quintana, H. (1986). Astron. J., **91**, 217.
Jarvis, J. F., and Tyson, J. A. (1981). Astron. J., **86**, 476.
King, C. F. and Ellis, R. S. (1985). Ap. J., **288**, 456.
Kirshner, R. P., Oemler, A., and Schechter, P. L. (1978). Astron. J. **83**, 1549.
Koo, D. C. and Kron, R. (1980). Proc. Astron. Soc. Pacific, **92**, 537.
Koo, D. C. (1981). Ph.D. Dissertation, Univ. California, Berkeley.
Koo, D. C. (1986). in "The Spectral Evolution of Galaxies", eds. C. Chosi and A. Renzini (Dordrecht: Reidel), p 419.
Kron, R. G. (1980). Ap. J. Supp., **43**, 305.
Kron, R. G. (1982). Vistas in Astronomy, **26**, 37.
Larson, R. B., Tinsley, B. M., Caldwell, C. N. (1980). Ap. J., **237**, 692.
McCarthy, P. J., Spinrad, H., Djorgovski, S., Strauss, M., van Breugel, W., and Liebert, J. (1986). Bull. Am. Astron. Soc., **18**, 903.
Meier, D. L. (1976). Ap. J., **207**, 343.
Ostriker, J. P. and Heisler, J. (1984). Ap. J., **278**, 1.
Partridge, R. B., and Peebles, P. J. E. (1967). Ap. J., **147**, 868.

Partridge, R. B., and Peebles, P. J. E. (1967). Ap. J., **148**, 377.
Pence, W. (1976). Ap. J., **203**, 39.
Peterson, B. A., Ellis, R. S., Kibblewhite, E. J., Bridgeland, M., Hooley, T., and Horne, D. (1979). Ap. J., **233**, L109.
Sandage, A. (1961). Ap. J., **133**, 355.
Sandage, A. (1975). in "Galaxies and the Universe", Stars and Stellar Systems IX, A. S. Sandage M. S. Sandage, and J. Kristian, eds. (Chicago, Univ. Chicago Press), p 761.
Seitzer, P., Tyson, J. A., and Butcher, H. (1987). Astron. J. (in press).
Shanks, T., Fong, R., Ellis, R. S., and MacGillivray, H. T. (1980). MNRAS **192**, 209.
Shull, J. M., and Silk, J. (1979). Ap. J., **234**, 427.
Silk, J. and Norman, C. (1981). Ap. J., **247**, 59.
Silk, J. and Wyse, R. F. G. (1986). in "Structure and Evolution of Active Galactic Nuclei", G. Giuricin, et al., eds. (Reidel), p 173.
Sunyaev, R. A., Tinsley, B. M., and Meier, D. L. (1978). Comments Astrophys., **7**, 183.
Tinsley, B. M. (1977). Ap. J., **216**, 349.
Tinsley, B. M. (1978). Ap. J., **220**, 816.
Tinsley, B. M. (1980). Ap. J., **241**, 41.
Tinsley, B. M. (1981). Ap. J., **250**, 758.
Tinsley, B. M. and Gunn, J. E. (1976). Ap. J., **203**, 52.
Toomre, A. (1977). in "Evolution of Galaxies and Stellar Populations", eds. B. M. Tinsley and R. B. Larson (Yale Observatory, New Haven).
Tyson, J. A. (1983). Proc. IAU 78, in "Astronomy with Schmidt-Type Telescopes", M. Capaccioli, ed. (Reidel, p. 489).
Tyson, J. A. (1984). Pub. Astron. Soc. Pacific, **96**, 566.
Tyson, J. A. and Jarvis, J. F. (1979). Ap. J., **230**, L153.
Tyson, J. A. and Seitzer, P., Weymann, R., and Foltz, C. (1986). Astron. J., **91**, 1274.
Tyson, J. A. (1986a). Astron. J., **92**, 691.
Tyson, J. A. (1986b). J. Opt. Soc. Amer. - A, **3**, 2131.
Tyson, J. A. and Seitzer, P. (1987). Preprint
Valdes, F. (1982). Proc. SPIE, **331**, 465.
Valdes, F., Tyson, J. A., and Jarvis, J. F. (1983). Ap. J., **271**, 431.
White, S. D. M., Frenk, C. S., and Davis, M. (1983). Ap. J. **274**, L1.
Whitford, A. E. (1975). in "Galaxies and the Universe", Stars and Stellar Systems IX, A. S. Sandage M. S. Sandage, and J. Kristian, eds. (Chicago, Univ. Chicago Press), p 159.
Windhorst, R. A. (1984). Ph.D. Thesis, University of Leiden, Netherlands.
Wolfe, A. M., Turnshek, D. A., Smith, H. E., and Cohen, R. D. (1986). Ap. J. Supp., **61**, 249.
Yoshii, Y. and Takahara, F. (1987). Preprint.

RESULTS FROM A SAMPLE OF 1000 FIELD GALAXIES

Edwin D. Loh
Joseph Henry Laboratories
Princeton University
Princeton, New Jersey 08544
USA

ABSTRACT. The angular correlation function for galaxies with redshifts z between 0.15 and 0.85 has been measured. The data are fitted to a spatial correlation function of the form $\xi(r_p) = (r_0/r_p)^\gamma (1+z)^{-(3+\epsilon)}$, where r_p is the proper distance, and ϵ is a measure of the evolution of the clustering. These data and other local measurements imply $r_0 = (4.8 \pm 0.3)h^{-1}$ Mpc and $\epsilon = -0.5 \pm 0.5$ for an Einstein-deSitter universe ($\Omega = 1$). If instead $\Omega = 0$, then $\epsilon = -1.3 \pm 0.5$; i. e., clustering was stronger in the past. Thus the data are consistent with the assumptions that $\Omega = 1$ and that physical associations of galaxies neither expand nor contract ($\epsilon = 0$).

1.   INTRODUCTION

I report on a program to compile a catalog of a magnitude-limited, complete sample of field galaxies with coarse redshifts. Thus far, data on 1000 galaxies have been reduced; the median redshift is 0.5 and the error $\delta z/z$ is about 0.1 at $z = 0.5$. With this well-defined sample, the goal of the project is to study the geometry of the universe, the evolution of field galaxies, the correlation function, and secondarily the stellar content of the Galaxy. The techniques and first results on cosmology have been published (Loh and Spillar 1986a and 1986b). Here I report on a new result from the 1000 galaxy sample, a measurement of the correlation function.

2.   PHOTOMETRIC REDSHIFTS

The technical advance that enables these data is the photometric method for measuring redshifts, which Baum (1962) invented 30 years ago and has now been used with charge-coupled detectors (Loh and Spillar 1986a). One measures a coarse spectrum by means of 2-dimensional photometry with filters for each of about 300 objects in a picture. The spectral range from 380 to 930 nm is split into six spectral bands. One measures the redshift with the apparent wavelength of the 400 nm break and identifies the color type of the galaxy by the shape of the spectrum.

In the present implementation of the photometric method, we minimize the χ^2 between the observed data and the colors of a fiducial object over a set of fiducial objects, which are 11 types of galaxies at various redshifts and many types of stars. The fiducial

R. G. Kron and A. Renzini (eds.), Towards Understanding Galaxies at Large Redshift, 197–201.
© 1988 by Kluwer Academic Publishers.

colors are computed by integrating the measured spectra and the measured spectral response of the system. The object is identified as the fiducial with the lowest χ^2. This implementation has several advantages over eyeball estimates or fits to a composite color such as $U - J - F + N$ (Koo 1987). (1) The technique is impartial. (2) Parts of the spectrum with high signal carry more weight in determining the redshift. Therefore, the redshift determined this way is biased by evolution to a lesser degree than that determined by the $U - J - F + N$ method. For example, for elliptical galaxies at high redshift, the fit, determined by the brighter part of the spectrum, is unaffected by enhanced UV light, but the $U - J - F + N$ color is.

The principal data to support the validity of the photometric redshifts is a comparison (Loh and Spillar 1986a) with the spectroscopic redshifts (Dressler, Gunn, and Schneider 1985) of the cluster 0024+1654 at $z = 0.4$. This sample contains 15 red galaxies and 21 blue galaxies that resemble nearby field galaxies in their correlation between OII strength and color.

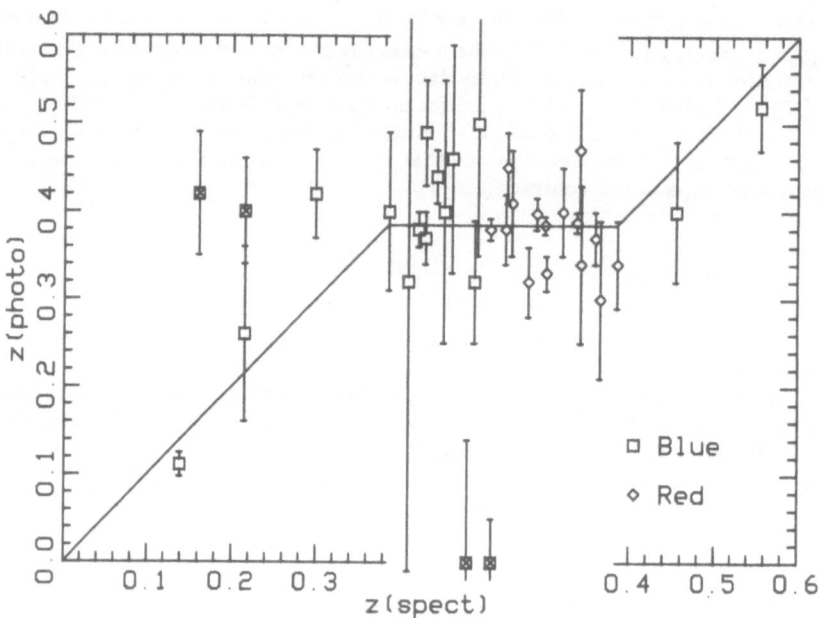

Fig. 1—Comparison between photometric and spectroscopic redshifts. The galaxies that are found spectroscopically to be cluster members are plotted at $z_{\text{spect}} = 0.385$ for clarity. Four points with discrepant redshifts are plotted with solid symbols.

From Fig. 1 one concludes that the comparison is quite good. The parameter x, defined to be $(z_{\text{photo}} - z_{\text{spect}})/\sigma_{\text{photo}}$, shows no bias either for the mean redshift or for the estimated redshift error σ_{photo}: $\langle x \rangle = -0.3 \pm 0.3$ and $\langle x^2 \rangle = 1.0 \pm 0.4$ for the red galaxies and $\langle x \rangle = 0.0 \pm 0.3$ and $\langle x^2 \rangle = 1.3 \pm 0.4$ for the blue galaxies.

Doubts about the photometric redshifts—a possible difference between cluster and field galaxies, biases that appear at higher or lower redshifts, biases that might become apparent with higher S/N levels, the frequency of objects that are not represented among the fiducials—these should be clarified presently with the completion of the comparison with the spectroscopic survey of field galaxies by Koo and Kron.

3.   CORRELATION FUNCTION

The two-point angular correlation function $w(\vartheta, z_0)$ for a redshift bin centered at z_0 is measured directly (Fig. 2), and the correlation length r_0 is deduced from it. To find $w(\vartheta, z_0)$, each galaxy with redshift z_1 for which $|z_1 - z_0| < \Delta$ is correlated with all galaxies with redshift z_2 for which $|z_2 - z_1| < \delta$. For the three bins centered at $z_0 = 0.25, 0.50$, and 0.75; $\Delta = 0.10, 0.15$, and 0.15, and $\delta = 0.10, 0.25$, and 0.25, respectively. The parameter δ is chosen to be large enough that redshift errors do not dilute the correlation and not needlessly large that the correlation function is diluted by the projection of uncorrelated galaxies.

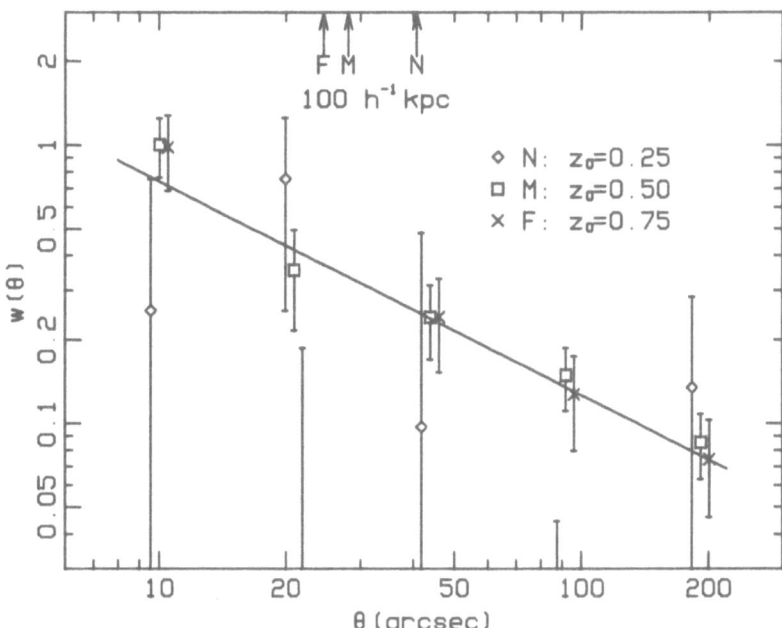

Fig. 2—Angular correlation function $w(\vartheta, z_0)$ for the subsamples at $z_0 = 0.25$ (N), 0.50 (M), and 0.75 (F). The error bars are drawn with Poisson statistics. The angles that subtend $100h^{-1}$ Mpc for the three subsamples are shown for an Einstein-deSitter universe. The line shows the best fit for the amplitude for the M subsample.

200

A similar luminosity range is used for all redshifts, the above three redshift bins as well as other magnitude-limited samples at $z \ll 1$. In rough terms, galaxies 1.7 mag fainter than the Schechter M^* are excluded. This same cut was used in the measurement of the volume element (Loh and Spillar 1986b), and the cut is explained precisely in that reference.

The correlation length r_0 is related to the angular correlation function by $w(\vartheta, z_0) = I_\gamma (r_0 D)^\gamma / [2\delta(\vartheta z_0)^{\gamma-1}]$, where $D(z; \Omega, \epsilon) = (z/r)^{1-1/\gamma} E^{1/\gamma} (1+z)^{1-(3+\epsilon)/\gamma}$ reflects the geometry and evolution, $E = \sqrt{[\Omega(1+z)^3 - (\Omega-1)(1+z)^2]}$ for a universe dominated by pressureless matter, I_γ is a constant, and r is the comoving coordinate. In the following $\gamma = 1.77$ is assumed. Since the geometry and evolution are not known *a priori*, the quantity $r_0 D$ (not r_0) is derived from the data, and this is shown in Figure 3.

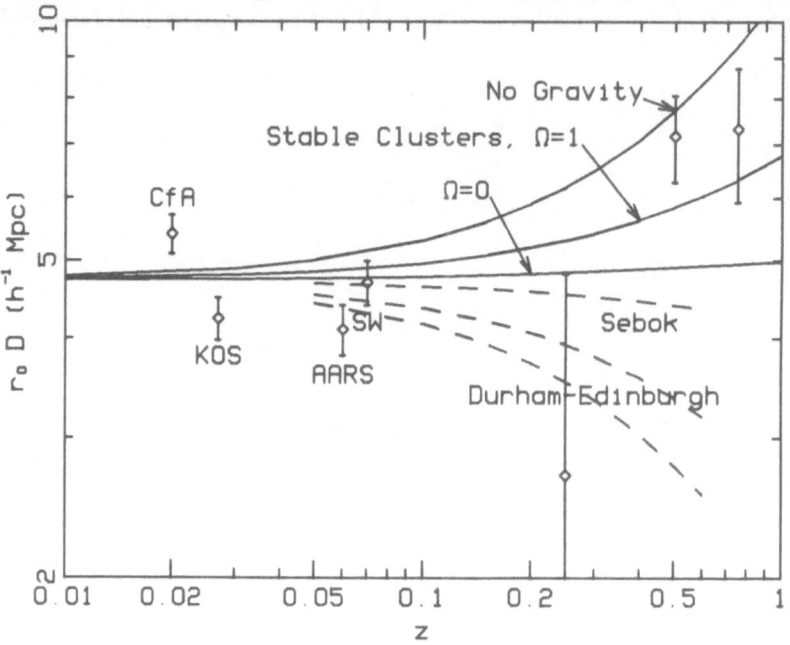

Fig. 3—Correlation length $r_0 D$ *vs.* redshift z. The points at $z = 0.25, 0.50$, and 0.75 show the present data. Data from other redshift catalogs are labeled CfA, KOS, and AARS. Also shown are several measurements derived from galaxy counts, Shane-Wirtanen (Groth and Peebles 1977), labeled SW, Sebok's data with his model for luminosity evolution, and Durham-Edinburgh data with two different models for evolution. Models for dynamical evolution and cosmology are shown as solid lines. The model labeled 'Stable Clusters, $\Omega = 1$' assumes that $\Omega = 1$ and that clusters, once formed, are stable ($\epsilon = 0$). The model labeled '$\Omega = 0$' assumes $\Omega = 0$ and clusters are stable. To fit the data with $\Omega = 0$ requires that the correlation be stronger in the past; i. e., that clusters are dissolving. The model labeled 'No Gravity' assumes that $\Omega = 1$ and that the clusters expand with the Hubble expansion as if they are not bound by gravity.

Two studies of galaxy counts to comparable distances but without redshifts, by Sebok (1986) and the Durham-Edinburgh group (Stevenson *et al.* 1985), find weaker correlation than the present data (Fig. 3), but Koo and Szalay (1984) with similar data show that luminosity evolution allows a wide range for r_0. Crucial to interpreting the data is the conversion of the angular correlation to a correlation length. This involves the depth of the survey, and without redshifts, requires a fine model for the luminosity function and the luminosity evolution. It is difficult to see how the present measurement of r_0 is erroneously too high, since mixing misclassified stars or galaxies with incorrect redshifts with the galaxies can only lower the correlation, and to reduce r_0 by a factor of 2 requires that the redshifts be too high by the same factor, in strong disagreement with Figure 1. Possibly the galaxy counts are overinterpreted: the models for luminosity evolution may be wrong, or some more distant galaxies are diluting the correlation.

The dependence of $r_0 D$ on redshift reveals the cosmology and the dynamical history. For that, the data at $z \ll 1$ are crucial, but the chief difficulty is to find a nearby sample comparable to the present one. One of our five fields has a cluster in it, and because the cluster was known *a priori*, not found by chance, I have excluded that field. Three nearby redshift samples for which r_0 has been measured are the CfA survey (Davis and Peebles 1983), the KOS survey (Peebles 1979), and the AARS (Bean *et al.* 1983). The CfA survey includes the Virgo cluster; the others cover leaner regions. I assume that the average of the CfA sample and the KOS and AARS samples mimics our survey. This is a major uncertainty and requires further work with larger samples both nearby and at high redshifts.

If the density parameter Ω is 1, then $\epsilon = -0.5 \pm 0.5$. Thus these data are consistent with the model that physical associations of galaxies neither expand nor contract and the universe has the Einstein-deSitter geometry. If instead $\Omega = 0$, then $\epsilon = -1.3 \pm 0.5$; *i. e.*, clustering was stronger in the past.

These preliminary results invite more work. If the local sampling variations can be assessed and the errors of the distant samples are reduced, then one has a constraint tying the evolution of clustering and the geometry of the universe. If the evolution of clustering can be understood theoretically, then one has a new measurement of geometry. This measurement of geometry is independent of the redshift-volume test (Loh and Spillar 1986b) in its systematic errors and its assumptions, but its random error is a factor of two larger.

4. REFERENCES

Baum, W. A. 1962, in *Problems of Extragalactic Research, IAU Symposium 15* (New York: Macmillan), p. 390.

Bean, A. J. *et al.* 1983, *M. N. R. A. S.*, **205**, 605.

Davis, M. and Peebles, P. J. E. 1983, *Ap. J.*, **267**, 465.

Dressler, A., Gunn, J. E., and Schneider, D. P. 1985, *Ap. J.*, **294**, 70.

Groth, E. J., and Peebles, P. J. E. 1977, *Ap. J.*, **217**, 385.

Koo, D. C. 1985, *A. J.*, **90**, 418.

Koo, D. C., and Szalay, A. S. 1984, *Ap. J.*, **282**, 390.

Loh, E. D. and Spillar, E. J. 1986a, *Ap. J.*, **303**, 154.

Loh, E. D. and Spillar, E. J. 1986b, *Ap. J. Lett.*, **307**, L1.

Sebok, W. L. 1986, *Ap. J. Suppl.*, **62**, 301.

Stevenson, P. R. F., Shanks, T., Fong, R., and MacGillivray, H. T. 1985, *M. N. R. A. S.*, **213**, 953.

DEEP IMAGING OF FIELD GALAXIES AT SHORT WAVELENGTHS

S. R. Majewski *
Yerkes Observatory
P.O Box 258
Williams Bay, Wisconsin 53191
U.S.A.

ABSTRACT. A deep CCD survey of field galaxies to U ≥ 25.5 verifies to fainter magnitudes the steep slope in the ultraviolet galaxy counts found by Koo (1986) for 20 ≤ U ≤
22. The sensitivity to the epoch of galaxy formation of still fainter surveys at short
wavelengths is discussed.

1. INTRODUCTION

Deep surveys of field galaxy counts have for the most part avoided very short optical
wavelengths (≤4000Å). Low atmospheric transparency, low detector quantum
efficiencies, and problems associated with differential refraction between object fields
and guide stars have conspired to make this part of the spectrum difficult for faint
galaxy work. Given that the near ultraviolet is sensitive to the light from young massive
stars and emission from hot gasses typically associated with star forming regions, this
part of the spectrum is the most dynamic in terms of galactic evolution. In addition, the
sky is darker at these shorter wavelengths, an advantage when working at low signal-
to-noise. With the advent of blue sensitive CCDs, this exciting part of the spectrum is
now more accessible to study and can place valuable constraints on models of galaxy
formation and evolution.

2. DEEP U-BAND COUNTS

The deepest galaxy survey in the ultraviolet to date is the photographic work of Koo
(1986). For the magnitude range 20 ≤ U≤ 22 he finds a steep slope in the counts of

---

* Visiting Astronomer, Kitt Peak National Observatory, National Optical Astronomⁱ Observatories, operated by the
Association of Universities for Research in Astronomy, Inc., under contract with the National Science Foundation.

R. G. Kron and A. Renzini (eds.), Towards Understanding Galaxies at Large Redshift, 203–207.
© 1988 by Kluwer Academic Publishers.

dlogA/dm = 0 68, whicn is higher than expected for even a non-expanding Euclidian universe (0 6) This result is certainly suggestive of some form of galaxy evolution since typical counts in the B band yield slopes of around 0 5 (which in itself is steeper than that predicted by non-evolutionary models) As might be expected, Koo finds that fainter galaxies are bluer, a result consistent with work at longer wavelenghts (Kron 1980, Shanks *et al* 1984; Tyson 1987), however the color-magnitude effect is greatest in U-B.

Motivated in part by the results of Koo (1986), a new survey has begun in colla- boration with D. Koo and R. Kron to obtain deep CCD U-band images of field galaxies with the purpose of pushing the counts to even fainter limits. The larger quantum efficiency of CCDs results in greater efficiency in telescope time not only because a given magnitude limit can be reached in less time, but also because the shorter exposures mean that given fields can be observed in the U band farther from the meridian without worrying about differential refraction. A run with RCA CCD#1 at the prime focus of the Mayall 4-m in June, 1986 yielded five hours of integration in the U band on a field at the North Galactic Pole and 3.5 hours in a field in Hercules. Integrations of from 30-60 minutes were also obtained in each of the B, V, R, and I bands to determine the color distributions (to a necessarily brighter magnitude limit).

In any study requiring identification of sources very close to the sky background fixed-pattern noise in the CCD is a concern. In order to minimize the effects of fixed-pattern noise we have used the method of Seitzer *et al.*(1987) whereby the many individual frames comprising a single net image are "dithered" with respect to one another by random, few-arcsecond shifts. In addition, the individual exposures were taken in "short scan mode", a technique in which the CCD camera is regularly stepped in the focal plane by one pixel (30 micron) shifts while the accumulating charge packets are coincidentally transferred in the opposite direction by one pixel. In this way each final *exposure* pixel is averaged over many *CCD* pixels, thereby reducing the effects of pixel to pixel variations. It was found that the short scan technique was also extremely helpful in reducing fringing effects, even in the red where stronger emission from night sky lines makes the problem significantly worse. Typical scans were 32 pixels in the blue and 64 in the red

Preliminary results from our Galactic Pole data confirm the findings of Koo to still fainter magnitudes. Specifically, we find that the dlogA/dm slope is still very steep at 22 ≤ U ≤ 23, of the order of that found by Koo. Because of the very much smaller field size of the CCD fields, our data are highly subject to statistical fluctuations at brighter magnitudes. Therefore we cannot directly confirm the Koo result for U ≤ 22 Fainter than U ~ 23 the slope of the counts seems to fall off, but it is too early to say whether this is real or the result of incompleteness. In our Galactic Pole U-band image we are seeing galaxies to U ≳ 25.5. In addition, we find that the very faintest galaxies seem to have overall bluer colors.

3. CONSTRAINTS ON THE EPOCH OF GALAXY FORMATION

While almost all galaxy count surveys, including our own, indicate some luminosity evolution at short wavelengths, the implications of the data for a picture of galaxy formation and evolution are subject to a wide range of interpretations due to a number of uncertainties in models and lack of redshifts. It is not unreasonable to say that each of the major elements (e.g. the initial mass function, the history of star formation, the luminosity function, the ultraviolet spectra of galaxies, distribution of spectral types, environmental effects) comprising most evolutionary models is still unsatisfactorily determined. Given that redshifts are still difficult to obtain for "normal" galaxies at cosmologically significant magnitudes, can faint count surveys say anything about the evolution of galaxies that is reasonably free from the above ambiguities?

Without the use of models and the benefit of redshifts, deep galaxy counts at short wavelengths can place important constraints on the epoch of galaxy formation if we can find a feature in the counts and colors that is a signature of a given redshift. The most promising test may be a search for the effects of the Lyman continuum being redshifted into a particular band. Because galaxies presumably are dark shortward of the Lyman limit, the slope of the counts should be seen to drop suddenly when the depth of the survey is such that significant numbers of galaxies at the critical redshift would otherwise contribute. For the U band this critical redshift is 2.5 and for the B band it is 3.3. The task then is to search the U band counts for a rapid decline in slope, and monitor the U-B colors to see if they suddenly become much redder. If such a feature is found, it would imply that the contribution of $z \sim 3$ galaxies to the counts is substantial. If the feature is not obvious, we will at least be able to place an upper limit on the number of $z \sim 3$ galaxies exhibiting a stellar continuum, a severe constraint for "bright phase" models at this redshift. Note that there is no reason for this test to be limited to the U band, but of course redder bands require still fainter magnitude limits.

In the table below are listed the B band differential galaxy counts one would expect as a function of redshift assuming $q_0 = 0.5$ and $H_0 = 50$ km s^{-1} Mpc^{-1} for one of our CCD fields (area = 4×10^{-3} deg^2). The counts were calculated using a Schechter function with the parameters suggested by Felton (1985), $M_{BT}^* = -21.56$, $\alpha = -1.25$, $\varphi^* = 1.5 \times 10^{-3}$ Mpc^{-3}. The calculations include just the bandwidth term of the K-correction, 2.5 $\log(1+z)$, which applies to energy distributions that are flat in f_λ. This is not strictly consistent with the use of the Felton normalization since φ^* includes early type galaxies. Locally these have large K_U and therefore may not be expected to contribute to deep counts at short wavelengths. The effect would be to introduce a scale factor which is likely to be around 0.75, based on the fraction of ellipticals in the local field. On the other hand, evolutionary effects will influence the colors and intrinsic magnitudes of early type galaxies if seen near the cut-off age of large-scale star formation; in the redshift range of interest here it may make them significantly bluer (Tinsley 1977). Also, dust has been neglected since its effect on the counts is not well under-

stood. However, ultraviolet data on blue star-forming galaxies by Hartmann *et al.*
(1984) and late type galaxies by Code and Welch (1982) suggests that dust may only play
a minor role in shaping the galactic energy distributions.

Differential galaxy counts per magnitude per 4×10^{-3} deg^2

z B_{lim} = 28	27	26	25	24	23	22	21
0.0 – 0.25 10.2	8.1	6.4	5.2	4.0	3.1	2.5	1.8
0.25 – 0.75 51.9	40.9	32.1	24.5	17.7	11.2	5.0	1.1
0.75 – 1.25 54.6	41.9	30.5	19.3	8.9	2.0	0.1	
1.25 – 1.75 44.5	31.7	19.2	7.9	1.4			
1.75 – 2.25 33.3	20.6	9.0	1.7	0.1			
2.25 – 2.50 12.7	6.6	1.9	0.2				
2.50 – 2.75 10.5	4.9	1.1					
2.75 – 3.25 15.2	5.5	0.7					
3.25 – 3.75 9.4	2.2	0.1					
3.75 – 4.25 5.3	0.7						
total 247.6	163.1	101.0	58.8	32.1	16.3	7.6	2.9

We see that this no-evolution simple model predicts that a survey reaching to B_T =
27.0 will contain approximately 12 galaxies per CCD field with the Lyman limit
redshifted into the U band, and will have several of these as bright as B_T = 26.5. The
equivalent depths in the U band are U = 26.45 and 25.95 respectively. Our work to date
shows that such limits are accessible, and we estimate that these magnitudes would be
reached in an integration of order 20–25 hours. The individual high-redshift galaxies
would be identifiable as those objects with very red U–B colors.

The number of accessible high-redshift galaxies tabulated above can be viewed as a
conservative estimate on many accounts. Ultraviolet spectra derived by Pence (1976),
Coleman *et al.*(1980) and Code and Welch (1982) from satellite photometry, and recent
balloon work by Donas *et al.*(1987), show that the energy distributions of late type
galaxies exhibit a significant *rise* at wavelengths $\leq$ 3000Å.

Also, the numbers were calculated using a *non-evolving* energy distribution. In
light of the various evidence supporting at least some evolution, the numbers would be
an underestimate whether the form of the evolution is a brightening or a bluing with
redshift, or a combination of both. If star formation was greater in the past, the
ultraviolet flux of a galaxy will resemble an increasingly earlier stellar continuum with
lookback time.

Finally, nebular emission in Lyα and other lines may contribute prominently to the
rest frame ultraviolet flux of young galaxies. Estimates figure the Lyα flux to be large
(Partridge and Peebles 1967, Meier 1976), but observations of high-redshift galaxies (cf.

Spinrad *et al.* 1985) and local blue star-forming galaxies (cf. Hartmann *et al.* 1984) exhibit a range of emission characteristics, with those at the high end associated with radio galaxies and therefore not necessarily representative of "normal" galaxies.

The effect of each of these qualitative refinements to the simple model presented is to enhance the contrast of galactic spectral energy distributions on either side of 912Å and boost the contribution of high-redshift galaxies to the net counts. This serves to make the proposed effect of a feature in the galaxy counts more dramatic. However, it should be pointed out that even if interpretation of the shape of deep U counts is ambiguous in these terms, we can still rely on the *positive* signal of bona fide high-redshift galaxies (those with very red U-B) to place constraints without the use of sophisticated models.

I would like to thank Rich Kron and Dave Koo for innumerable valuable discussions and help with this report. Also, George Jacoby and Roger Lynds (NOAO) were extremely gracious in helping us work with the virgin scanning CCD system on the 4-m. We would also like to thank the NOAO downtown staff, especially Frank Valdes and Jeannette Barnes, for help with the reduction software. This work was supported under NSF grant AST-8314232.

References

Code, A. D., and Welch, G. A. 1982, Ap.J. 256, 1.
Coleman, G. D., Wu, C., and Weedman, D. W. 1980, Ap.J. Supp. 43, 393.
Donas, J., Deharveng, J. M., Laget, M., Milliard, B., and Huguenin, D. 1987, A&A. 180, 12.
Felton, J. E. 1985, Comments Astrophys. 11, No. 2, 53.
Hartmann, L. W., Huchra, J. P., and Geller, M. J. 1984, Ap.J. 287, 487.
Koo, D. C. 1986, Ap.J. 311, 651.
Kron, R. G. 1980, Ap.J. Supp. 43, 305.
Meier, D. L. 1976, Ap.J. 207, 343.
Partridge, R., and Peebles, P. 1967, Ap.J. 147, 868.
Pence, W. 1976, Ap.J. 203, 39.
Seitzer, P., Tyson, J. A., and Butcher, H. 1987, A.J. (in press).
Shanks, T., Stevenson, P. R. F., Fong, R., and MacGillivray, H. T. 1984, MNRAS 206, 767.
Spinrad, H., Filippenko, A., Wyckoff, S., Stocke, J. T., Wagner, R. M., and Lawrie, D. G. 1985, Ap.J. 299, L7.
Tinsley, B. M. 1977, Ap.J. 211, 621.
Tyson, J. A. 1987, A.J. (submitted).

A DEEP REDSHIFT SURVEY OF FIELD GALAXIES

David C. Koo
Space Telescope Science Institute
3700 San Martin Drive
Baltimore, MD 21218   U.S.A.

Richard G. Kron
Yerkes Observatory
University of Chicago
Williams Bay, WI 53191   U.S.A.

ABSTRACT. A spectroscopic survey of over 400 field galaxies has been
completed in three fields for which we have deep UBVI photographic
photometry. The galaxies typically range from B = 20 to 22 and
possess redshifts z from 0.1 to 0.5 that are often quite spiky in
distribution. Little, if any, luminosity evolution is observed up to
redshifts z ≃ 0.5. By such redshifts, however, an unexpectly large
fraction of luminous galaxies has very blue intrinsic colors that
suggest extensive star formation; in contrast, the reddest galaxies
still have colors that match those of present-day ellipticals.

1. INTRODUCTION

Over the past six years, we have been undertaking a deep redshift
survey of field galaxies. Our original and still primary goal is to
measure directly, with redshifts and multicolor photometry, the evolution in the luminosity function and color distributions of field
galaxies. Adding the use of spectral features that relate to age,
abundance, and star-formation rate (such as emission line strengths,
the amplitude of the 4000Å break, or the relative strength of various
absorption lines), we track evolution through several other parameters and thereby better understand the nature of any detected
changes.   Furthermore, our high quality imaging data yield limited,
but still useful, information on the morphology and sizes of distant
galaxies; both the angular positional data and redshifts probe evolution in the clustering behavior of galaxies; and finally, with an
understanding of distant galaxies, even the geometry of the universe
is in principle, measureable (cf. talk by Loh).   At this time, our
data are still being reduced, but we are able to report some preliminary results that do place interesting constraints on the clustering
and evolution of field galaxies up to redshifts z ~ 0.5.

R. G. Kron and A. Renzini (eds.), Towards Understanding Galaxies at Large Redshift, 209–212.
© 1988 by Kluwer Academic Publishers.

2. OBSERVATIONS

The spectroscopic data have been acquired with the Cryogenic Camera
spectrograph system on the 4m telescope at Kitt Peak National Observatory. The detector is a TI 800 x 800 CCD chip. By using a mask
with multiple slits at the focal plane, about 10 objects can be
observed simultaneously over a 5 arcmin field of view. Our typical
setting uses a 15Å FWHM spectral resolution at 4Å per pixel; a
wavelength range of ~3000Å centered near 6000Å; and exposures as
short as 10 min for very bright galaxies to several hours for very
faint objects.   Between 5 to 10 of these 5 arcmin fields were
observed in each of three 0.3 deg^2 regions of the sky where we
already have deep 4m plates taken with UBVI filters: Selected Area
57 (1305+30), SA 68 (0015+15), and Hercules No. 1 (1720+50).
     The main reason our redshift survey has taken so long has been
our attempt to achieve completeness, not in the sense of measuring
redshifts of all galaxies to a given magnitude limit over a specified
area, but rather in having subsamples in both magnitude and area that
are representative so that the selection function is well defined,
i.e. unbiased by color or strength of spectral features. This sample
can then be legitimately used for statistical analysis and for direct
comparisons to our models of galaxy spectral evolution. For galaxies
that were faint or which did not have strong spectral features,
repeat exposures for several years were often required. At present,
we have measured redshifts for over 400 field galaxies, but only 300
of these constitute the statistically complete sample. Another similar and yet complementary project is the AAT faint galaxy survey of
Ellis, Broadhurst, and Shanks (Ellis 1987, talk by Ellis), which
differs from ours in that we have selected our sample using red
(6100Å) magnitudes rather than blue; their spectra cover a wider
wavelength range from 3700Å to 10,000Å with higher resolution; we
possess multicolor photometry whereas their galaxies have mainly blue
photometry; and their ~200 galaxies are divided among five fields,
thus averaging 40 galaxies per field, while we have about 100 in each
of SA 68 and Hercules and over 200 in SA 57.

3. LARGE-SCALE CLUSTERING

One of the most striking results is displayed in Figure 1. Very
strong fluctuations can be seen in the redshift distribution of
galaxies in SA 57, over scales of up to 100 Mpc. These include not
only strong peaks but also very large regions in which the densities
of galaxies are low. For example, in the gap centered on redshift z
= 0.15, about 15 galaxies are expected and yet at most 1 galaxy is
found. As an example in the opposite sense, i.e. strong clustering,
over 50% of all galaxies over the entire 0.3 deg^2 with B < 20
congregate at the same redshift of 0.125! Our other two fields also
show fluctuations, but not as dramatically (Koo, Kron, and Szalay
1987; see also Ellis 1987).

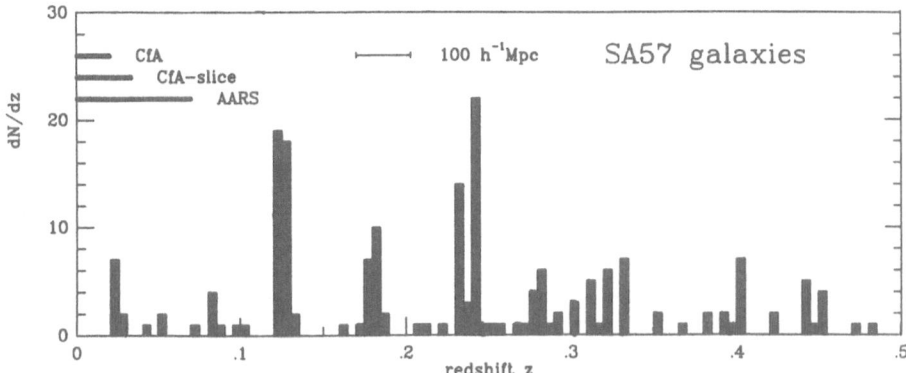

Figure 1. The distribution in redshift bins of 0.005 of the 204 galaxy redshifts at the North Galactic Pole over an area of 0.3 deg^2. Note the extreme clumpiness. CFA shows the depth of the B = 15 Center for Astrophysics redshift survey; the slice reaches B = 15.5; AARS shows the depth of the B = 17 Anglo Australian Redshift Survey.

4. FIELD GALAXY EVOLUTION

Of more interest to this workshop are the clues to galaxy evolution. In general, the preliminary results of our red-selected survey (Koo and Kron 1987) are in good agreement with those from the blue-selected one of the faint AAT survey (Ellis 1987). Some of the data is shown in Figure 2a. The model predictions given in Figure 2b fully account for photometric errors and the decreasing effective area of the redshift survey at fainter magnitudes and are otherwise similar to those of Bruzual and Kron (1980) that give good fits to faint galaxy counts and colors.

One surprise was that, on average, the distribution of higher redshifts is most consistent with little or no luminosity evolution among galaxies observed to redshifts z ~ 0.5. Evidence for this is partially seen in the near equality between the total observed numbers for z > 0.25 and those of the no-evolution model. There is some evidence for red galaxies being brighter in the past, but clustering (note clump at z ~ 0.4) makes this a marginal result. More secure is the evidence that luminous galaxies with extremely blue (Im type) colors were more common at higher redshifts, perhaps by a factor of two (Fig. 2b). This evidence for color evolution is substantiated by the presence of strong emission lines in many of the faint galaxies (Ellis 1987). In contrast to this trend, the reddest field galaxies have colors consistent with those of ellipticals found today.

Another clear result is that the intrinsically faint galaxies at low redshifts predicted by the models are not observed, and since these do not involve any evolution, the input parameters of our models must be in error at some level. We hope to improve the models by, e.g., using different shapes of the luminosity function for

212

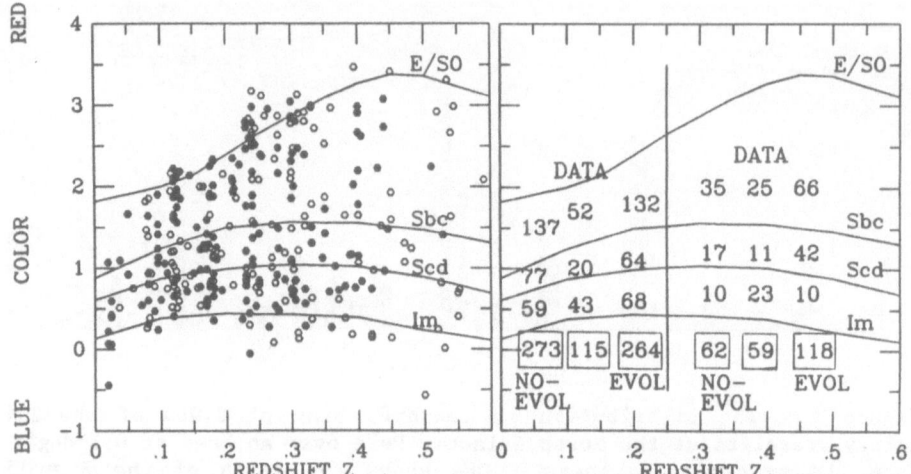

Figure 2 (a) Color (U − F) versus redshift for ~230 field galaxies in
SA 57 and SA 68, where U is our ultraviolet band (3600Å) and F is our
red band (6150Å); filled circles are those constituting the statistically complete sample. The tracks taken by various types of galaxies
are also shown. (b) Observed numbers (statistically complete sample
only) versus predictions of models with mild and without evolution in
the regions bounded by the Sbc and Scd galaxy types and divided at
z = 0.25. The boxed numbers are totals.

galaxies of different colors (Sandage et al. 1985). As suggested by
Ellis (1987), however, perhaps our picture of evolution, and thus our
model, needs more substantial revision.

We acknowledge the support of Kitt Peak National Observatory,
Department of Terrestrial Magnetism (Carnegie Institution of
Washington), and NSF Grants AST 81-21653 and AST 83-14232.

5. REFERENCES

Bruzual, G. A. and Kron, R. G. 1980 *Ap. J.*, **241**, 25.
Ellis, R. S. 1987 in IAU Symp. No. 124, **Observational Cosmology**, eds.
    A. Hewitt, G. Burbidge, and L. Z. Fang (Reidel, Dordrecht)
    p. 367.
Koo, D. C. and Kron, R. G. 1987 in IAU Symp. No. 124, **Observational
    Cosmology**, eds. A. Hewitt, G. Burbidge, and L. Z. Fang (Reidel,
    Dordrecht) p. 383.
Koo, D. C., Kron, R. G., and Szalay, A. S. 1987 in 13th Texas
    Symposium in Relativistic Astrophysics, ed. M. P. Ulmer (World
    Scientific, Singapore) p. 284.
Sandage, A., Binggeli, B., and Tammann, G. A. 1985 *A. J.*, **90**, 1759.

NEW LIMITS ON THE SURFACE DENSITY OF Z = 5 PRIMEVAL GALAXIES

C.J. Pritchet and F.D.A. Hartwick
Physics Department
University of Victoria
Victoria, B.C., CANADA V8W 2Y2

ABSTRACT. We have searched for faint emission line objects using a narrowband imaging technique. From our observations we conclude either (i) that primeval galaxies with $4 \lesssim z \lesssim 6$ are fainter than R = +27.5, or (ii) that such objects have a surface density $\lesssim 0.1$ arcmin^{-2}.

1. INTRODUCTION

A variety of techniques have been proposed for detecting primeval galaxies (PG's), and the reader is referred to a superb review of the current observational situation by Koo (1985). Here we describe a search for Ly α emission objects at z ≃ 5. Our technique is to image a small area of sky through a series of narrow-band filters; emission line objects might be detected in a filter tuned to the wavelength of an emission line, but would be weak or absent in other filters (including broadband filters). Clearly emission line objects detected at 6000 $\lesssim \lambda \lesssim 8500$Å are candidates for primeval galaxies at $4 \lesssim z \lesssim 6$ if the observed emission line is Ly α.

2. OBSERVATIONS AND DATA REDUCTION

All imaging observations were obtained at the prime focus of the Canada-France-Hawaii 3.6m telescope (CFHT) using an RCA 320 × 512 CCD detector. The median seeing for the observations was ~ 1 arcsec. The field observed was a 2.2 arcmin × 3.3 arcmin area centered on the radio galaxy 3C217 (Laing et al. 1978, Spinrad and Djorgovski 1984).

Observations were obtained through commercially-available narrowband filters with typical bandwidths 100–130 Å, and through filters emulating the Johnson B, V and Kron-Cousins R, I systems.

Narrow-band observations consisted of three 2400 s exposures through each of 9 filters (7500–9000 Å) for our 1984 January run, and four 4600 s exposures through each of 4 filters (6300Å, 6800Å, 7500Å, 8100Å) for our 1986 March run. Multiple exposures through the same filter were shifted by ~4 arcsec to enable fringe removal. Data reduction was performed as described in Pritchet and Hartwick (1987).

R. G. Kron and A. Renzini (eds.), Towards Understanding Galaxies at Large Redshift, 213–216.
© *1988 by Kluwer Academic Publishers.*

3. SEARCH FOR EMISSION LINE OBJECTS

In our 1984 January data approximately ten faint candidate emission
line objects were discovered that were > 3σ above sky in one filter,
but weak or absent in other filters. In a two night observing run in
1985 February we attempted to obtain followup spectroscopy of these
candidates using the Cryocam aperture plate spectrograph on the KPNO 4m
telescope. No emission lines or continuum were visible in any of the
slitlet spectra. We estimate the limiting magnitude for our Cryocam
data to be $R_{lim} \approx 28.3$ for a source whose luminosity is concentrated in
a single emission line. The quoted limit is for a source with diameter
~ 2.5 arcsec.

Our 1986 March observations also failed to confirm any of our PG
candidates. The limiting magnitude for these observations is R > +27.5
at λ6800; this limit is an equivalent limiting <u>broadband</u> magnitude that
would be observed for a <u>pure</u> emission source. Since this limit is considerably (~ 1 mag) fainter than obtained in 1984 January, in what
follows we shall refer only to this limit.

4. DISCUSSION

In Figure 1 we present our observed limits on the numbers of emission
line objects in a plot of surface density versus R magnitude. The
placement of our data point in this diagram assumes <1 object per CCD
field, and is plotted for our deepest filter (λ6800). In addition, we
also plot the limit obtained from our Cryocam observations ($R_{lim} \approx 28.3$
for an observed area 272 arcsec^2).

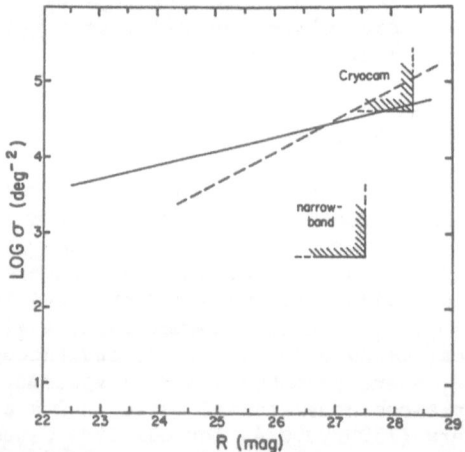

Figure 1. Comparison of our observational limits with theoretical
predictions. The solid and dashed lines represent the predictions of
models (see text). The sense of the limits is to exclude PG's whose
properties lie to the upper left of the plotted points.

Our constraints are plotted for unresolved or marginally-resolved PG's. Our limits for objects of characteristic size ~ 10" are about 2 mag brighter than plotted in Fig. 1 for our narrowband observations, and ~ 1 mag brighter for our Cryocam observations. We have also assumed that the _entire_ redshift range of PG's $\Delta z(PG)$ has been sampled by our narrow-band observations.

Also plotted in Fig. 1 is a solid line that represents the predictions of R magnitudes and surface densities from a variety of detailed models of PG's by Meier (1976). The plotted line is representative of models with collapse times ~ 10^9 y and cosmologies ranging from $\Omega = 0.06$ ($z_f \simeq 10$) on the right to $\Omega = 2$($z_f \simeq 3$) on the left. The locus of these models by Davis (1980) (as modified by Koo 1985) is plotted as a dashed line in Fig. 1.

From Fig. 1 it can be seen that significant limits have been placed on the surface density of PG's, even with the small imaging area of the CFHT CCD (~ 7 arcmin^2). Our limiting magnitudes for unresolved objects are similar to those obtained by Cowie (1986) in an imaging and long-slit spectroscopic search for faint emission line objects.

Our observations appear to conflict with a class of PG model which reaches maximum luminosity at z = 5. There are, however, many possible avenues of reconciliation between theory and observation, and these have been enumerated by many authors (e.g. Davis and Wilkinson 1974).

1. PG's may reach maximum luminosity at z $\gtrsim$ 5.

2. The Ly α luminosity of PG's may be much lower than expected (e.g. Hartmann, Huchra, and Geller 1984), although this possibility seems rather unlikely in the low metallicity environment of a PG.

3. The collapse of protogalaxies may be very slow (as, for example, advocated for the Milky Way by Yoshii and Saio 1979), resulting in very low PG luminosities, and an extended range in redshift over which PG's can be seen.

4. Objects at z $\gtrsim$ 4 may be obscured by a wall of absorption produced by overlapping foreground galaxies (Ostriker and Heisler 1984).

5. In a cold dark matter-dominated Universe, dwarf galaxies form before large protogalaxies; the maximum luminosity of protogalaxies corresponds to the maximum luminosity of its component dwarf galaxies, which may be reached well before the collapse of the protogalaxy within which the dwarf galaxies reside. PG's would then be extended, low surface brightness objects, and would be difficult to observe.

This work was supported by the Natural Sciences and Engineering Research Council of Canada through operating grants to the authors.

REFERENCES

Cowie, L. 1987, private communication.
Davis, M. 1980, in Objects of High Redshift, ed. G.O. Abell and P.J.E. Peebles (Dordrecht: Reidel), p. 57.
Hartmann, L.W., Huchra, J.P., and Geller, M.J. 1984, Ap.J., 287, 487.
Koo, D.C. 1985, in The Spectral Evolution of Galaxies, ed. C. Chiosi and A. Renzini (Dordrecht: Reidel), p. 419.

216

Laing, R.A., Longair, M.S., Riley, J.M., Kibblewhite, E.J., and
    Gunn, J.E. 1978, M.N.R.A.S., 183, 547
Meier, D.L. 1976, Ap.J., 207, 343.
Ostriker, J., and Heisler, J. 1984, Ap.J., 278, 1.
Pritchet, C.J., and Hartwick, F.D.A.H. 1987, Ap.J., 320, in press.
Spinrad, H., and Djorgovski, S. 1984, Ap.J. (Letters), 285, L49.
Yoshii, Y., and Saio, H. 1979, Pub. A.S. Japan, 32, 339.

CLUSTERS OF GALAXIES

Guido Chincarini
Universita' di Milano e
Osservatorio Astronomico di Brera
via Brera 28
Milano , Italy

ABSTRACT. The all sky catalogue of clusters of galaxies prepared by Abell,Corwin and Olowin is now completed and the first catalogue of distant clusters by Gunn,Hoessel and Oke has been published. I stress, in relation to such catalogues, the need of using numerical simulations to estimate their completeness , understand the observations and achieve unbiased informations on cosmology and cluster evolution as a function of cosmic time. It is suggested that an early interaction between the galaxies and the intracluster medium may enhance star formation.

INTRODUCTION

Clusters of galaxies have been a main subject in extragalactic research for about 50 years. In the last 20 years,however,we have seen a change of emphasis. On the one hand we gained evidence that not all clusters are fully relaxed so that the understanding of the dynamical evolution became of primary importance and on the other hand we realized that evolutionary phenomena are not so far in the past and that indeed they are at presently observable values of redshift. The study of the evolution of galaxies as a function of the cosmological time is one of the most challenging and fascinating problems of modern observational cosmology and,as expected,one of the most controversial.

Naturally the study of evolutionary phenomena implies the understanding of the clusters of galaxies as observed in different epoch and in particular at the present cosmological time. The understanding of the physics and dynamics of clusters of galaxies is also a fascinating topics of its own. For a detailed discussion a usefull list of references the reader is referred to the excellent and comprehensive review written by Sarazin (1986). Clusters

R. G. Kron and A. Renzini (eds.), Towards Understanding Galaxies at Large Redshift, 217–225.
© 1988 by Kluwer Academic Publishers.

of galaxies have been used also as markers of the large
scale structure,Tully (1986) , Bahcall and Soneira (1984
) , see also references in the review paper by Chincarini (
1987). In a previous Erice summer school (at that time we
were closer to a school than to a meeting or workshop) ,
Chincarini (1980) discussed some of the cluster
parameters.

In what follows I will sketch very briefly the present
situation on catalogues of clusters of galaxies (the
starting step for any type of research) and a possible step
to better understand their completeness (and eventually
evaluate evolutionary effects) by simulations. At the end
I will put forward the idea that at large z the clumpy
structure of the intracluster gas and density of galaxies
may favor the probability for star formation , and related
phenomena , in galaxies via a possible enhanced interaction
between the galaxies and the gas.

THE CATALOGUE OF CLUSTERS OF GALAXIES

Before his sudden death (October 1983) George Abell had
undertaken , in collaboration with H. Corwin , the revision
of the northern catalogue of galaxies and the preparation of
its extension to the southern emisphere. As a tribute to
the memory of George Abell , Corwin and I asked a
contribution to the N.S.F. for the completion of the
catalogue and for the collaboration of Dr. Olowin. Corwin
and Olowin completed the catalogue so that the all sky
catalogue of rich clusters of galaxies is now available.
The catalogue lists 4076 rich clusters of galaxies each
having at least 30 members within the magnitude range m(3)
to m(3)+2. The complete catalogue will be essential to the
astronomical community for a variety of studies ; it wil be
particularly usefull to the european astronomical community
since the main observing facilities are in the southern
emisphere. The catalogue will be available on magnetic tape
from the NASA Data Center and from the Strasbourg Data
Center. Detailed information on it can be found in Abell
and Corwin (1982) and Abell,Corwin,Olowin (1988).

The catalogue lists clusters up to a redshift of about z=0.2
it is rather incomplete in the range 0.2 < z < 0.4. It is
complitely blind at z>0.4. Indeed very distant clusters are
not detectable in the ESO/UK Schmidt survey. To understand
the completeness of catalogues of galaxies,the possibility
of detecting distant clusters of galaxies in deeper surveys
and detect , if present , evolutionary phenomena , we
decided to approach the study and analysis of distant
clusters of galaxies by simulating what we expect to observe

on the basis of what we know today in extragalactic
astronomy and observational cosmology. This work , carried
out in collaboration with Cappi,Conconi, Manoussoyanaki e
Vettolani (see also Cappi et al. 1987) does not take into
account dynamical evolution and only preliminary tests have
been done to simulate the effects of possible evolution of
galaxies. As pointed out by Gunn et al (1986) ,however,
dynamical evolution may play a very important role: at
large z the collapse time of a cluster of galaxies could be
larger then the Hubble time , that is a not too dense
cluster may not yet be formed,figure 1.

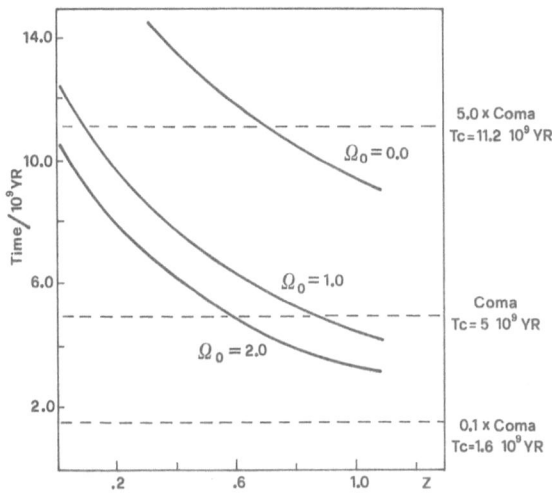

Figure 1. The lookback time as a function of redshift. The
collapse time for some chracteristic cluster densities is
also marked. Ho=50.

The aim of the simulations is to determine the probability
of detecting a cluster in a sample complete to a limiting
magnitude . Since our main interest is for CCD images at
very faint limiting magnitudes from large ground-based
telescopes and/or data obtained with deep exposures and the
wide field camera of the Space Telescope,we disregarded the
separation between stars and galaxies. The effect of
stellar contamination is estimated to be unessential for the
present purpose and in any case it could be easily added to
the algorithm. We did not take into account the response
and resolution of the detector to limit the computer time of
the simulations and analysis. This,however, could be easily
added to our software. For the momemt we must keep in mind

that the real situation will be somewhat worse of the one we will describe,that is our detection probability is somewhat optimistic.

The simulations are conceptually rather simple.   Using our present   knowledge   in   cosmology,galaxian distribution, clusters of galaxies and the spectroscopic and photometric characteristics of   galaxies for   the various types of the Hubble sequence, we create a non cluster field of galaxies (background) , clusters of galaxies of various galaxian content and richness located at various z and a cutoff in the limiting magnitude.   After defining a search area we determine both analitically and by numerical simulations the signal to noise based on the galaxy statistics (counts in the search area)as a function of cluster type and z.

In this context it is clear that different realizations of the same cluster located at a preselected z cause a dispersion in the signal to noise ratio , an effect which is certainly present also in nature and wich must be taken into account if we want to determine a realistic probability of detection.

Using a search area of about 1 - 2 core radii and defining:

Nc cluster objects / area
Nb back-forground objects / area
Na Number of areas
S/N signal to noise ; $S/N = Nc/\sqrt{Nc+2Nb}$

The probability that one area has at least N objects is:

$P(N) = 1$ for $N < $ or $= Nb$

$P(N) = 1- (SUM\ P'(i)) ** Na$; $i=0$, $N-1$;for $N > Nb$

where $P'(i) = Nb ** i * \exp (-Nb) / i!$

The probability of detection is the SUM for $0 < i < Nb*Na$ of the products of the probability that a cluster is in an area with i background objects and of the probability that in the background there is not an area with at least $Nc + i$ objects (spurious cluster) is

$P(Detect) = SUM (P'(i)*P(Nc+i))$; $i = 0$, $Nb*Na$

and this is the relation we shoud use (derivation by Conconi) for the theoretical expectation.

Such probability has been computed for a cluster at z=0.6 as a function of the S/N and compared to 3000 simulations obtained for clusters of various richness and type located in the redshift range between 06-07.

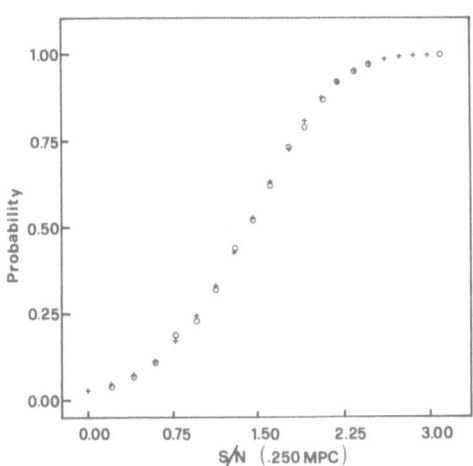

Figure 2. The probability of detection as a function of the signal to noise ratio. Crosses computed for z=0.6 , open circles derived from 3000 simulations.

The agreement is excellent , figure 2 , and the curve is independent of z. The S/N as a function of the redshift is given in figure 3 for a cluster of richness class 2 (R2) rich in spiral content (T3) .
In the simulations we account for the morphology of the galaxies,their color and we apply all the corrections related to the variation of the spectroscopic and photometric parameters as a function of z. To have a realistic background we extrapolated the luminosity function of non cluster galaxies at large redshifts.

By folding the relations obtained above : that is dispersion in S/N , S/N as a function of z and probability of detection as a function of S/N we obtain the percent of detectable clusters as a function of the redshift. In figure 4 we illustrate the resulting curve obtained for a cluster of richness class 2 , intermediate galaxy content in an exposure to the limitimg magnitude J1 = 23.0. For a cluster of richness class 4 (R4)and rich in spiral galaxies (T3) the curve would maintain roughly its shape and shift to the right (larger redshifts) of about 0.15. For the same cluster,R4 and T3,in an exposure reaching the magnitude J1 = 25,the curve would shift in z of about 0.5.

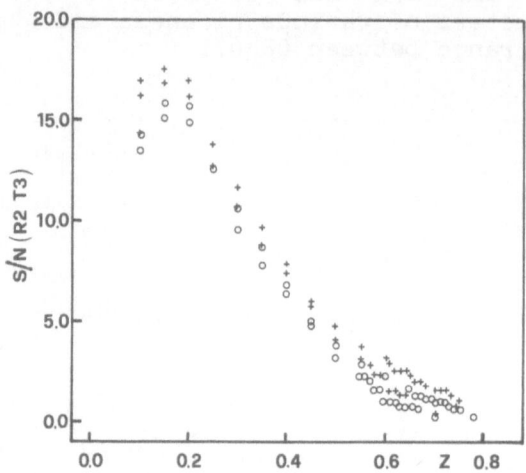

Figure 3. The signal to noise ratio as a function of the redshift for a cluster of richness class 2 and rich in spiral galaxies derived using various simulations.

The simulations tell us known facts in a quantitative and significant way : at very large redshifts we will detect only very rich clusters (or in some cases more clusters seen in projection). Other simulations in which we used the evolutionary model C by Bruzual show that evolution enhances cluster detection . We are now testing this effect using the models computed by Buzzoni (these proceedings). I also emphasize that a good understanding of the statistics in connection with a catalogue of distant clusters in which the biases are known (assumed that at some point we will have to our disposal a large set of observations) may allow informations on the collapse time and cosmological constants.

Gunn,Hoessel and Oke (1986) undertook some years ago a fundamental search for distant clusters of galaxies. Simulations should fully explain their figure 9 and make us understand statistically how significant are the anomalies (evolution or not) observed in some of the distant clusters of galaxies. A point which has been stressed also by Koo in this conference.

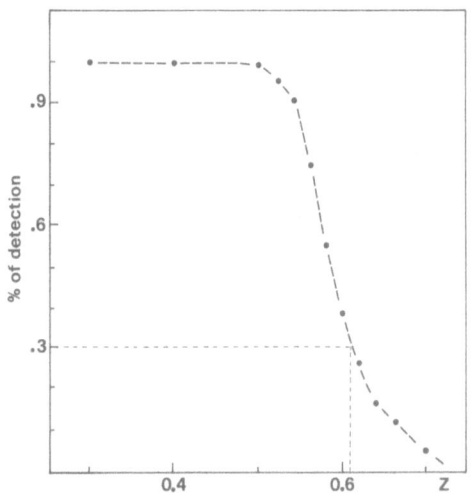

Figure 4. Percent of cluster detection as a function of redshift , see text.

IS DYNAMICAL EVOLUTION AND HOT-GAS/GALAXY INTERACTION PLAYING A ROLL?

The clusters of galaxies we observe locally show an optical morphology which goes from a rather symmetric (quasi isothermal) telescopic appearance to an irregular one. The cluster morphology is correlated with other cluster parameters as galaxy content , velocity dispersion , richness etc ; see the detailed review by Sarazin (1986). A similar behaviour has been observed in the X-ray morphology ; for an excellent review see Forman and Jones (1982). If we consider the gross features of clusters of galaxies we find that the X-ray morphology correlates fairly closely with the optical morphology. An operative way to interpret the morphological differences among clusters of galaxies is to refer to the cluster density as the fundamental cluster parameter and adopt the view that while dense clusters had the needed time to reach a sort of equilibrium the most irregular clusters (which are also the less dense) could not relax in the Hubble time. That is even locally we observe different phases of dynamical evolution , a concept which is supported by computer simulations.

It has been shown,furthermore,that a strong interaction
exists between the interstellar gas of cluster galaxies and
the hot intracluster gas. In other cases the halo gas of a
neighbour galaxy may cause a similar effect. Observational
evidence has been given by various authors,for a review on
HI depletion see Haynes,Giovanelli and Chincarini (1984).
That such interaction is detectable at various wavelengths
has been shown , among others ,by Kothanyi et al. (1983)
and by Chincarini and de Souza (1986). These authors
detected , at a faint level , also the H (alpha) emission in
the galaxy NGC 4438 and the shock model proposed would cause
, after cooling , turbolence and star formation.

Looking back in time we may have a) an enhancement of these
phenomena and b) a certain selection of these phenomena
because of the higher probability of detecting rich or
unusual rich clusters.

The enhancement of the intercation activity , or the
magnification of it , is probably independent of an x-ray
evolution in the sense that we have no evidence of larger
x-ray emission for more distant clusters. We know,on the
other hand,that in the early phases of cluster evolution the
space distribution of the galaxies and of the gas must be
very clumpy (see for instance Cavaliere et al , 1986 , and
references therein). If such clumpiness or related
phenomena enhances the phenomena due to the interaction
described above , then we may be able to explain , at least
in part , the phenomena of youth we may detect in distant
clusters of galaxies, since at large z we would expect some
enhancement (bursts ?) of star formation.

What I suggest is that the interaction Galaxy -ICM may be a
function of cosmic time possibly via dynamical evolution and
subclustering. The phenomenon is then regulated by
dynamical evolution and statistics. It seems to me that
this phenomenon must also be understood if we want to
understand the observations of distant clusters of galaxies.

REFERENCES

Abell,G.O.and Corwin,H.G.,1983,in EARLY EVOLUTION OF THE
UNIVERSE AND ITS PRESENT STRUCTURE,page 179, Eds. G.O.Abell
and G.Chincarini , published by D.Reidel

Abell,G.O.,Corwin,H.G. and Olowin,R.,1988, in preparation

Bahcall,N. and Soneira,S.N.,1984,Astroph.J. 277 , 27

Cappi,A.,Chincarini,G.,Conconi,P.,Manoussoyanaki,I.   and
Vettolani,G., 1987 , The Messenger , N.48 , June

Cavaliere,A.,Santangelo,P.,Tarquini,G. and Vittorio,N.   ,
1986 , Astroph.J., **305** , 651

Chincarini,G.,1980,in   X-RAY   ASTRONOMY,page   197,Eds.
R.Giacconi and G.Setti , published by D.Reidel

Chincarini,G.,1987,in OBSERVATIONAL COSMOLOGY,page 275, Eds.
A.Hewitt et al.   (I.A.U.   Symposium China),published by
D.Reidel

Chincarini,G. and de Souza,R.,1985,Astron. and Astroph.,
153 , 218

Forman,W.   and   Jones,C.,1982, in Ann.   Rev.   Astron.
Astroph. Vol.20, page 547, published by Annual Reviews Inc.

Gunn,J.E.,Hoessel,J.G. and Oke,J.B.,1986, Astroph.J. **306** ,
30

Haynes,M.P.,Giovanelli,R.   and   Chincarini,G.L.,1984,   in
Ann.Rev.Astron.Astroph.   Vol.22, page 445, published by
Annual Reviews Inc.

Kotanyi,C.G.,van Gorkom,J.H.   and Ekers,R.D.   , 1983 ,
Astroph.J.Letters, **273** , L7

Sarazin.C.L.,1986,Reviews   of   Modern   Physics,Vol.58,N1.
January

Tully,R.B.,1986,Astroph.J. **303** , 25

AKNOWLEDGEMENTS

I appreciate the kindeness of the Editors for letting me
submit the paper after the deadline , Dr. Buzzoni for
technical help and Mr. Garignani for the figures.

CLUSTER GALAXIES AT HIGH REDSHIFT

James E. Gunn
Peyton Hall
Princeton University
Princeton, NJ 08544
Visiting Associate, Palomar Observatory

Alan Dressler
Mount Wilson and Las Campanas Observatories
813 Santa Barbara Street
Pasadena, CA 91101

ABSTRACT. Modern CCD instrumentation has made it possible to obtain photometric and spectroscopic data on normal galaxies in clusters at redshifts up to nearly unity. In this paper we discuss a survey for such clusters and preliminary spectroscopic results for galaxies in a sample of 7 clusters near redshift 0.5, and another at z=0.75. The data clearly indicate strong evolution of a sizeable fraction of the cluster population, in the sense that there were many more "active" galaxies then than at the present epoch. The physical mechanism responsible for the activity is as yet uncertain, but we discuss what we consider a likely candidate. The evolution of the "passive", red population is sufficiently slow that strong constraints are placed on the rate of continuing star formation or, alternatively, on the formation epoch.

1. SURVEYS FOR HIGH-REDSHIFT CLUSTERS

It is self-evident that if one wishes to study the galaxies in high-redshift clusters one must first find the clusters, and it was to this end that Oke and Gunn began in 1970 the survey which has led to the discovery of most of the clusters discussed here. The survey is described in some detail in Gunn, Hoessel, and Oke (1986), but we will review its main properties here and describe further work in progress which should allow a much more quantitative investigation of the clustering properties at high redshift.

R. G. Kron and A. Renzini (eds.), Towards Understanding Galaxies at Large Redshift, 227–237.
© *1988 by Kluwer Academic Publishers.*

The survey was done in several stages, beginning with a IIIa-J survey on the 48-inch Schmidt telescope at Palomar; this survey produced 159 clusters over a survey area of 71.5 square degrees, most of which had redshifts between 0.2 and 0.3. This was extended over a smaller area, 42.5 square degrees, with IIIa-F plates, which produced an additional 63 clusters, mostly with redshifts in the range 0.3-0.4. We then surveyed 11.3 square degrees with a red-sensitive image tube with the 200-inch telescope and found 76 clusters with a wide range in redshift, but most in the interval 0.5-0.7. The image tube plates were not of very high quality, and when hypersensitization techniques became efficient and reliable enough we moved to the Kitt Peak 4-meter and worked with both IIIa-F and IV-N plates, surveying 5 and 24.5 square degrees, respectively, yielding 18 and 102 clusters, mostly with redshifts between 0.6 and 0.9. We have reshifts for at least one galaxy in 154 of the survey clusters; the highest redshift so far found (actually, for two clusters, probably in the same supercluster), is 0.92.

We have thoughout this exercise attempted to move the search band to the red as the mean survey redshift increased, in a (largely successful) effort to keep the rest wavelength of the images roughly constant at about 5000 Å. In this way we avoid differential K-correction biasing, though of course we are still subject to bias caused by any gross differential evolution.

The number counts indicate that the survey is in some sense reasonably complete in the relevant search areas to a redshift of about 0.5, above which the counts drop significantly below the predictions for constant comoving density. Many, if not most, theories for the formation of structure on large scales, and indeed the dynamical parameters of present-day clusters themselves, argue for relatively recent formation, so that it is not unexpected that the counts should fall off to higher redshift. The visual searches used in the survey, however, are so subjective that to conclude anything physical from the number counts would be sheerest folly.

What is clearly needed at this point is an objective survey with quantitative imaging data, and we (Gunn, Hoessel, Oke, Postman, and Schneider), with the aid of the new four-shooter CCD camera on the 200-inch telescope (Gunn et al 1987) began last year a new survey with this goal in mind. The camera is used in a scanning ("TDI") mode, in which the telescope is moved in synchronism with the charge on the chips to produce a continuous "tapestry" on the sky. The survey goes to about $m_r = 25$, and should permit identification of clusters of Abell (1958) richness 1 to redshifts of unity or a bit beyond, from catalogs produced by automatic object finding and identification software. The cataloguing routines are in hand and are now undergoing extensive testing; the survey itself, which will cover six square degrees, is about two-thirds complete, and we expect to publish the first results within a couple of years.

2. THE EVOLUTION OF CLUSTER GALAXIES

The study of galaxy evolution with large lookback times is dominated by two difficult issues. The first is a technical matter. Even the most luminous galaxies are faint ($m_r > 19$) at significant lookback times ($z \geq 0.5$), and so spectrophotometric observations of average galaxies challenge our present telescope and detector technology. The second issue is the selection of objects in an unbiased way in order to assemble a representative sample of galaxies at the remote epoch. It is far too easy to chase only exotic objects whose very peculiarity has brought them to our attention. Though observations of such objects may be fascinating and revealing, they may tell us little about the evolution of an typical galaxy like our own.

Early theoretical studies and modelling (see, for example, Tinsley 1980 and Bruzual 1983) led one to expect that there would be little of interest to see in the way of spectrophotometric changes in normal galaxies over lookback times corresponding to redshifts of 1/2 or so. It is fair to say that that has, in the main, proved to be the case, but there is a significant number of objects, twenty percent or so of the total, for which this is apparently not so. This was seen early in the pioneering photometric observations of Butcher and Oemler (1978,1984) which indicated a dramatic evolution of cluster galaxies in only the last 5 Gyr; it was that work which led us in 1981 to begin a program of spectroscopic observations of a sample of galaxies in distant clusters. We began with the PFUEI CCD camera/spectrograph (Gunn and Westphal 1981) on the 200-inch telescope and when the new four-shooter CCD spectrograph became available began using it. The latter exhibits a quantum efficiency on the sky of about 25% at peak and allows us to obtain useful low-resolution spectra of normal galaxies even up to $z = 0.8$ with exposure times of less than 10 hours. An aperture mask with of order 10 small slits produces that many spectra at one setting, and enables, but by no means makes easy, the study of typical galaxies at cosmologically significant lookback times.

Observations of galaxies in clusters provide a practical though imperfect solution to the problem of selection effects. Because of their high densities and predominance of early type galaxies, clusters do not offer us a sample of average galaxies in typical environments, but they do offer us a volume limited sample of galaxies under similar conditions over a large range of epoch. Though there may have been some question at first as to whether distant clusters were, by selection, all extraordinarily rich or dense, more complete surveys like that described above have identified clusters that appear, based on their space densities and luminosity functions, to be the ancestors of at least the richer present-epoch clusters like Coma or Hercules. The reader should bear in mind one strong caveat in all of this, and that is that if the cluster population itself evolves strongly, clusters which look like Coma at z = 0.5 may not look like Coma today. Until the issue of cluster evolution is settled, we have no ready answer to this problem, but the direction of evolution, namely to richer objects, almost certainly increases the strength of our conclusions. We will return to this question of the representativeness of the cluster sample later.

Our present sample consists of 7 intermediate redshift clusters in the range $0.37 < z < 0.55$ and 4 high redshift clusters with $0.65 < z < 1.0$. We typically have photometry of 100-300 galaxies with $18.5 < m_r < 23.5$ in fields of size $5' \times 5'$. For the intermediate redshift sample we have collected usable spectra of 20-50 galaxies with typical magnitudes of $20 < m_r < 22$ in each of these fields but we are are just beginning to accumulate a reasonable number of spectra for the high-z sample. We also have available a low-z sample of roughly 1000 spectra of galaxies in clusters with $z < 0.06$, obtained by Dressler and Shectman.

3. THE FRACTION OF ACTIVE GALAXIES IN HIGH REDSHIFT CLUSTERS

In our intermediate-redshift sample of galaxies in seven clusters, which represents a relatively homogeneous data set, we have redshifts for 236 objects. Of these 236, 163 are cluster members and 73 are "field galaxies". This averages to 15-30 members per cluster over an area which, at this redshift, corresponds to about two square megaparsecs.

Although Butcher and Oemlers' claim of significant evolution was based on a distinction between "red" and "blue" galaxies, our data lend themselves better to a separation into what we call "passive" and "active" galaxies. "Active" galaxies are those with signs of recent star formation or an active nucleus. These are to be contrasted with "passive" galaxies, those with a K-giant spectrum typical of an old stellar population, rest frame $B - V \simeq 0.9$, and little or no sign of star formation within the previous 5 Gyr. At the present epoch these are the spectral characteristics of E or S0 galaxies which now account for about 95% of the galaxies in the inner regions of dense, concentrated clusters. As we will see, by $z \simeq 0.5$ this fraction has dropped to about 70%.

The spectra of the "active" population can itself be divided into those with and without emission lines. Those with emission lines include galaxies with long-term, relatively steady star formation like spirals, "starburst" galaxies with greatly enhanced star formation rates (SFRs), and high-excitation spectra typical of active nuclei (AGNs). Examples of these are shown in Dressler, Gunn, and Schneider (1985). The other common type, first noted in our study of 3C295 (Dressler and Gunn 1983) is a basically old stellar spectrum, perhaps about 0.2 mag bluer in rest frame B-V than a typical passive galaxy, with strong Balmer *absorption lines* and little or no emission. We call these "E+A" spectra because they are well matched by adding the spectra of A stars to an elliptical-type (passive) spectrum. We have also called these "post-starburst galaxies", based on our interpretation that many of these galaxies had a significant increase in the SFR which subsided about 1 Gyr before the epoch of observation.

We have calculated the active fraction in each cluster based on both the spectral and photometric data in the following way. We divide the g-r histogram for each field studied into a red, yellow, and blue region. The contamination by field galaxies obviously increases substantially as one moves from red to blue. For each zone we use the spectral data to predict the cluster-to-field ratio, and then multiply this fraction by the total number of galaxies in the bin to the approximate magni tude limit sampled by the spectroscropic sample. This gives us a predicted cluster membership of red, yellow, and blue galaxies. We then use the spectral data to predict the active fractions in each bin based on the active-to-passive ratio for the spectroscopic sample. By dividing the predicted number of cluster members by the predicted number of active galaxies we derive the active fraction f given in Table 1. This procedure corrects for the bias we introduced by preferentially selecting bluer galaxies for spectroscopic study, although in practice this bias was almost entirely offset by the fact that the majority of these are not cluster members.

TABLE 1: The Intermediate Redshift Sample

Cluster	z	f (%)	E+A	em+AGN
9HFCL27	z = 0.38	27 ± 15	3	1
9HFαβ	z = 0.39	27 ± 8	2	10
Cl0024+24	z = 0.40	22 ± 9	6	1
3HFCL2	z = 0.42	34 ± 17	1	4
3C295	z = 0.46	39 ± 18	3	3
Cl0016+16	z = 0.54	31 ± 10	7	5
16HFββ	z = 0.54	37 ± 15	3	5

From Table 1 we see that the active fraction averages 30% and it is easily shown that the deviations are consistent with counting statistics. It therefore appears that with this still small sample of 7 clusters we have found a significant fraction of active galaxies which, within the errors, does not vary from cluster to cluster. It is also true but far less than obvious that *types* of active galaxies are not statistically significantly different from cluster to cluster. Note that the fraction of active galaxies is large even for the often-cited cluster Cl0016+16, which was claimed by Koo (1981) to be a *counterexample* to the Butcher–Oemler effect of a greater fraction of blue galaxies at high redshift. Koo suggested that this distant cluster was like the low-redshift Coma cluster, but our discovery of a large population of E+A galaxies demonstrates that this is not the case. While it is true that there are few *very blue* galaxies in this cluster, as Koo reported, this difference may not be very significant. If, for example, many galaxies in a given cluster experience a simultaneous but temporary increase in star formation activity, the blue population could rise for a few Gyrs. Afterwards, the cluster may return to a redder color distribution but still harbor signs of a very active period. Thus the accident of the epoch of observation may make two phases of the same activity appear different and unconnected. Thus the apparent differences in the type of activity from cluster to cluster in Table 1, while still not significant with this rather sparse data set, may well be real and telling us something about the history of the event(s) which are responsible for the activity.

How does this fraction of active galaxies of 30% compare to the populations of present-epoch clusters of similar richness and density? The ~ 16% fraction of emission-line galaxies is significantly higher than the value of 7% found by Dressler, Thompson, and Shectman (1985) in a sample of about 1000 galaxies in low-z clusters. The distribution of luminosity and average surface brightness for this low-z sample studied by Dressler and Shectman, is very similar to the characteristics of the high-z sample, but the area survey extends to larger radii in each cluster. Dressler has redone the analysis with an area similar to that surveyed for the distant clusters and finds an emission-line frequency of a bit less than 5%. Thus an increase of at least a factor of three in the frequency of emission-line galaxies at $z \simeq 0.5$ is implied.

The E+A case is more dramatic. Dressler identified 20 candidate objects from about 1000 low signal-to-noise spectra from the Dressler-Shectman sample, and then obtained better spectra for these with the du Pont telescope at Las Campanas Observatories. He found only 3 or 4 that were comparable to the high-z E+A galaxies, a fraction of less than 1%! Therefore, an order of magnitude increase is found in the frequency of E+A galaxies in high-z clusters. While it is true that there are examples of such galaxies in both clusters and the field at the present epoch, we stress that they are relatively rare (and also tend to be much lower-luminosity systems than the ones studied here).

Thus it seems clear that Butcher and Oemler were on the right track when they identified a change in cluster populations since $z \simeq 0.4$. Although it is too early yet to determine unambiguously the physical nature of this difference, there is a strong indication in our work, as well as studies by Couch and Sharples (1988), reported on at this workshop by Ellis, and Lavery and Henry (1986), also described here by Henry, that starbursts in some of these cluster galaxies are partly or wholly responsible for the change in spectral characteristics. We came to this conclusion for the E+A galaxies by a rough modeling of the Balmer absorption lines and continuum color (Dressler and Gunn 1983, Dressier 1987) and recently Couch and Sharples have made more detailed models of the same sort. It is difficult to model a system with such relatively red colors and such strong Balmer lines, and essentially the only way any of the workers have found is to make somehow a spectrum whose red end is dominated by K giants and whose blue light is dominated by A main-sequence stars, just as the qualitative appearance of the spectrum suggests. This can be accomplished with real galaxies either with a very peculiar initial mass function, or, more naturally, with a normal mass function and a burst, which (if the mass function is more–or–less normal) involves a fair fraction ($\sim 20\%$) of the mass and occurred long enough ago that the O and B stars are gone. The evolutionary tracks for such systems in the color/Balmer equivalent width plane are shown in Dressler (1987) and exhibited in a slightly different form in the Couch and Sharples paper.

Although it was our original interpretation that most of the objects with moderately strong emission-lines were likely to be normal spirals (e.g., Dressler, Schneider, and Gunn 1985), it is plausible that a significant number of these are starbursts caught in the act rather than spirals forming stars vigorously and *continuously*. Although their numbers are comparable with the E+A objects and the emission-line phase must be much shorter, they will be correspondingly brighter during that phase, so one will be looking farther down the galaxy mass function where there are many more objects.

What might trigger such an increase in star formation? We have advanced the idea that a sudden increase in the SFR may be triggered in a gas-rich galaxy if it falls into the dense, high pressure intracluster gas for the first time. While ram pressure may serve to sweep out the warm intercloud medium, the dense, cold clouds may be relatively unaffected by this and instead induced to collapse by the sudden and significant increase in pressure over the normal interstellar values, which can amount to a factor of a *thousand* for a galaxy penetrating the inner regions of a cluster. We have some evidence, admittedly circumstantial, from the observations that tends to support this model: (1) the active galaxies appear to preferentially populate an annular zone approaching but always avoiding the cluster center; (2) the typical velocities of active galaxies are about 50% higher, on average, than those of passive galaxies. Additional support for this hypothesis may be available in the analysis of high-resolution images of such galaxies that will be possible with the Hubble Space Telescope.

Consideration of an alternate hypothesis, that the starbursts might be triggered by mergers or tidal encounters, has prompted us to inspect visually all the galaxy images on our CCD frames and classify them as to whether they are isolated, have close companions, or look tidally disturbed. This classification (blind with respect to the colors or spectra of the galaxies) revealed no increase at all in the fraction of galaxies with companions or disturbed shapes for those with active spectra. One might not be surprised in the case of E+A galaxies since a merger might be well calmed after a Gyr or more, but we expect that some reasonable fraction of the emission-line galaxies if they are in fact the same creatures observed closer to the event, should have shown an increased incidence. The poor statistics of our small sample prevent us from ruling out such a model, but it is certainly clear that our data provide no support, as it probably should were this the correct interpretation. It also seems unlikely that this hypothesis can account for the spatial distribution of the objects, since the incidence of encounters is highest at the center, which the active galaxies avoid.

Nor can we rule out the possibility that such evolution is independent of the cluster environment and instead reflects a time evolution of galaxies in general. Detailed comparison of cluster samples to representative field samples, like the one being assembled by Kron and Koo, should be decisive on this point. In this regard the spatial distribution of the active systems in the clusters is of crucial importance; if it is statistically demonstrable that the active galaxies inhabit an annular zone, it says either that environment plays a crucial role or that the phenomenon is strongly related to morphological type, since there are strong morphological type gradients in clusters (Dressler, 1980).

In summary, we have strong evidence that cluster galaxies do evolve. In each of our 7 clusters at $z \simeq 0.5$ we see a significant increase in the fraction of galaxies in which recent star formation has been important. The interpretation of this result remains in question, but perhaps not too far beyond our reach.

Finally, the issue of selection effects deserves further attention. Koo, at this workshop, has argued that the reported increase in blue or active galaxies does not necessary imply that these galaxies are actually located *in the core* of the cluster. His principal concern is that interlopers in the vicinity of but not actually in the core are included because the redshift range covered by the velocity dispersion admits a volume hundreds of times larger than the core volume. This effect becomes more serious with increasing redshift. It is straightforward to show that this is not an important effect for the kind of sampling we have been doing. Even a volume hundreds of times larger would provide negligible contamination if the surrounding volume had the average field density because the average density contrast of the central regions of clusters is typically 10^4.

Koo has argued, however, that it is likely to be the case that rich clusters are imbedded in superclusters in which the density is much higher. If take a value of $n \sim 10^{-1}$ luminous galaxies per cubic megaparsec from present-epoch examples like the Coma and the Local Supercluster, and a thickness of the supercluster of about 5 Mpc (see, for example, the "cone-diagrams" of deLapparent, Geller, and Huchra 1986), it is clear that our 2 Mpc^2 areas will typically have about 1 supercluster galaxy superposed on the central part of the cluster. Even if the supercluster is seen edge-on (a chance occurrence of about 10% probability unless selection effects are important) the contamination is $\sim$ 5 interlopers. This is still small compared to the 20-30 active galaxies, but it may be large enough to be important. Therefore, it is important to know if such edge-on alignments have been preferentially selected. This is surely not the case for 5 out the 7 clusters in our sample. Four of these are from the Hoessel, Oke, and Gunn catalog, for which the critical selection criterion is the density contrast of a region of radius of 1-2 core radii, i.e., a fraction of a square megaparsec. Contamination by supercluster galaxies is totally unimportant in such small areas, as evidenced by the fact that nearly the entire core population of these clusters is red, passive galaxies, far different from the supercluster population. Attention was drawn to the 3C295 cluster by the very strong radio source, so this case too is immune from such a bias. As for the other two clusters, they were selected on photographic plates covering a large area so that a supercluster projection bias could have played a role, though they, like the Hoessel-Oke-Gunn clusters, were almost certainly found by virtue of the contrast of their central regions to the field. Koo himself has noted that Cl0016+16 has practically no blue population in the central region, so it seems unlikely that there is any serious contamination from a bluer supercluster population. Cl0024+24, which, interestingly enough, has the largest population of very blue, emission-line galaxies of any of our sample, may be the only good case for supercluster contamination.

While it is true, as Koo points out, that many of the first clusters studied may be unusual in one way or another (e.g., richness, color, density) it seems to us important that these selection biases appear to vary from cluster to cluster but the "active fraction" we find does not. We think this in itself is a strong argument for the universality of the effect and the relative unimportance of the selection bias.

4. THE EVOLUTION OF THE 4000 ANGSTROM BREAK

It is not clear how germane are the results just discussed to the general issue of galaxy evolution. It is possible that the Butcher-Oemler effect is quite environment specific and may tell us little about the history of star formation in the more typical galaxies not in rich clusters. This limitation has encouraged us to address the more general question of the history of star formation in the "passive" galaxies. Because these are likely to be among the "oldest" (in the sense of having the least late star formation) of any galaxies regardless of environment, they should be good cases in which to try and estimate the critical epoch of star formation when these oldest galaxies converted much of their gas into stars.

The diagnostic we are using for this study is the 4000 Åbreak amplitude introduced by Spinrad (1980,1986) and recently studied by Hamilton (1986). A new study by Dressler and Shectman (1987) shows, for a sample of about 800 galaxies in present-epoch clusters, that the break amplitude is a weak function of galaxy luminosity (as claimed by Hamilton) but a strong function of star formation history. These are exactly the desired characteristics for the task at hand. Our distant cluster sample provides a good set of galaxies in which to investigate this issue because it is, in some ways, a volume limited sample free of the more troublesome selection effects. We have shown (see Dressler 1987) that the maximum break amplitude does not change significantly in cluster galaxies from the present epoch to the redshift 0.5 sample. This alone implies that there has been little significant star formation for some 10 Gyr in $\sim 70\%$ of our cluster galaxies. Just for a point of comparison, we can parameterize this result in terms of Bruzual's (1983) models of the evolution of coeval stellar populations. For example, with $H_0 = 50$, $q_0 = 0$, and an epoch of formation $z_f = 4$, this implies a model of $\mu \geq 0.3$, i.e., an exponential decay in the star formation rate with 30% of the star formation occurring within 1 Gyr of formation. In such a model, 80% of the stars have formed by $z = 2$. This has obvious implications for models of the growth of structure in the early universe, especially if they predict a good deal of galaxy formation at a relatively late epoch. A parameterization of the observations that is less model-dependent would, of course, be desirable.

We have a small amount of data which, if representative, provides an even stronger constraint. These are the 15 spectra of cluster members in 13HFKPα, a rich, concentrated cluster at $z = 0.75$ that we are studying. As discussed in Dressler (1987), these appear to have a genuine drop of about 10% in the maximum break amplitude relative to the $z = 0$ sample. One might be concerned that aperture effects are responsible since the distant galaxies are studied using apertures that cover more of the galaxy than for the low-z sample. However, we are reassured by the fact that the change in the break amplitude shows up between the intermediate-z sample and the high-z cluster, over which the change in covered area is negligible. The Bruzual model which is appropriate for this amount of evolution in the break amplitude is $\mu = 0.7$. In such a model 90% of the stars have formed 2 Gyr after formation, which, for the cosmological parameters given above, corresponds to $z = 2.3$. At least 30% of the galaxies in 13HFKPα seem to fit into this category. This result, similar to what has been found for the reddest field galaxies by Lilly and Longair (1984) and Djorgovsky $et.\,al.$ (1987) and implies an early and rapid formation of at least some galaxies.

5. SUMMARY

Cluster galaxies do evolve. There is a higher fraction of active galaxies in clusters at redshits around 0.5, although these clusters are still dominated by passive galaxies. Judging from the amplitude of the 4000 Åbreak, even the passive galaxies show evidence for evolution by $z = 0.75$.

Cluster-to-cluster variations in the types of active galaxies may be significant, but could be a only an accident of the epoch of observation of transient populations. For example, many of the AGN and active starburst galaxies may signal an epoch of activity (like the infall of a subcluster of gas-rich galaxies), which is later seen as an increase in E+A galaxies. We have suggested that E+A galaxies are post-starburst galaxies, perhaps produced by ram-pressure induced star formation. Our data do not give any support for the model that mergers or tidal encounters are responsible for the increased star formation, though they may not be able to rule it out. A model in which such activity is independent of environment can be tested by comparing cluster and field samples.

In terms of future observations, we are turning more of our attention to the difficult task of getting spectra for galaxies in the high-z clusters. We are hopeful that evidence for galaxy evolution will be evident in *all* of the galaxies in such clusters, and that this, combined with long-awaited images from HST, which will allow a reasonably detailed study of the morphology of the galaxies themselves, will help in deciphering the nature of the evolution we have observed.

REFERENCES

Abell, G. O. 1958, *Ap. J. Suppl.* **3**, 211.

Bruzual A. G. 1983, *Ap. J.* **273**, 105.

Butcher, H., and Oemler, A. 1978, *Ap. J.* **219**, 18.

Butcher, H., and Oemler, A. 1984, *Ap. J.* **285**, 426.

Couch, W., and Sharples, R. M. 1988, preprint; see also R. Ellis, this volume.

de Lapparent, V., Geller, M. J., and Huchra, J. P. 1986, *Ap. J. (Lett.)* **302**, L1.

Dressler, A. 1980, *Ap. J* **236**, 351.

Dressler, A. 1987, in *Nearly Normal Galaxies* ed. S. M. Faber, (New York: Springer-Verlag) p. 265.

Dressler, A., Gunn, J. E., and Schneider, D. P. 1985, *Ap. J.* **294**, 70.

Dressler, A., and Gunn, J. E. 1983, *Ap. J.* **270**,

Dressler, A., and Shectman, S. A. 1987, *A. J.* in press.

Dressler, A., Thompson, I. B., and Shectman, S. A. 1985, *Ap. J.* **288**, 481.

Djorgovski, S., Spinrad, H., and Dickinson, M. 1987, *Ap. J.* submitted.

Gunn, J. E., Hoessel, J. G., and Oke, J. B. 1986, *Ap. J.* **306** , 30.

Gunn, J. E., Carr, M., Danielson, G. E., Lorenz, E., Lucinio, R., Nenow, V., Schneider, D., Smith, J. D., Westphal, J., and Zimmerman, B. 1987, *Optical Engineering* **26**, 779.

Gunn, J. E. and Westphal, J. A. W. 1981,*Proc SPIE* **290**, 16.

Hamilton, D. 1985, *Ap. J.* **297**, 371.

Koo, D. C. 1981, *Ap J. Lett* **251**, L75.

Koo, D. C. 1988, this volume.

Lavery, R. J., and Henry, J. P. 1986, *Ap. J. Lett.* **304**, L5; see also Henry, J. P., this volume.

Lilly, S. J., Longair, M. S. 1984 *M.N.R.A.S.* **211**, 833.

Spinrad, H. 1980, "Spectroscopy and Photometry of Faint Galaxies" in G. O. Abell and P. J. E. Peebles, eds.: *IAU Symposium 92, Objects of High Redshift* pp. 39-48 (Reidel, Dordrecht).

Spinrad, H. 1986, *P. A. S. P.* **98**, 269.

Tinsley, B. M. 1980, *Ap. J* **241**, 41.

DISTRIBUTION OF COLORS IN THE Z = 0.24 CLUSTER II Zw 1305.4+2941

D. Trevese
Astronomical Institute, University of Rome
Via G.M. Lancisi, 29
00161 Roma, Italy

D. Nanni, A. Vignato, D.C. Koo, and R.G. Kron

ABSTRACT. The color distribution and spatial distribution of galaxies in a rich condensed cluster at z = 0.24 is analysed. We find no evidence for any population of blue galaxies within the central region.

1.   INTRODUCTION

In this contribution we present multicolor photometric data for a field containing II Zw 1305.4+2941, a rich Bautz-Morgan type I cluster at z = 0.24. The principal motivation for the study is the determination of the blue fraction of the galaxy population of the cluster. To do this we follow the recipe of Butcher and Oemler (1984), but we elaborate their procedure to make the result more reliable in this case. The most important difference is the use of the distribution of galaxies in a color-color diagram to recognize and isolate galaxies in the foreground. A more complete description of this work will appear elsewhere (Koo et al. 1987). In that paper we also consider fits to the cluster profile, the cluster luminosity function, and the similarity of the cluster shape and orientation to that of the cD.

2.   OBSERVATIONS

This study is based on a complete photometric catalogue of all objects in an 18.6 x 18.6 arc min area surrounding the cluster (Koo et al. 1986). We also make use of a faint redshift survey of SA 57 (Koo and Kron 1987): there is an overlap of 50 galaxies in our catalogue. 17 of which turn out to be at the

R. G. Kron and A. Renzini (eds.), Towards Understanding Galaxies at Large Redshift, 239–244.
© *1988 by Kluwer Academic Publishers.*

cluster redshift. The catalogue is limited at sufficiently bright magnitudes that the separation of stars by image profile is reliable. The catalogue contains the photographic colors U-J, J-F, and F-N, which are similar in photometric behavior to U-B, B-V, and R-I, respectively. We also use the F magnitude $\lambda_{eff} \sim 6150$, which at z = 0.24 corresponds to restframe between B and V. Figure 1 shows the color-magnitude plot for all of the non-stellar objects in the field. At the cluster redshift z = 0.24, the limit F = 21 corresponds closely to $M_V = -20$ for galaxies of typical color (we adopt the same cosmological parameters as Butcher and Oemler, $H_0 = 50$ km sec^{-1} Mpc^{-1} and $q_0 = 0.1$).

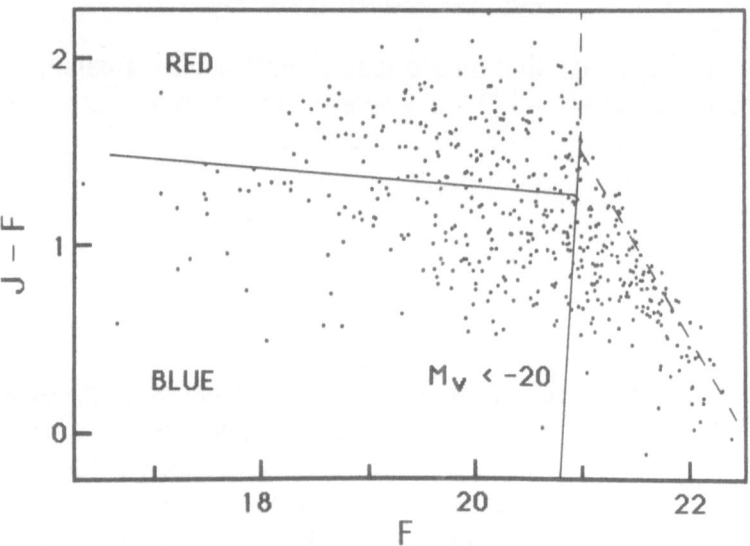

Figure 1. Color-magnitude distribution for the catalogue. The dashed line gives the completeness limits J = 22.5 or F = 21. The essentially vertical line gives the absolute magnitude limit for galaxies at z = 0.24, and the essentially horizontal line defines the division between blue and red.

3. REDSHIFT CUT

The most important technical problem in the study of distant clusters is the contamination of the cluster membership by the projected field. The characteristic energy distributions of galaxies allow redshifts to be determined photometrically under certain conditions (Koo 1985), which should provide an efficient way to address this problem. The 50 galaxies in

our field with spectroscopic redshifts are representative of the whole
catalogue, except that they tend to be brighter than average and their
photometric errors are correspondingly smaller. We have investigated how
well our photometry can discriminate redshift intervals based on this
empirical training set. We have decided to use just the U-J versus J-F
diagram (also because the N photometry has larger errors), and we are content to isolate statistically the foreground, z < 0.15, without claiming
determination of individual redshifts. Figure 2 shows the empirical redshift
cut that is intended to divide groups at z = 0.125 from the Zwicky cluster at
z = 0.24. Note this this empirical approach differs from our earlier modeling
of colors as a function of redshift (Koo et al. 1984). According to this
selection by U-J and J-F colors to eliminate the foreground, we construct for
the following analysis a working subcatalogue. This so-called FAR catalogue
contains only the galaxies which fall to the right of the line in Figure 2, and
only galaxies which if at z = 0.24 would have M_V < -20, as determined by an
appropriate color transformation from the J and F magnitudes.

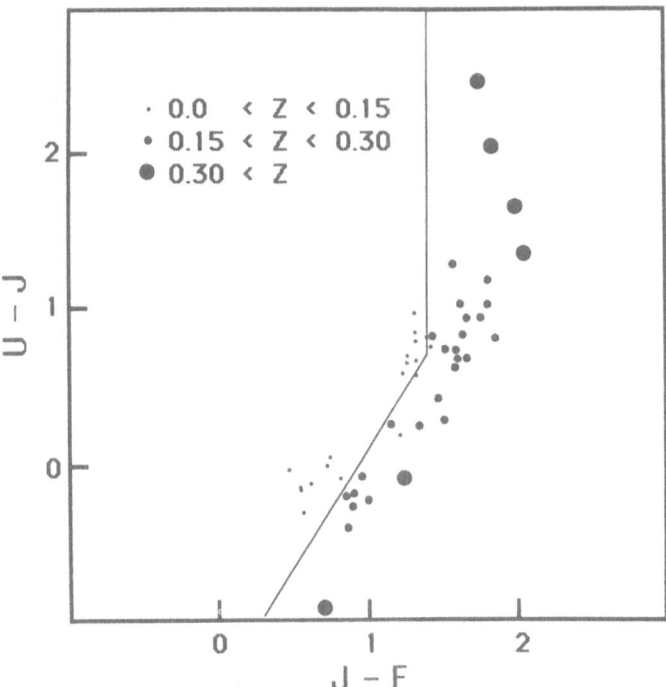

Figure 2.   Color-color diagram for galaxies with spectroscopic redshifts, as coded. The line is
an empirical division between the low-redshift systems (mostly near z = 0.125) and the
remainder, mostly at the cluster redshift.

242

4. ANALYSIS

Our analysis of the FAR catalogue is directed at determining the fraction of
blue galaxies as defined by Butcher and Oemler (1984). We adopted the cD
galaxy as the cluster center, and computed the surface density in radial bins.
The surface density appears to approach an asymptotic value of 0.52 galaxies
per square arc min at ~ 7.5 arc min radius, which we adopt as the background
and the cluster extent. (The redshift survey does however find galaxies at
$z = 0.24$ well outside of this region, including for example the peculiar, luminous galaxy discussed at this conference by S. Majewski.) The growth curve
for the cluster with this background subtracted was then computed in order
to determine the radii enclosing 20%, 30%, and 60% of the total population.
These values are R_{20} = 1.2 arc min, R_{30} = 1.7 arc min, and R_{60} = 4.3 arc
min, respectively. The concentration index C = $\log(R_{60}/R_{20})$ = 0.55, easily
qualifying this cluster as condensed. The determination of the blue fraction
is to be done within R_{30} (Figure 3). The line in Figure 3 gives the division
between nominally blue galaxies and nominally red galaxies according to
Butcher and Oemler (1984); it has a finite slope in order to account for the
color-luminosity relation for elliptical and S0 galaxies. It is clear from
Figure 3 that there is no substantial blue population in the cluster center,
even before correction for the background! The usual way of representing the
background-corrected color distribution is by histograms as in Figure 4,
which shows a strong trend in the blue-to-red ratio as a function of radial
distance. We derive for the blue fraction parameter f_b = -0.01 ± 0.07.

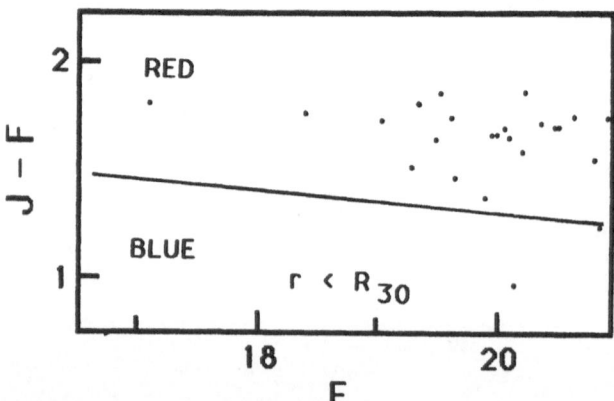

Figure 3.   Same as for Figure 1, except only galaxies within a radius of 1.7 arc min of the cD
are plotted, and only those with colors indicating $z > 0.15$.

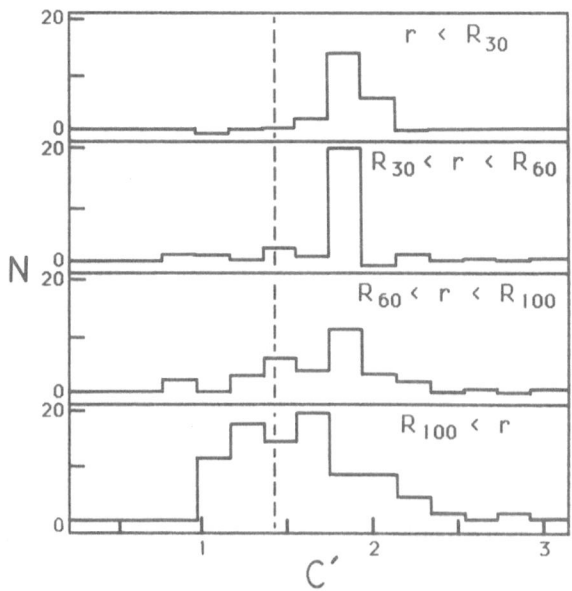

N

C'

Figure 4.   Color histograms in various radial zones, corrected for the background (bottom panel). The abscissa $C' = (J-F) + 0.05 (F - 17)$ sharpens the red peak by accounting for the color-luminosity correlation for intrinsically red galaxies. The vertical dashed line at $C' = 1.46$ gives the division between blue and red galaxies.

5. DISCUSSION

We have repeated the analysis for the blue fraction parameter with some variations to test its sensitivity. If the procedure is repeated in the same way using the original catalogue, i.e. if the redshift cut of Figure 2 is not made, then $f_b = 0.17 \pm 0.11$. Two of the galaxies contributing to this value have spectroscopic redshifts $z = 0.125$; correcting for these gives $f_b = 0.10 \pm 0.10$. If the elongated structure of the cluster is taken into account by considering elliptical rather than circular zones, we derive $f_b = 0.01 \pm 0.08$ without the redshift cut, and $f_b = -0.05 \pm 0.08$ with the redshift cut. These differences are expected because the elliptical-zone analysis rejects more of the low-density area. The conclusion is that the derived blue fraction is indeed moderately sensitive to the method of analysis, and had we not recognized the foreground groupings at $z = 0.125$, we would have found a positive f_b. The important point is that as the level of information about which galaxy is really a cluster member decreases, the net error systematically goes in the direction of increasing the blue fraction because the field is bluer than the central regions of condensed clusters. The observational spread will increase with redshift because the contrast of the cluster with respect to the projected field is lower. Photometric studies of clusters at large redshift are thus expected to display a trend towards larger f_b, quite apart from evolutionary effects. As a result, observational errors and observational selection must be more critically assessed before one can confidently interpret the apparent trend of f_b with redshift.

Spectroscopic studies have revealed populations of galaxies in distant clusters that are unknown in nearby clusters (e.g. Dressler, Gunn, and Schneider 1985). This result is certainly interesting, but we are very far from having established a systematic trend with redshift of any spectroscopic property. This is because the number of clusters is still so few, because there is evident dispersion in the various properties, and most importantly because there could be serious bias in the selection of clusters for spectroscopic work. Contributions at this conference by Chincarini and by Koo review some of the problems of selecting clusters at high redshift.

REFERENCES

Butcher, H. and Oemler, A., Jr. 1984, Ap. J., **285**, 426.

Dressler, A., Gunn, J.E., and Schneider, D.P. 1985, Ap. J., **294**, 70.

Katgert, P., Thuan, T.X., and Windhorst, R.A. 1983, Ap. J., **275**, 1.

Koo, D.C. 1985, A. J., **90**, 418.

Koo, D.C., and Kron, R.G. 1987, in IAU Symposium 124, Observational Cosmology, eds. A. Hewitt, G. Burbidge, and L.Z. Fang (Dordrecht:Reidel), p. 385.

Koo, D.C., Kron, R.G., Nanni, D., Trevese, D., and Vignato, A. 1984, in Clusters and Groups of Galaxies, eds. F. Mardirossian, G. Giuricin, and M. Mezzetti (Dordrecht:Reidel), p. 159.

Koo, D.C., Kron, R.G., Nanni, D., Trevese, D., and Vignato, A. 1986, A. J., **91**, 478.

Koo, D.C., Kron, R.G., Nanni, D., Vignato, A., and Trevese, D. 1987, submitted.

SPECTROSCOPIC OBSERVATIONS OF GALAXIES IN A370

Y. Mellier
Observatoire de Toulouse
14, avenue E. Belin
31400 Toulouse
France

1. INTRODUCTION

The observation of distant clusters of galaxies is the best way to understand the evolution of galaxies in clusters. Since about 5 years, the photometric and spectroscopic studies of these systems have largely confirmed that the blue galaxy excess first mentionned by Butcher and Oemler [1] is common in distant rich clusters of galaxies. However, the origin of this effect is always unclear. Moreover, the first spectroscopic observations of Dressler and Gunn [2,3] have shown that: (1) this effect present a large discrepancy from cluster to cluster either quantitatively or in their spectral content (AGN or normal spirals). (2) In addition to the blue galaxies, distant clusters have a large extend of spectral types as the Post-Star-Burst galaxies (E+A) or S0 with UV-excess [4]. The understanding of the evolution of galaxies in clusters remains a challenge: are these blue galaxies result from galaxy/gas interactions or galaxy/galaxy interactions due to close encounters ? Are they young normal galaxies falling inward the cluster center or old cluster members which have undergone star-burst formation? Although, spectral indications inform us on possible scenarios, not enough data are presently available to firmly conclude on the evolutionary processes responsible of this spectral content in clusters of galaxies. A larger sample of clusters with very good data is still needed.

A few years ago, the Toulouse Observatory has developed an on-line multiaperture spectroscopic system called PUMA for faint object spectrograph (FOS) of the 3.60m C.F.H. Telescope and E.S.O. [5]. It is well suited to study distant clusters of galaxies since it is possible to obtain both photometry of the field and the spectroscopy of about 20 objects simultaneously during the same night. This work which has been described in a serie of papers [6] summarize the first observations on the distant cluster of galaxies A370.

R. G. Kron and A. Renzini (eds.), Towards Understanding Galaxies at Large Redshift, 245–250.
© *1988 by Kluwer Academic Publishers.*

2. THE PUMA SYSTEM

The multiaperture spectroscopic system PUMA was developed at the Toulouse Observatory. It consists in three parts: the first one is fixed on the FOS focal reducer and corresponds to a mask holder which acts as an entrance slit assembly. The second is the flexible software for an on-line selection of the objects for which a spectrum will be requested. The last part is an electromechanical machine which punch the mask (PUnching-MAchine) just before the spectroscopic observations.

The general procedure during the night on the telescope can be described as follow: after a short CCD exposure of the field, the CCD frame is reduced such that all the objects present in the field are detected. An automatic and/or interactive selection is done which optimizes the place available accounting for the length of the spectra on the CCD. The coordinates of these objects are transformed in the corresponding ones on the focal plane of the Cassegrain focus and transfered to the puncher which can then realize the mask (holes or slits). The mask is finally positionned on the focal plane with an accuracy better than 10µm before starting the spectroscopic exposure.

It should be mentionned that during all the procedure described above, the telescope is always guiding on the field such that it is possible to use this time for photometric exposures.

Presently this system allows to obtain about 60 spectra (3h-exposure) plus the BVR CCD photometry of the field in 3 half-nights. The limiting magnitude we obtained is R=20.8 or B=22.9 (objects for which we obtained the redshift, at least).

3. OBSERVATION OF A370

3.1. The data

The photometric and spectroscopic data have been collected during three observing runs at the C.F.H.T. (September 85 and November 86) and E.S.O. (November 86). We have obtained 84 spectra in a field of 5'x7', of which 53 are clusters members and 47 with good radial velocities (±200 Km/s). This sample is now complete to R=19.5 and represents the largest one on a distant cluster of galaxies. Moreover, the observations provides the BVR CCD photometry, so thus 330 objects have been measured with a completeness limit of R=21.

3.2 Dynamical analysis of the cluster

Abell 370 (z=0.374) is a very rich cluster of galaxies which looks like the Coma cluster (z=0.023): same richness, dominated by two bright central galaxies and a high X-ray luminosity [7]. In this point of view, A370 is a "sosie" of Coma.

Using the 47 radial velocities, we derive the velocity dispersion inside the cluster: σ_{los}=1500 Km/s. From this value, we deduce its virial mass: M_{vir}=1.5 10^{15} M_0 or a mass-to-light ratio $(M/L)_R$=60, not

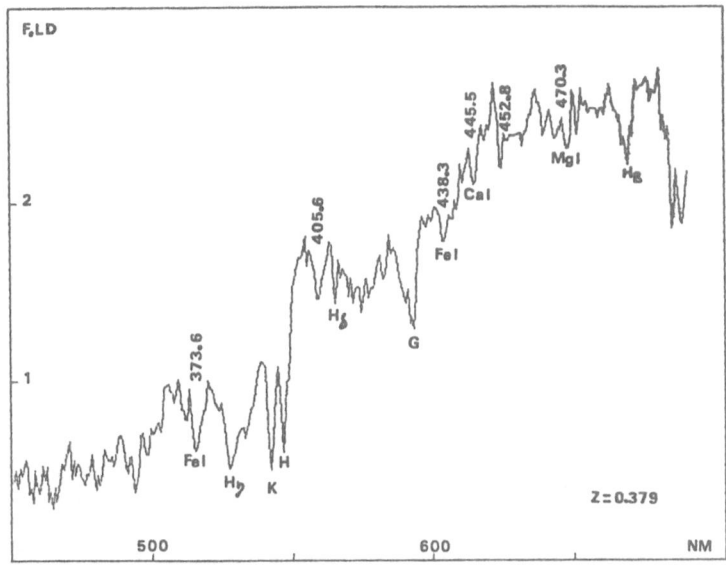

Figure 1: *Spectrum of the brightest galaxy of the cluster A 370 (No. 20, cD type) with a redshift of z = 0.379.*

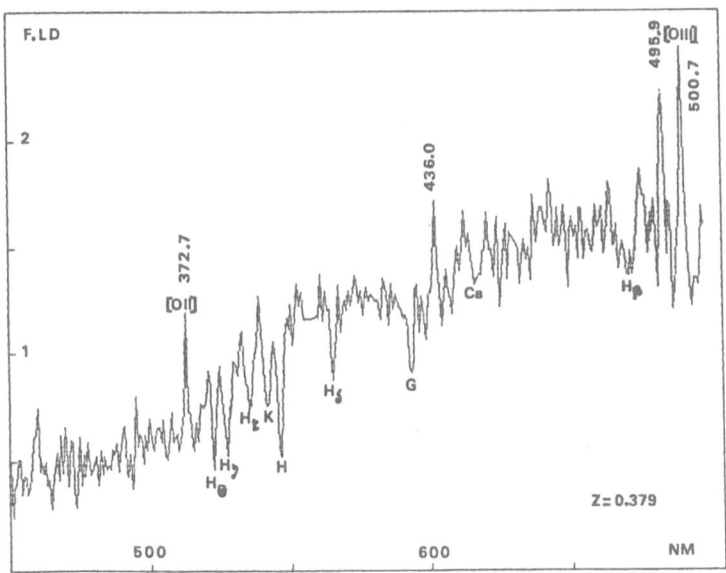

Figure 2: *Spectrum of a blue galaxy of the cluster identified as an irregular type (No. 41, z = 0.379). Note the emission lines typical of H II regions and the strong Balmer absorption lines.*

far from the M/L measured in nearby clusters.

It is possible to estimate the dynamical stage of a cluster of galaxies from the measurement of the luminosity and color segregations present in these systems: two body relaxation tends to concentrate the massive objects in the core of a gravitational system. This leads to a mass segregation which can be observed from the luminosity segregation. On the other hand, it is well known that elliptical and lenticular galaxies are in the dense regions of clusters. In so far these galaxies are the reddest members, one can observe a color segregation as a proof of a morphological segregation. We checked these segregations are present in A370 using the mean galaxy-galaxy distance in magnitude bins or in color bins, following the method described by Capelato et al [8]. We then observe a strong luminosity segregation and a color segregation which appear both on the photometric and the spectroscopic sample (which only contents clusters members): the mean galaxy/galaxy distance decreases as the luminosity increases (luminosity segregation) or as the (V-R) increases (color segregation). Such segregations are commonly observed in nearby clusters of galaxies. Hence in this point of view, A370 in not different of nearby clusters. More generally, on the basis, of its virial mass and of these segregations, this cluster has the same properties than closer ones which probably means that it is roughly in the same dynamical evolution stage. Moreover, accounting for the two boby relaxation time scale the presence of luminosity segregation in clusters at z=0.4 strongly suggests that this phenomenon already occurs in substructures which are formed during the cluster formation [9].

3.3 Spectral observation of the galaxies of A370

The data reduction procedure has been described elsewhere [10]: for each spectrum, the spectral energy distribution was compared with different synthetic spectra obtained with the reference model of Guiderdoni and Rocca- Volmerange [11]. The spectral identification in the range 4500 to 7500Å allows us to classify the galaxies in three groups: ellipticals or S0s (see Fig. 1), red spiral-like spectra (Sa and Sb) and blue spiral-like spectra (Sc, Sd and Irr). Indeed, this spectral classification must be taken with caution since it is has not been observationaly established that the morphological type really corresponds to the integrated spectrum given here. A better approach is to only consider two classes: the star-forming galaxies (red spiral-like or blue spiral-like spectra) and the passive ones (ellipticals or lenticulars). We also search for the presence "E+A" galaxies.

We did not found any "E+A" galaxy in our sample. However, the main result derived from this classification is that A370 has a high fraction of star-forming galaxies (50%)- "red" or "blue": if we assume that these spectra really correspond to spiral galaxies (in term of morphological type) this is a very high spiral fraction. For instance, the Coma cluter only has 3% of spirals in the same magnitude range and within the same radius. Moreover, a plot of the spiral fraction versus the richness or the X-ray luminosity of A370 shows that it drastically escapes from the correlations observed in rich clusters below z=0.1 .

Among this star-forming galaxies, we observed 13 Butcher-Oemler objects for which a redshift was obtained: 8 of them are cluster members (Fig. 2). so, the blue galaxy fraction in this sample 8/53=15% instead of the 0.21% measured by Butcher and Oemler from their photometric sample [12]. These blue galaxies show strong Balmer absorption lines, with sometimes [OII]λ3727 emission lines indicative of ongoing star formation. Note that no AGNs were detected in this sample.

4. EVOLUTION OF THE SPECTRAL CONTENT IN A370

From the spectroscopic sample presented here, we have seen that A370 has 50% of star-forming galaxies. Whatever their morphological types are, this is a high fraction of active galaxies which is not observed in dense nearby clusters of galaxies in this magnitude range (star-forming galaxies observed in Coma by Bothun and Dressler are much less luminous [13]). Hence, the questions are: why is there such a high star-forming galaxies in A370 ? What are the differences between "red" and "blue" star-forming galaxies ? Indeed, galaxies observed at z=0.4 probably have a larger fraction of gas than present-day galaxies and the star formation must be more effective, even in lenticulars or ellipticals. These high gas fraction galaxies could be the red star-forming galaxies we observed whereas the blue ones are those which have undergone a burst of star formation. A possible origin of these bursts is given in Bothun and Dressler [13] which propose that blue galaxies are triggered by the interaction with the Intra-Cluster-Medium (ICM). If distant galaxies really have a higher gas content than now, one easily understand why the Butcher-Oemler effect is not observable at low redshift. This model implies that it must be observed a strong correlation between the blue galaxy fraction and the X-ray luminosity. The other interpretation in term of galaxy/galaxy interactions is also compatible with observations. However, if this assumption is correct, one has to find close companions near blue galaxies.

In fact, it is not possible to answer to these questions with the present data. Ram-pressure-stripping as well as galaxy/galaxy interactions are both compatible with star-forming galaxies spectra we present here. The understanding of the mechanism responsible of the Butcher-Oemler effect need high resolution imaging both in optics and in the X-ray range. No strong correlation is observed between the blue galaxy fraction and the X-ray luminosity and the presence of a companion near blue galaxies has to be confirmed on a statistical sample before any conclusion.

5. CONCLUSIONS

From a very large photometric and spectroscopic sample of galaxies in the field of A370, some different aspects of the cluster evolution have been presented:
- the dynamical stage of A370, is not so far from nearby ones, showing the same luminosity and color segregations although it is observed 5 Gyrs before. This suggest that these mechanisms are fast and already

occurs in substructures during the cluster collapse.

- A370 presents a high star-forming galaxy fraction (50%) composed of two populations: a red one and a blue one, this latter beeing constituted of the blue Butcher-Oemler objects. From the spectroscopic sample of cluster member, we found that the blue galaxy fraction is of 15%. These star-forming galaxies could be galaxies with a higher gas fraction than nearby ones.

- no firmly conclusions can be given on the origin of these blue galaxies: ram-pressure-stripping and galaxy/galaxy interaction are possible mechanisms. one has to wait for high resolution imaging to conclude about this problem.

Finally, I mention that during our first observation of A370 at the CFHT in September 1985, we have discovered a giant luminous arc in the center of this cluster [14, 15]. From our first spectroscopic data of the east part of this arc it appears that this structure does not present any emission lines as it is expected in star formation regions. In view of this spectrum, we suggest that this arc could be the result of a gravitational lensing by the cluster center on a background galaxy at z=0.59. A model has been developed [15] which is enable to represent the arc produced in such a lensing case. The fit with the data is very good although a simple multi-point-mass model has been used for the deflector. This reinforces the lensing hypothesis. However, the signal-to-noise of the spectrum is low and this model has to be confirmed by future spectroscopic observations all along the arc structure.

REFERENCES

[1]: Butcher, H., and Oemler, A. 1978, Astrophys. J. 219, 18

[2]: Dressler, A., and Gunn, J.E. 1982, Astrophys. J. 263, 533

[3]: Dressler, A., and Gunn, J.E. 1983, Astrophys. J. 270, 7

[4]: MacLaren, I., Ellis, R.S., and Couch, W. 1986, STScI preprint n[0] 130

[5]: Fort, B., Mellier, Y., Picat, J.P., Rio, Y., and Lelièvre, G. 1986, S.P.I.E. Vol. 627, "Instrumentation in Astronomy" VI, p. 321

[6]: Mellier, Y., Soucail, G., Fort, B., and Mathez, G. 1987, submitted to Astronom. Astrophys.

[7]: Henry, J.P., Soltan, A., and Briel, U. 1982, Astrophys. J. 262, 1

[8]: Capelato, H.V., Gerbal, D., Mathez, G., Mazure, A., Salvadore Solé, E., and Sol, H. 1980, Astrophys. J. 241, 521

[9]: Roos, N., and Aarseth, S.J. 1982, Astron. Astrophys. 114, 41

[10]: Soucail, G., Mellier, Y., Fort, B., Picat, J.P., and Cailloux, M. 1987, Astron. Astrophys. (in press)

[11]: Guiderdoni, B., and Rocca-Volmerange, B. 1987, Astron. Astrophys. (in press)

[12]: Butcher, H., and Oemler, A. 1984, Astrophys. J. 285, 426

[13]: Bothun, G.D., and Dressler, A. 1986, Astrophys. J. 301, 57

[14]: Soucail, G., Fort, B., Mellier, Y., and Picat, J.P. 1987, Astron. Astrophys. Letters 172, L14

[15]: Soucail, G., Mellier, Y., Fort, B., Hammer, F., and Mathez, G. 1987, Astron. Astrophys. Letters (in press).

NEAR-INFRARED OBSERVATIONS OF DISTANT CLUSTER ELLIPTICALS

S. E. Persson

Mount Wilson and Las Campanas Observatories

813 Santa Barbara St.

Pasadena, CA 91101, USA

ABSTRACT. Visual-to-infrared colors of 56 red-selected, first ranked, cluster ellipticals out to a redshift of 0.92 are discussed. Color versus redshift diagrams display substantial scatter at all redshifts blueward from a well-defined ridge line that is a good match to the energy distribution of an old gE galaxy. Beyond $z = 0.6$ there are few red galaxies. Selection effects that operate even in this sample are evaluated via simple combinations of starbursts plus an old galaxy. The "first-ranked" object is not necessarily relevant for a determination of q_o in the Hubble diagram. An age test based on the reddest galaxies at high redshift looks promising.

1. INTRODUCTION

Study of the stellar content of gE galaxies at high redshift can be significantly improved by extending the wavelength coverage of the spectral energy distribution (SED) to the K band (2.2 μm). The original motivation for such work was to draw a Hubble diagram at a wavelength where luminosity evolution and the effects of residual star formation are minimized (Grasdalen 1980). From a practical standpoint the peak of the SED moves *into* the K band with increasing z, making distant red galaxies easier to observe in the infrared. The overall SED contains information on the color evolution of the stellar population, and the expectation was that through detailed comparison of the SEDs with those of model evolving stellar populations, the models could be used to calibrate the degree of luminosity evolution (at any wavelength) over lookback times of several Gyr (Tinsley 1972a,b, 1976). Serious complications (viz., galaxy cannibalism) were recognized in 1975, and it is now appreciated that at least one other parameter, such as the light distribution, is needed to calibrate the "standard" candles. Improved understanding of galaxy evolution, and not a direct measurement of q_o will probably be the ultimate result.

Most of the infrared data on distant galaxies obtained to date pertains to luminous radio sources, mostly selected from the 3CR catalog (Lilly and Longair 1982, 1984; Lilly, Longair, and Allington-Smith 1985; Lebofsky and Eisenhardt 1986; Eisenhardt and Lebofsky 1987). All of the infrared data on these objects have been assembled into a Hubble diagram at K by Spinrad and Djorgovski (1987), who claim to derive q_o to ±0.2.

R. G. Kron and A. Renzini (eds.), Towards Understanding Galaxies at Large Redshift, 251–257.
© *1988 by Kluwer Academic Publishers.*

However, Lilly and Longair (1984) have turned the problem around and used the infrared Hubble diagram to derive the luminosity evolution at K (one magnitude at $z = 1$) for their sample. A basic result from the radio-selected samples is that the visual-to-infrared colors of radio galaxies do not follow a simple color-evolution line; considerable scatter sets in at redshifts larger than 0.4 or 0.5. It is easy to fit Bruzual (1983) models to the colors of these objects, and to thus claim detection of color evolution. It is possible, however, that bursts of star formation and enhanced radio emission (e.g., brought about by mergers; Heckman *et al.* 1986) may compromise their use as standard candles, even at K. It is important therefore to also investigate the SEDs of *optically* selected high z gE galaxies, the problem being to find them in the first place. In this paper we present a preliminary report of visual-to-infrared colors of such a sample. A complete report, by Persson, Oke, Porter, and Matthews, will appear in the Astrophysical Journal.

2. OBSERVATIONS

2.1 The Distant Cluster Sample, SEDs, and Redshifts

Gunn, Hoessel, and Oke (1986; henceforth GHO) have presented a new sample of 418 distant clusters found on red and IV-N plates. Working as far into the red as possible increases the probability of finding high redshift gEs because they are much redder than those at low redshift. SEDs and redshifts for over a hundred of the brightest cluster galaxies in the sample were obtained by Gunn and Oke using the Multichannel Spectrometer and the PFUEI spectrograph on the Palomar 5-m telescope. These data provide enough resolution that reliable redshifts could be obtained by visually matching the observed SED and spectral features with those of an average brightest cluster gE (Yee and Oke 1978). For the more distant and fainter cluster galaxies redshifts were obtained from PFUEI slit spectra. Most of the galaxies are first ranked in both metric and total luminosity.

2.2 Infrared and Optical Observations

A subset of the optical sample was selected, without regard to optical color, for observation in the infrared. Our approach was to measure the visual-to-infrared *colors* as precisely as possible, aiming for uncertainties less than $\sim \pm 0.1$ mag. All the observations were done in dark time on the 5-m telescope, and all the objects were centered in the 5" diameter infrared aperture with the aid of an integrating television system. Red and infrared magnitudes were measured simultaneously with an auxiliary optical detector mounted on the infrared photometer. The advantages of simultaneous photometry are discussed by Eisenhardt and Lebofsky (1987). This technique yields precise and well understood colors that link the visual to the infrared. The standard J, H, and K infrared passbands were used and the red filter was that of Thuan and Gunn (1976).

Optical colors of the cluster galaxies in the infrared sample were obtained from images taken with the Four-Shooter Cassegrain CCD camera on the 5-m telescope. The passbands used were the g, r, and i bands defined by Schneider *et al.* (1983). The software photometric aperture matched that used in the infrared. The data set thus consists of g-r, r-i, and r-J, H, and K colors; all have been corrected for galactic reddening which is small.

3. RESULTS

3.1 Color-redshift Diagrams

The g-r colors are well correlated with the r-H and r-K colors and we use g-H here, as the H-band data are the most complete and accurate. Figure 1 presents the basic results: (1) There is a fairly well-defined upper envelope to the distribution of color with redshift. (2) For each color the upper envelope is moderately well followed by the smooth curve which represents the expected colors of a standard gE galaxy, consisting of a Yee and Oke (1978) SED for the visual, tied into the infrared SED of α Tau (data from Strecker *et al.* 1979), so as to reproduce the V-K color (3.35) of an average first-ranked gE at zero redshift. (3) There is considerable blueward scatter away from the upper envelope at all redshifts, with roughly 25 % of the sample significantly bluer than the ridge line.

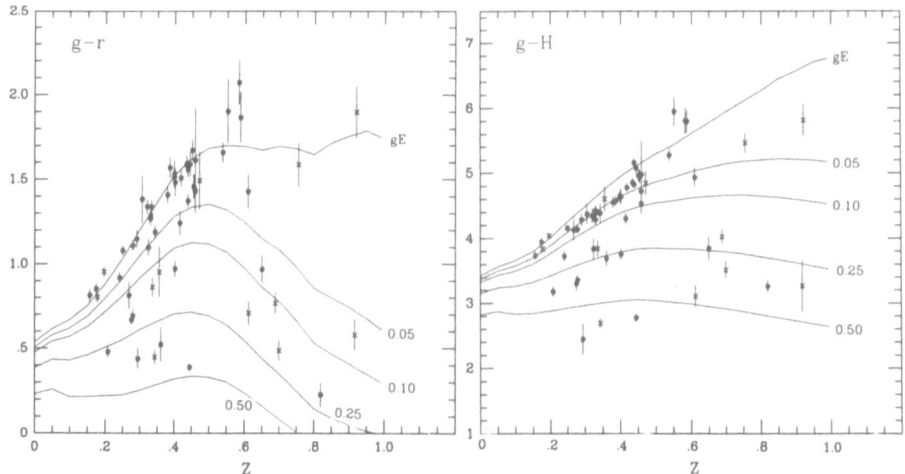

Figure 1. The g-r and g-H color-redshift diagrams for red-selected first-ranked cluster galaxies. The error bars are 1 σ. The crosses denote galaxies selected in a passband near 8000 Å; the others were selected at 6700 Å. The upper smooth curve represents the expected color dependence for an old stellar population, and the lower curves include a young cluster contribution (SWB type I); the numbers give the fraction of the light at V due to the aging starburst.

Although not explicitly shown here, it is the case that if a galaxy is relatively blue in one color, then it is blue in all the colors. The color deviations can be extremely pronounced, with the range exceeding 2.5 mag in g-H. The light from these galaxies is obviously not coming from an old, red, stellar population. (4) At z $\sim$ 0.6 the distribution changes from one having a well-defined upper envelope to one in which there are mostly only blue galaxies. At z $\gtrsim$ 0.7 one galaxy lies close to the "no evolution" line. It also has a normal 4000 Å break parameter (Gunn 1988; this conference). (5) The basic character of the color-redshift plots is qualitatively similar to that found for samples that are mostly radio-selected from the 3CR catalog. This result strongly suggests a common physical origin.

3.2 Possible Interpretations

Two explanations for the blue galaxies are possible. First, we may be seeing color evolution in the usual sense that a relatively blue object is a young gE that has formed all of its stars in a single burst. The scatter then indicates a large range in formation epoch. If true, it may be possible to model the color and luminosity at any redshift. Alternatively, some or all of the very blue objects could contain an underlying old red population, upon which is superposed a strong burst of star formation. A range in the strength or age of the burst would naturally account for the scatter, and no systematic deviation from the ridge line would be expected. Rather strong selection effects might still be present, *even at GHO's survey wavelengths*, and it is not at all clear that the brightest galaxy at 8000 Å is necessarily the most luminous, averaged over the lifetime of the cluster. The most extreme example is the very blue object at $z = 0.92$. It has a bright $\lambda 3727$ line, and is surely undergoing a burst of star formation.

3.3 Evaluation of Selection Effects

To evaluate the magnitude of the starburst selection effect, we assume a simple combination of light from a starburst of some age superimposed on an old population. For the present exercise the details of the input components do not matter much, as the observed scatter in g-H is more than a magnitude. Nevertheless, the aging starburst component must have an SED that properly includes the light from red supergiant stars which begin to contribute not long after the burst starts, and AGB stars that contribute for many GYr. Persson *et al.* (1983) studied these effects by measuring the visual-to-infrared colors of young and intermediate age clusters in the Magellanic Clouds. Serious disagreement between the V-K colors of young clusters and the models of Struck-Marcell and Tinsley (1978) that did not include red supergiants was noted. We have constructed synthetic SED's of these same clusters from the uvgr colors of Searle, Wilkinson, and Bagnuolo (1980; SWB) and the V-K colors.

Different age cluster models were added in varying proportion to the Yee-Oke gE SED to produce synthetic color-redshift curves. It was found that the bluest galaxies could not be matched with any fraction of cluster light older than a few Gyr. Figure 1 shows four examples using SWB type I cluster light (age $\sim 10^7$ yr). The model galaxy fluxes in the GHO IV-N survey bandpass are brightened by 0.09, 0.17, 0.44 and 1.03 mag at a redshift of 0.7. (The K-band flux is brightened by 0.34 mag for the latter model.) The essential result is that it is easy to account for the bluest galaxies in figure 1 if the starburst is young enough, and the extra flux may be sufficient to cause their spurious selection at 8000 Å as the time-averaged brightest object in the cluster.

4. DISCUSSION

It seems plausible that many of the gEs in the optically selected sample are affected to some degree by starbursts. This is consistent with the fact that nearly all of the radio-selected galaxies observed in the infrared and spectroscopically display exceedingly strong $\lambda 3727$ emission, some at a level that dwarfs what is known locally by orders of magnitude (see review by Spinrad 1986). We conclude that aging and/or ongoing starbursts and their associated selection effects are important in both samples and deserve close attention.

While the use of a luminous radio galaxy to locate a putative standard candle for infrared photometry is certainly appealing, the systematic bias is potentially serious. The evolving models pioneered by Tinsley and carried forward by Bruzual (1983, 1986, and this conference), can only give a baseline solution, as they do not adequately treat the very cool and luminous AGB stars, which could contribute half the bolometric light over a large fraction of the age of a gE galaxy. (The theoretical framework has been discussed by Renzini (1981, 1986), and Renzini and Buzzoni (1986); see also Wyse 1985). Attempts to detect (and then model) these stars in integrated light are difficult, because they have virtually the same colors as the reddest giants in the stellar population. Lilly and Gunn (1985) interpreted their near-infrared photometry of galaxies in the 0024+1654 cluster in these terms. Recent studies of the Giants in Baade's Window by Frogel and Whitford (1987) and Frogel (1988) indicate just how lacking our library of infrared stellar SEDs really is.

Until the selection effects and the role of the AGB are understood, we find little justification for comparing 3CR galaxies to old, first-ranked gE's in a Hubble diagram.

4.2 An Age Constraint

Hamilton (1985), Spinrad (1986), Windhorst *et al.* (1986), and O'Connell (1988) have pointed out that old and red galaxies at high redshift can be used to set a constraint on the age of the universe. Two ingredients are needed: large samples of red-selected galaxies to establish the distribution of visual-to-infrared color and break parameter in the redshift range of 0.7 to say 1.0, and model stellar populations that include both AGB stars and a metal abundance distribution. Suppose one allows 1 Gyr for the galaxy to assemble itself, a conservative 4 Gyr for the population to age to such an extent that it appears "old" in all respects (despite the AGB). Adding the lookback time to $z = 0.8$, one arrives at an upper limit for H_o of 50 km/sec/Mpc for a formation redshift of 10, $q_o = 0.5$, and $\Lambda = 0$. Larger values of H_o are allowed for $q_o = 0$, but both a larger population age and a lower (more realistic) formation redshift force H_o to smaller values.

4.3 Conclusion

An increased sample of red-selected clusters between $z = 0.6$ and 1.0 would be of great importance in studying the color distribution of cluster galaxies, and the apparent transition from mainly red to mainly blue objects. A combination of high precision visual-to-infrared colors plus break parameter data for several objects in each cluster could lead to ages for the oldest galaxies. If only one galaxy in the cluster were undergoing a starburst, then the serious selection bias could be evaluated or avoided. Alternatively, if all or most of the galaxies in a cluster turned out to have similar visual-to-infrared colors, and also followed a familiar luminosity function, then one could conclude that quiescent color evolution had been found. The development of near-infrared array detectors will significantly aid in this difficult but decisive test.

REFERENCES

Bruzual, G. 1983, *Ap.J.*, **273**, 105.

Bruzual, G. 1986, in *Spectral Evolution of Galaxies*, ed. C. Chiosi and A. Renzini (Dordrecht:Reidel), p. 263.

Eisenhardt, P. R. M. and Lebofsky, M. J. 1987, *Ap.J.*, **316**, 70.

Frogel, J. A., and Whitford, A. E. 1987, *Ap.J.*, in press.

Frogel, J. A. 1988, this volume.

Grasdalen, G. L. 1980, in *IAU Symposium No. 92, Objects of High Redshift*, ed. G. O. Abell and P. J. E. Peebles, (Dordrecht:Reidel), p. 269.

Gunn, J. E. 1988, this volume.

Gunn, J. E., Hoessel, J., and Oke, J. B. 1986, *Ap.J.*, **306**, 30.

Hamilton, D. 1985, *Ap.J.*, **297**, 371.

Heckman, T. M., Smith, E. P., Baum, S. A., van Breugel, W. J. M., Miley, G. K., Illingworth, G. D., Bothun, G. D., and Bulich, B. 1986, *Ap.J.*, **311**, 526.

Lebofsky, M. J. and Eisenhardt, P. R. M. 1986, *Ap.J.*, **300**, 151.

Lilly, S.J. and Gunn, J. E. 1985, *M.N.R.A.S.*, **217**, 551.

Lilly, S. J. and Longair, M. S. 1982, *M.N.R.A.S.*, **199**, 1053.

Lilly, S. J. and Longair, M. S. 1984, *M.N.R.A.S.*, **211**, 833.

Lilly, S. J., Longair, M. S., and Allington-Smith, J. R. 1985, *M.N.R.A.S.*, **215**, 37.

O'Connell, R. W. 1988, this volume.

Persson, S. E., Aaronson, M., Cohen, J. G., Frogel, J. A., and Matthews, K. 1983, *Ap.J.*, **266**, 105.

Renzini, A. 1981, *Ann.Phys.*, **6**, 87.

Renzini, A. and Buzzoni, A. 1986, in *Spectral Evolution of Galaxies*, ed. C. Chiosi and A. Renzini (Dordrecht:Reidel).

Schneider, D. P., Gunn, J. E., and Hoessel, J. G. 1983, *Ap.J.*, **268**, 476.

Searle, L., Wilkinson, A., and Bagnuolo, W. G. 1980, *Ap.J*, **239**, 803.

Spinrad, H. 1986, *Pub.A.S.P.*, **98**, 269.

Spinrad, H., and Djorgovski, S. 1987, in *IAU Symposium No. 124, Observational Cosmology*, ed. G. Burbridge (Dordrecht:Reidel).

Strecker, D. W., Erickson, E. F., and Witteborn, F. C. 1979, *Ap.J. Suppl.*, **41**, 501.

Struck-Marcell, C., and Tinsley, B. M. 1978, *Ap.J.*, **221**, 562.

Thuan, T. X., and Gunn, J. E. 1976, *Pub.A.S.P*, **88**, 543.

Tinsley, B. M. 1972a, *Ap.J. (Letters)*, **173**, L93.

Tinsley, B. M. 1972b, *Ap.J.*, **178**, 319.

Tinsley, B. M. 1976, *Ap.J.*, **203**, 63.

Windhorst, R. A., Koo, D. C., and Spinrad, H. 1986, in *Galaxy Distances and Deviations from Universal Expansion*, ed. B. F. Madore and R. B. Tully (Dordrecht:Reidel), p. 197.

Wyse, R. 1985, *Ap.J.*, **299**, 593.

Yee, H. K. C. and Oke, J. B. 1978, *Ap.J.*, **226**, 753.

LYMAN–ALPHA GALAXIES AT LARGE REDSHIFTS

S. Djorgovski
Astronomy Department and Palomar Observatory
California Institute of Technology, 105–24
Pasadena, CA 91125, USA.

ABSTRACT. Strong Lyα emission has been detected from several different types of distant galaxies, radio galaxies and their companions at $z \simeq 1.8$, and galaxy companions of quasars at $z > 3$. This line emission is a good way of detecting galaxies at very large redshifts, and a powerful probe of the physical conditions and processes in them. A subset of the Lyα radio galaxies have photometric, spectroscopic, and morphological properties unlike those of any previously known class of objects, and may be interpreted as galaxies in the process of formation. These and other observations agree very well with some modern theoretical ideas, which predict that galaxy formation is an extended process which may begin at large redshifts, but culminate at $z \sim 1-2$. If this is the case, we may have already detected primeval galaxies. Searches can be (re)designed to test this notion.

There is now abundant and varied observational evidence for galaxy evolution at large look-back times. Whereas the interpretations of the observed phenomena are still far from being final or even generally accepted, most workers in this field would agree that we have seen objects, events, and populations back there and then which do not exist here and now. Some of the evidence is statistical, e.g., the evolution of quasars and AGN's (Schmidt 1987; Koo & Kron 1988), or the behavior of faint galaxy counts in the field (Tyson, this conference), or near distant quasars (Tyson 1986). There are Butcher–Oemler–Dressler–Gunn effects (Dressler 1987, and references therein; Gunn, this conference) and the existence of "E+A" galaxies both in rich clusters and the field (Ellis 1987, and this conference).

Probably the most dramatic evidence for evolution is found in the studies of complete samples of radio galaxies, which reach up to $z \simeq 1.8$, or look-back times of $\sim 2/3$ of the Hubble time (Lilly & Longair 1982,1984; Lebofsky & Eisenhardt 1986; Eisenhardt & Lebofsky 1987; Spinrad 1986; Djorgovski et al. 1985a, 1988; Djorgovski 1987, 1988; etc.) The samples are well defined in the radio and practically complete, and in the absence of an important correlation between the radio powers and optical luminosities they can be used to study galaxy evolution in the optical bands (for the conflicting views on this possible bias, cf. Eisenhardt & Lebofsky 1987, or Spinrad & Djorgovski 1987 vs. Yates et al. 1986). Only the galaxies without detectable nuclear nonthermal light contributions are used, in order to probe the behaviour of stellar populations alone. Luminosities, colors, surface brightness, morphology, and spectra of these distant galaxies paint a consistent picture, in which stellar populations evolve actively in a declining sequence

R. G. Kron and A. Renzini (eds.), Towards Understanding Galaxies at Large Redshift, 259–274.
© 1988 by Kluwer Academic Publishers.

of massive starbursts, with e-folding times for an *average* SFR of about $1 - 1.5$ Gy. The physical mechanisms causing the starbursts probably include dissipative mergers of gas-rich galaxies (Djorgovski *et al.* 1987a), some kind of interaction between the radio-sources and their host galaxies (McCarthy *et al.* 1987b; Chambers *et al.* 1987), and it is prudent to assume that a variety of distinct processes may be at work. Traces of past mergers are discernible in many low-z radio galaxies (Heckman *et al.* 1986). The evolution of these powerful radio galaxies may be (and probably is) unrepresentative of elliptical galaxies in general, but that is something we really do not know yet. It is encouraging that we can follow significant evolutionary changes in at least some galaxy samples at high redshifts, and see the galaxies grow.

Beyond $z \sim 1.7$, Lyα line appears from the atmospheric O_3 cutoff, and becomes a potentially powerful tool to detect and study extremely distant galaxies. To date, all previously known galaxies from which we could detect the Lyα line did show a strong Lyα emission, and a number of additional galaxies were discovered through their strong Lyα. These objects fall roughly in four groups, as described below. More details can be found in the review paper by Spinrad (1988).

First, there are powerful 3C and 1-Jy radio galaxies which continue the observed trends of photometric, morphological, and spectroscopic properties from lower redshifts. These galaxies are relatively bright ($V \sim 22 - 23$ at $z \simeq 1.8$), generally compact or barely multimodal in shape, with projected isophotal sizes of $3 - 5$ arcsec (at $z = 1.8$, 1 arcsec corresponds to $\simeq 7$ kpc, in a $H_0 = 75$, $q_0 = 0.2$ Friedman cosmology; I will adopt these cosmological parameters throughout this paper, unless indicated otherwise). The observed mean surface brightness in these galaxies is indicative of a large luminosity density in the rest frame. Their Lyα emission lines have equivalent widths of some hundreds of Å (Spinrad *et al.* 1985), and the ionized gas is distributed in the same way as the stars. Examples of this

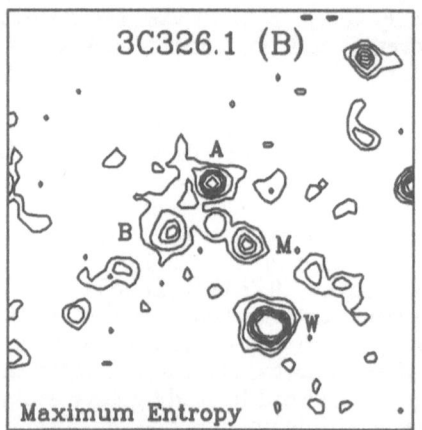 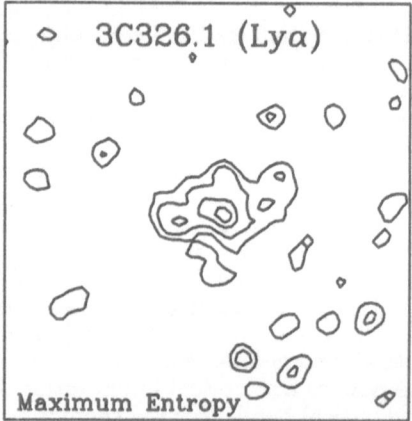

Figure 1: Contour maps of 3C326.1 ($z = 1.825$), in the broad-band B, which is free of any line emission (left), and a 60 Å narrow-band filter, centered on the redshifted Lyα (right). Both fields are 37.4 arcsec square, with North to the top, East to the left. The data were obtained with the KPNO 4-m telescope and a PF RCA CCD, and processed with a Maximum Entropy algorithm. The object notation in the B frame follows that by McCarthy *et al.* (1987a).

type include 3C 239, 241, 256, 454.1, and 1-Jy galaxy 1141+35. These galaxies are already well formed, even though they may be dominated by still relatively young stellar populations.

Another, more interesting group (a new class of objects?) are some $z \sim 1.8$ radio galaxies with remarkable properties. The first example in this category, 3C326.1, at $z = 1.825$, was discovered in the final stages of optical identifications of 3C sources by Spinrad and collaborators. The first results on this object were published by McCarthy *at al.* (1987a) and Strauss *et al.* (1988). This object has a relatively large size ($\sim$ 12 arcsec, or $\sim$ 90 kpc), and a very low continuum surface brightness with a lumpy appearance: three distinguishable condensations have V magnitudes of $\sim$ 23.5, 24.5, and $\geq$25 (Figure 1). However, narrow-band images and long-slit spectrograms reveal a Lyα-luminous cloud of ionized gas, of the same extent as the continuum images, and encompassing the radio source. This Lyα emission has a large velocity shear along the cloud's major axis ($\simeq$ 1000 km/s in

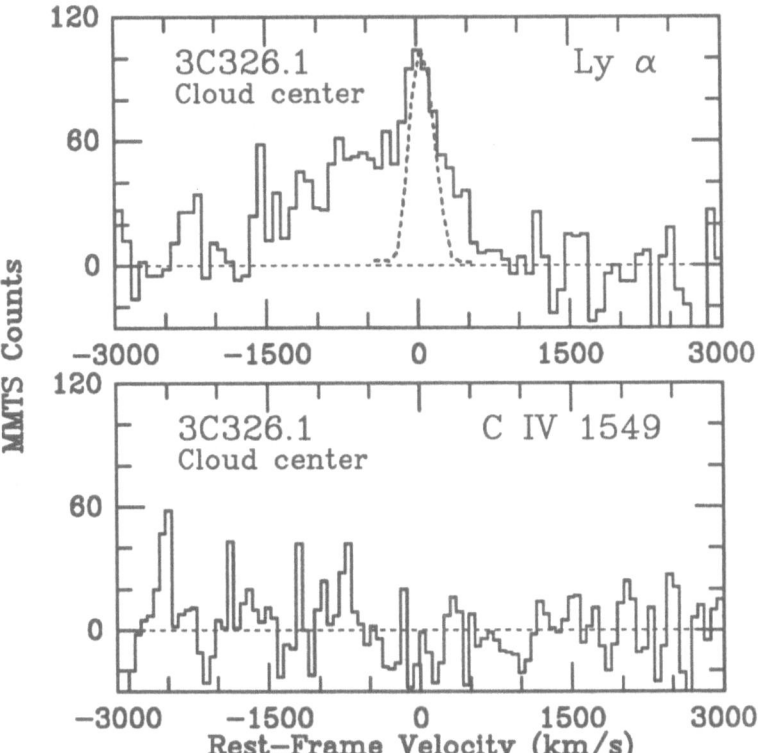

Figure 2: Zoom-in on the Lyα line (top) in the central part of 3C326.1, plotted against the rest-frame velocity; note the extensive blue wing. The data were obtained with the MMT Reticon spectrograph, in a 2×3 arcsec aperture; the dotted line indicates the instrumental resolution. The counts are in arbitrary units, but the total line flux from the entire Lyα cloud is $\sim 7 \times 10^{-14}$ erg/cm^2/s. The bottom panel demonstrates a complete absence of the C IV high-ionization line, from the same data set.

the rest-frame), and a comparable intrinsic velocity width at every point examined. There is also weak extended emission from CII] 2325 and CIII] 1909, but no C IV 1549, so that the overall ionization is fairly soft, but the abundances are not primordial. Our VLA maps show the radio source to be a $\sim$ 7 arcsec double without a detectable core at the level of $\sim$ 0.5 mJy. That, and the absence of any C IV 1549 emission, suggest that there is no hidden QSO-like nucleus which would ionize the gas. From the intensity of Lyα, and assuming a photoionization by young stars and the Salpeter IMF, we deduce the SFR of several hundred $M_\odot$/yr. However, it is likely that at least some ionization, and most of the observed kinematics, are provided by the collisions of infalling gas clouds and supernova shock waves.

3C326.1 is not unique: another object with very similar properties is 3C294 at $z = 1.779$, discovered since (Spinrad *et al.*, in preparation). The spectrum of the dominant ($\sim 24^m$) condensation in this object is shown in Figure 3, narrow-band Lyα image in Figure 4, and the velocity field of Lyα in Figure 5. The internal kinematics of 3C294 are even more spectacular than 3C326.1: the velocity field of Lyα extends over $\geq$ 15 arcsec and $\sim$ 2000 km/s, and has a distinctive structure in it. One important difference is the presence of high-ionization lines, including C IV 1549, and a radio core: some nonthermal optical activity is present. Moreover, the high-ionization lines also seem to be extended, but not at the same level as the Lyα. There are some other possible candidates for this class of objects: 3C194, possibly 3C68.2, and 3C222, but they require a further study.

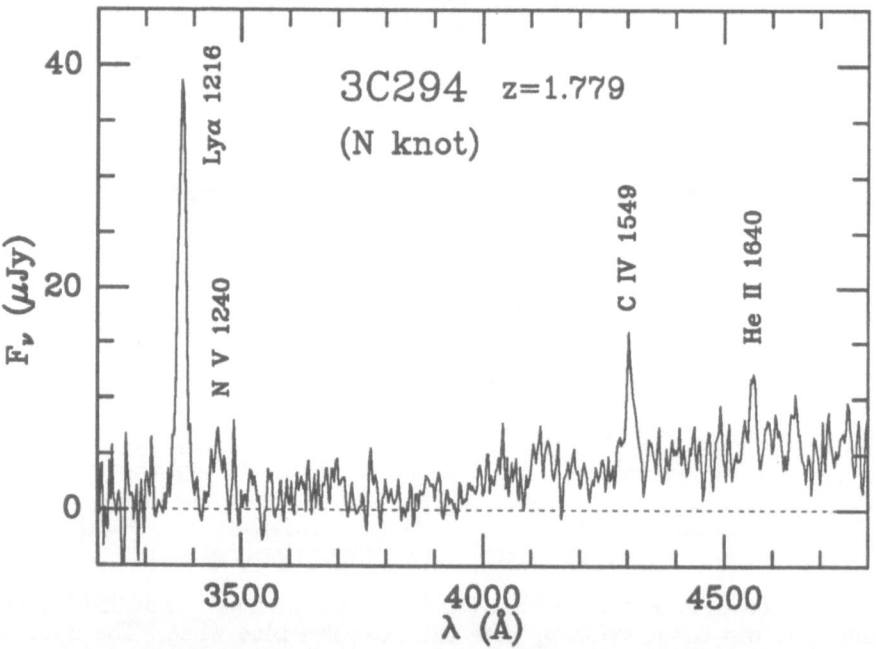

Figure 3: Spectrum of the northern knot of 3C294, obtained at the MMT, with a Reticon spectrograph, through a 2×3 arcsec aperture. The ionization appears to be lower in other regions of the object. The continuum rise redward of $\sim$ 5000 Å is due to the scattered light from the nearby foreground star.

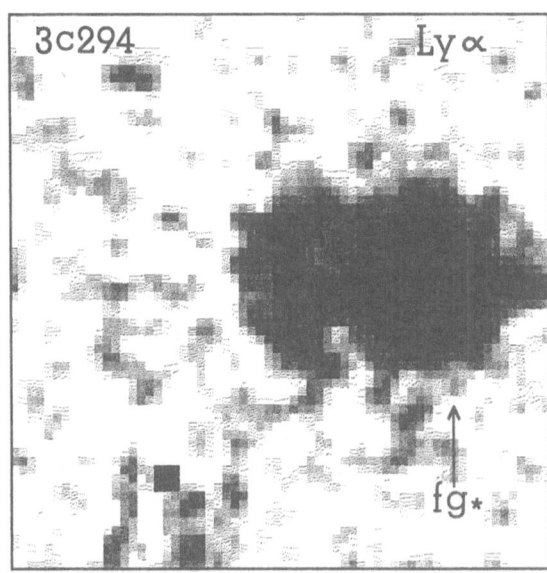

Figure 4: Image of 3C294, obtained in a 60 Å narrow-band filter, centered on the redshifted Lyα. The field is 37.4 arcsec square, with North to the top, East to the left. The data were obtained with the KPNO 4-m telescope and a PF RCA CCD. The bright object just West of the galaxy is a foreground star.

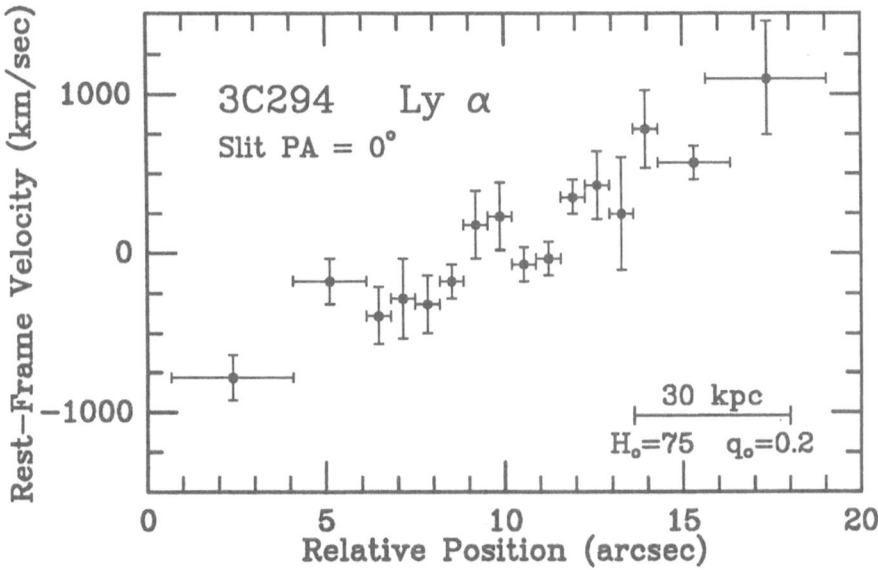

Figure 5: Velocity field of the Lyα line in 3C294, measured along the N–S axis. Long-slit CCD data were obtained with the Lick 3-m telescope and the Cassegrain spectrograph. The bar in the lower right indicates the physical scale along the slit.

Here we have a class of high-redshift galaxies, distinguished by a very strong, extended ($\geq$ 100 kpc) Lyα emission, with a generally low ionization, strong velocity shear and turbulence ($\geq$ 1000 km/s) , weak, probably stellar, continua, and large inferred star-formation rates (a few hundred $M_\odot$/yr). Their properties, save for the "low" redshifts of $\simeq$ 1.8, and the presence of well-developed radio sources, are almost exactly what textbook primeval galaxies are expected to look like. As will be argued below, these objections may be irrelevant. There is nothing quite like this known at lower redshifts: extended emission-line regions in powerful low-z radio galaxies (Robinson et al. 1987; Di Serego Alghieri, this conference; McCarthy 1988) are some two orders of magnitude less luminous, and probably have quite different ionization physics. Even the most luminous cooling flow filament systems at low redshifts (e.g., N1275; Fabian at al. 1984) have Lyα luminosities two orders of magnitude lower than the high-z galaxies described here.

The third group of distant Lyα galaxies are optically selected companions of 3CR sources. At this time, there are two possible cases, one in the field of 3C239 ($z = 1.78$), and one in the field of 3C256 ($z = 1.81$). The first one was selected through its optical colors, and the second one through narrow-band Lyα imaging. Both galaxies were followed spectroscopically, and repeated the Lyα emission within several hundred km/s of the radio source itself; both need further confirmation. We have seen other emission-line companions of $z > 1$ radio-galaxes in [O II] 3727 (e.g., near 3C267, at $z = 1.139$). Their equivalent widths of Lyα or [O II] are smaller than those of the 3C galaxies, but the lines are still very strong. This is important, because it demonstrates that strong Lyα (and other) emission at high redshifts is not an exclusive property of "monster" radio-sources, and may be used to find somewhat more normal galaxies at large look-back times.

The last category of Lyα galaxies at large redshifts are the companions of $z > 3$ quasars, selected through narrow-band imaging. There are 2 (3?) such objects so far, and they are the most distant (non-QSO) galaxies now known. The search method was defined in the paper by Djorgovski et al. (1985b): known high-z quasars are used as "beacons" pointing towards other galaxies. At low redshifts, the best place to look for a galaxy is next to another galaxy, and if the degree of sub-Mpc-scale clustering did not evolve dramatically (this is a very important assumption), looking for companion galaxies around distant quasars may be propitious. Moreover, in order to assure that any candidate companions are at the quasar redshift, they will have to have strong line emission, and Lyα is the natural choice. Emission line galaxies would stand out in a narrow band centered on the redshifted Lyα, as compared to the broad-band images.

The first time we tried this experiment, we discovered a galaxy companion of quasar PKS 1614+051, at $z = 3.215$ (Djorgovski et al. 1985b, 1987b). This object is separated in projection by 6.55 arcsec ($\sim$ 40 kpc) from the QSO, and the velocity difference between them is 300 – 700 km/s, depending on which emission lines are used; the geometry is thus typical of compact galaxy groups or clusters. The object clearly stands out in the narrow-band Lyα images. The broad-band red images, which also include the C IV line, but are not dominated by it, show a nucleated, but non-stellar (i.e., marginally resolved) shape of the companion, as well as two other faint objects next to the quasar. The line-corrected magnitude of the companion is $R \simeq 24$. The Lyα images show a nucleated emission, a faint bridge between the QSO and the companion, which then extends past the companion's nucleus, and possibly some slight enhancements of line emission corresponding to the other two faint objects. This morphology is suggestive of a tidal interaction, and perhaps we are witnessing an event which caused the QSO activity in this system. Whereas the QSO is a 1-Jy source, the companion is not detected down to the VLA limit (e.g., 30 μJy at 6 cm), an upper limit which corresponds to the rest-frame power which is a median for the low-z Seyferts. The spectra of the companion show strong C IV 1549 and C III] 1909 lines, but with a considerably lower ionization than the neighboring QSO. The carbon lines are slightly broadened ($\leq 600 - 800$ km/s), but the core of the Lyα line is narrow (velocity width ~ 160 km/s). The presence of the C IV line, and its width are suggestive of an active nucleus, which may contribute some light, both in the lines and continuum. The continuum is also clearly detected in our long-slit spectroscopic data between the carbon lines.

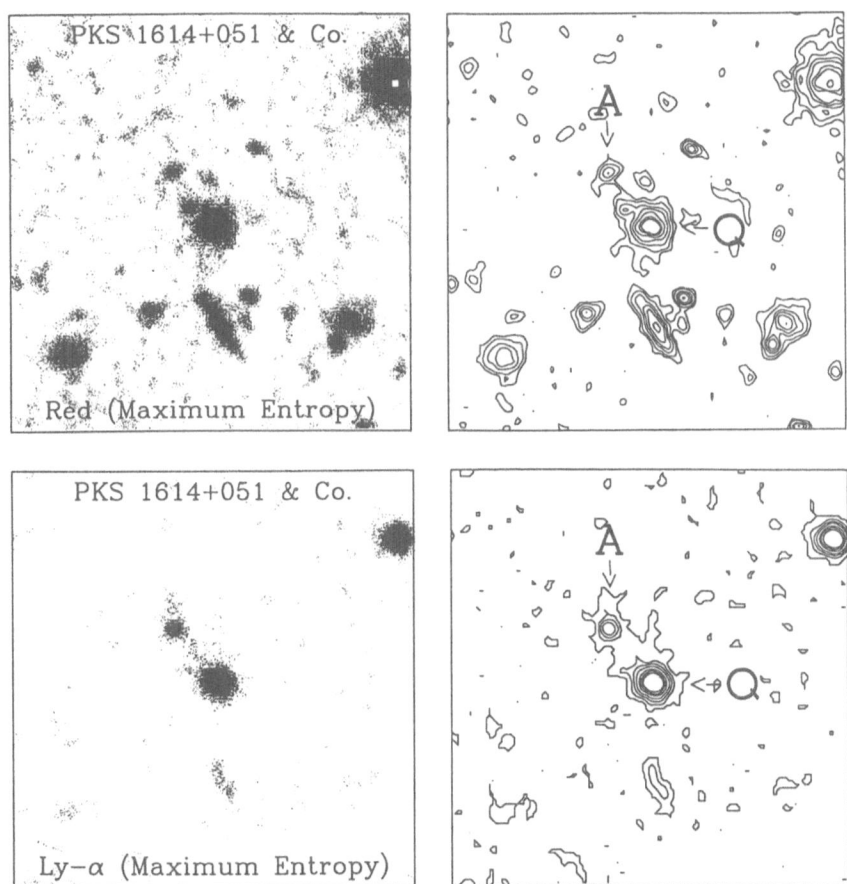

Figure 6: Images of the quasar PKS 164+051 ("Q") and its companion galaxy ("A"), at $z = 3.215$, from Djorgovski *et al.* (1987b). The broad-band R images (top) also contain the C IV 1549 emission line, but the line contributes $< 50\%$ of the total red light from the companion. The redshifted Lyα images (bottom) were obtained in a 70 Å narrow-band filter, centered on λ5139Å. Note the extended morphology of the Lyα gas. The field size is 38 arcsec square, with North to the top, East to the left. The data were obtained with the KPNO 4-m telescope and a PF TI CCD, and processed with a Maximum Entropy algorithm.

266

In the absence of a strong active nucleus and any radio emission from the companion, we conclude that this continuum is probably starlight. Hu & Cowie (1987) differ, but their conclusions seem to be based on a more limited data set.

Another pair of Lyα companion objects were discovered in the field of gravitational lens MG 2016+112 by Schneider *et al.* (1986, 1987), at $z = 3.273$; these may be two gravitationally lensed images of a single Lyα companion of MG 2016+112 (itself a triple image of a narrow-line AGN). These objects are much fainter than the companion of PKS 1614+051, and have no detectable lines other than Lyα, and thus a lower ionization. Their nature is still a mystery. All these objects must be very young on account of their large redshifts, independent of the redshift at which they *started* to form.

Our group checked some forty other QSO fields since then, and no objects as bright as the first one were found. We did detect, however, some very faint galaxies which seemed to show possible Lyα excesses, but they are too faint ($> 25^m$) for follow-up spectroscopy with the present generation of telescopes.   Other groups

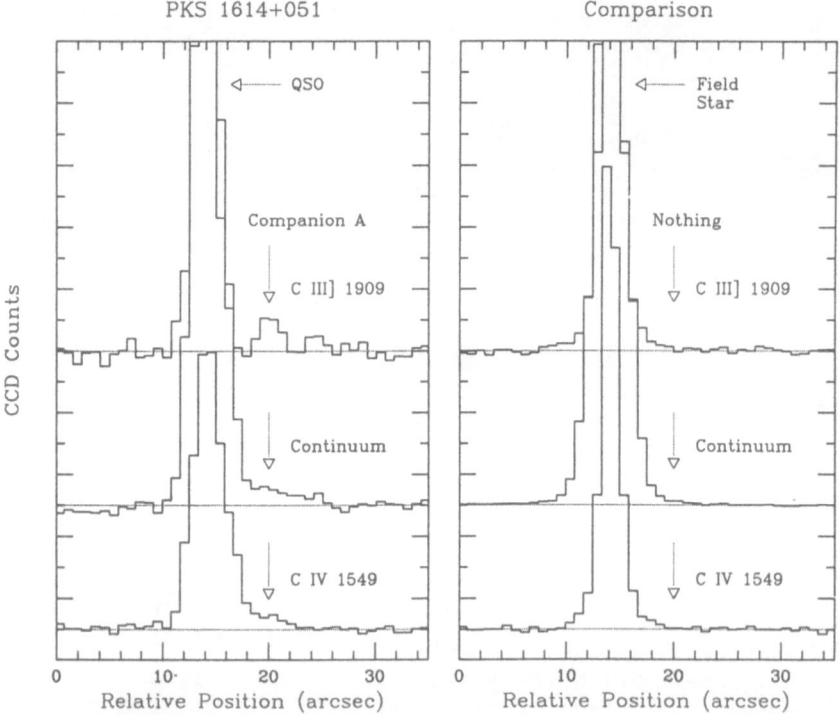

Figure 7: Presence of the optical continuum in the companion of PKS 1614+051 is demonstrated in these intensity cuts along the slit from our spectroscopic near-IR CCD data (Djorgovski *et al.* 1987b). The left panel shows the data on the QSO and the companion, and the right panel shows equivalent data for a comparison star. The carbon line emission seems to be nucleated, but the continuum between the lines appears to be diffuse; the line-corrected magnitude of the companion is $R = 24 \pm 0.5$. The data were obtained with the KPNO 4-m telescope, and the Cryocam spectrograph.

(Hu & Cowie 1987; Barthel *et al.*, priv. comm.; Hunstead & Johnston, priv. comm.) checked another dozen fields, also without definite results. This scarcity may have more than one interesting explanation: the small-scale clustering may have evolved drastically since $z \sim 3$; the companion galaxies are there, but are either shrouded in dust, or somehow deactivated by the neighboring QSO's (for a low-redshift version of a similar problem, see Filippenko 1985); or there simply were fewer galaxies back then. A more detailed discussion will be presented elsewhere (Djorgovski *et al.*, in preparation); the sensitivity of our search, and the resulting limits on the number-density of emission-line objects are similar to those by Pritchet & Hartwick (1987; and this conference), but with a different redshift range.

It is important to understand the source of these Lyα emissions. At least in the cases of most radio galaxies and companions, we know that there are almost certainly no UVX-strong AGN's in the vicinity, whose dilute radiation would ionize the gas. The Lyα gas is generally distributed in the same way as the stars, and some *in situ* ionization mechanism seems more likely. A good candidate is the photospheric radiation from young stars. We can test that, at least crudely, with the stellar population synthesis models (Bruzual 1983), which fit the colors and magnitudes of these objects by integrating the fluxes of ionizing photons generated by the massive young stars. The biggest uncertainties in this procedure are the IMF shape at the high end (for a lack of knowledge, we assume the Salpeter power-law), and the amount of internal reddening (we assume none, since the Lyα photons seem to be escaping easily). The conversion between the number of ionizing photons, and the number of released Lyα photons is uncertain, but a mental average over many reasonable photoionization models of extragalactic H II regions (e.g., by Stasinska 1982) suggests that about 10 ionizing photons are needed for each Lyα photon. This analysis is illustrated in Figure 8 for two different redshifts of galaxy formation, and a range of plausible evolution models. If galaxies commence forming at a large redshift (e.g., $z_{gf} \simeq 5$), the photoionization by young stars seems to fail by one or two orders of magnitude; but if one allows the start of galaxy formation to be at a lower redshift ($z_{gf} \simeq 2$), it may just be barely possible. Presence of any internal extinction would make the discrepancy even worse. The models are too simple, but there is a suggestion that some other ionization mechanism may be operating.

Collisional ionization provides a plausible solution. As mentioned before, morphology and kinematics in these galaxies are strongly suggestive of merging. The available orbital kinetic energies are typically $\sim 10^{60}$ erg, and depending on an unknown efficiency factor, dissipation in colliding gas clouds may support the observed line emission for $\sim 10^{8\pm1}$ yr, a time comparable to the dynamical crossing times in these systems. This process must be even more important in the initial collapse of protogalaxies, since vast amounts of potential energy, perhaps $\sim 10^{61}$ erg for a proto-gE/cD galaxy, have to be dissipated (Fish 1964). In most early work on the collapse of proto-ellipticals, it was assumed that the bulk of this energy is released through Compton cooling at large redshifts (Silk 1977). But if the collapse of relatively cold gaseous units occurs at lower redshifts, some other mechanism, e.g., the Lyα cooling, may become important. Finally, the large SFR's in these galaxies imply large supernova rates ($1 - 10$ yr^{-1}), which may release $\sim 10^{59-60}$ erg in the ambient gas during the galaxy free-fall or a dynamical crossing time, or duration of a typical starburst. Some part of this energy budget must go into powering the radio lobes, but there seems to be plenty left to maintain the observed line emission (cf. Silk & Shull 1979). Thus, the collisional ionization mechanisms are plausible suppliers of Lyα photons in these objects.

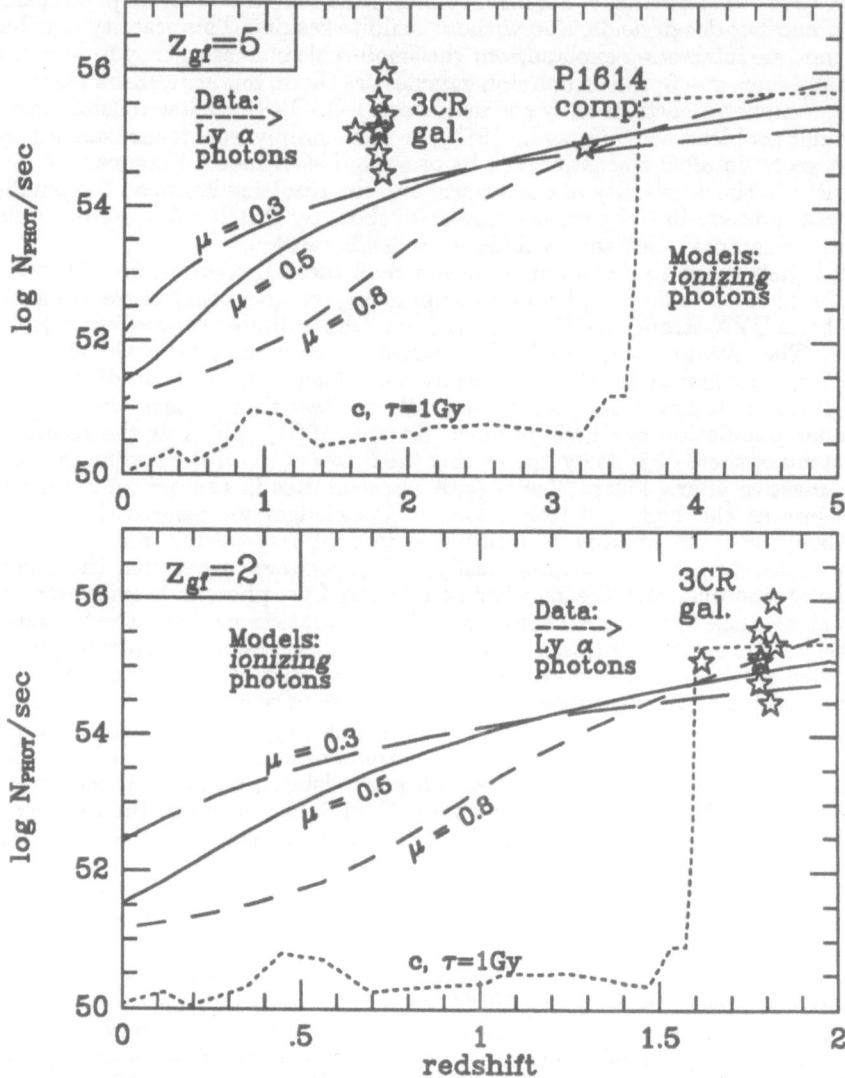

Figure 8: Evolution of the number-fluxes of ionizing photons for different population synthesis models (Bruzual 1983). The models have the Salpeter IMF, no mass infall or outflow, no internal extinction, and are scaled to have $M_V = -23$ at $z = 0$. The cosmology is a $H_o = 50$, $\Lambda_o = 0$, $\Omega_o = 0$ Friedman model. None of these assumptions affect our conclusions very much: numbers of ionizing photons and Lyα photons scale in similar ways. Two redshifts of galaxy formation are assumed: $z_{gf} = 5$ (top), and $z_{gf} = 2$ (bottom). The data on derived Lyα luminosities for distant galaxies are also shown. About 10 ionizing photons are needed to produce 1 Lyα photon if photoionization is the dominant mechanism; this may be marginally possible for $z_{gf} = 2$ models, but not for the $z_{gf} = 5$ models.

Prevalence of shock heating, combined with the UV radiation field, may help explain another important observed fact, namely the absence of significant amounts of dust. Several groups (Meier & Terlevich 1981; Hartmann *et al.* 1984, 1987; Deharveng *et al.* 1985) observed low-redshift starburst dwarfs, and found little or no Lyα emission, and there is a good anticorrelation of Lyα/Hβ line ratios with the oxygen abundance. Apparently, Lyα is mostly trapped by the dust in these dwarf galaxies, but certainly is not in the high-redshift giants which we observe. The survival of dust grains in an ISM subjected to supersonic shocks, and in a radiation field of many massive young stars, is a complex issue, and it is plausible that most grains evaporate, or do not even form, in such physical conditions. Some Lyα in the low-z dwarfs may also be trapped in the ambient gas, through resonant scattering. Here again, the objects which we see at high-z are different: their large velocity fields broaden the line so much that the overall cross section for the self-absorption of Lyα must be very small. It is not clear how long such a transparency phase would last, but a good guess may be that it is closely related to the dynamical and starburst time scales ($\sim 10^8$ yr). These growing/forming galaxies would then pop up from the general self-induced obscurity in one or more brief time windows, when they would be observable in all of their ionized splendor.

Before confronting these data with the modern theoretical views, let us briefly recall some other relevant observations. There is now a general agreement that QSO counts indicate a broad peak in the co-moving number density near $z \sim 2 - 2.5$, with a gradual falloff at larger redshifts (Schmidt 1987, and references therein). Damped Lyα absorption systems at $z \simeq 2$ can be interpreted as H I proto-disks, but they would have to be bigger than the disks of $z \simeq 0$ spirals (Wolfe 1986; Smith *et al.* 1986; Wolfe *et al.* 1987). This is not the only possible interpretation, and the absorbers may be normal-sized or small disks, but with a higher number density than at the present epoch, which would imply a substantial depletion by merging since $z \simeq 2$. Alternatively, the absorbers may be a new class of H I rich objects, without low-z counterparts. This is intriguing, since Tytler (1987) argues that *all* Lyα absorption systems are members of a single population, and there are no obvious, direct low-z counterparts for any of them. Deep galaxy counts by Tyson *et al.* (e.g., this conference) show a large numbers of faint galaxies down to $\sim 27^m$, whose suface density is indicative of a strong evolution. These faint galaxies are in the average perhaps too blue to be at $z > 3$, and too faint to be at $z < 1$, or so. The software used to compile the catalogs would split lumpy composite objects (e.g., like 3C326.1) into several fainter galaxies, and the standard analysis assumes a number conservation of galaxies. It is quite possible that a considerable fraction of them will merge by the $z \simeq 0$ epoch. There is another very intriguing observation (Miley 1987; Barthel 1987), that there seems to be a threshold in redshift (near $z \simeq 1.5$), for the onset of dramatical radio jet bending in radio-loud quasars. In many cases, the jet bending is correlated with an associated C IV absorption in the QSO spectrum. The jets at larger redshifts may be confined by gas clouds or neighboring dwarf galaxies, which then disappear through merging or get dissipated by the QSO radiation or winds by $z \simeq 1.5$. And finally, numerous primeval galaxy (PG) searches failed to reveal any luminous candidates at large redshifts, in particular at $z \sim 5$ and beyond (Koo 1986, and references therein).

Galaxy formation was traditionally viewed as a remote and well-synchronized event, ocurring at redshifts greater than those of the most distant QSO's known (viz., $z_{gf} \geq 4 - 5$). Formation of ellipticals and bulges at least was supposed to occur in a narrow redshift interval, and last about one free-fall time ($\sim 10^8$ yr).

Since vast amounts of binding energy had to be released, and most or all stars formed in a relatively short time interval, the PG's were supposed to be extremely luminous, albeit for a short time ("ten million Orions at a large redshift" — see Sunyaev *et al.* 1978). There were really no good reasons to believe all this. The early rapid formation models (e.g., Eggen *et al.* 1962) did well with the data available at the time for our Galaxy's halo, but their conclusions were subsequently generalized into a broader picture, and a rapid formation of spheroids became astronomical folklore.

On the theoretical side, there is now something of a paradigm shift on the subject of galaxy formation. For many years, numerical simulations of the evolution of large-scale structure had a quandary: the models which can best reproduce the topology of large-scale structure, galaxy clustering, etc., would require galaxy formation at "unacceptably low" redshifts of 1 or 2 (e.g., Davis *et al.* 1985; Frenk *et al.* 1985). Purely *ad hoc* biasing schemes were designed to combat this. In the framework of a still fashionable CDM scenario, a natural picture would have galaxies form *gradually*, over a large stretch of time, by stochastic merging and accretion of smaller fragments (Silk & Wyse 1986; Silk & Szalay 1987; Baron & White 1987). Whereas the first protogalactic lumps ("galaxlets") form at large redshift ($z_{init} \sim 10$), and can harbor quasars at $z \sim 3 - 5$, or radio sources at $z \sim 2$, the bulk of the present-day galaxies is assembled (that is, the accumulation peaks) at $z \sim 2 \pm 1$, or so. The merging process can stimulate starbursts in the colliding fragments, and the final stellar population would not be coeval. Formation of disks may proceed even slower. If this picture is correct, there never was such thing as *the* epoch of galaxy formation, and PG's would be a diverse class of mostly low-luminosity objects, to be found at a wide range of redshifts. The relevant galaxy formation time scales are then $\geq 10^9$ yr, rather than the free-fall time ($\sim 10^8$ yr). Because of the nonlinearity of look-back time as a function of redshift, this translates into a wide redshift interval, which we are already probing, independent of when the galaxies *started* to form.

This is very exciting, as it would mean that we have already seen luminous examples of PG's. Radio galaxies like 3C326.1 or 3C294 may be easily understood as just-collapsing systems, where one of the protogalactic lumps comes equipped with a radio transmitter. A gas-rich merger environment provides favorable conditions for the fueling of such incipient active nuclei. Similarly, systems like PKS 1614+051 or MG 2016+112 may be examples of luminous protogalaxies at an earlier stage of development, where one accretion core has developed a QSO-type nucleus: in the Silk & Szalay (1987) model this can happen in the protogalactic cores with a mass as low as $10^{8-9} M_\odot$. Their model also predicts the correct shape of the QSO luminosity function at $z \simeq 3$. The damped Lyα absorbers studied by Wolfe, Smith, *et al.* may be gas-rich protogalactic fragments or small disks, more numerous at $z \simeq 2$ than they are now. Faint galaxies counted by Tyson *et al.* may be visible images of such fragments at $z \sim 1 - 3$. Searches for luminous PG's at $z \geq 3$ or 5 failed because perhaps there were no luminous PG's that far away!

Another possibility is that these systems represent formation of compact galaxy *groups*. The observed velocity scales are too large even for gE/cD galaxies, and the virialization can only increase the final stellar velocity dispersion. Alternatively, some fraction of the observed motions may be hydrodynamically induced, e.g., by multiple supernova events, galactic winds from starburst cores, or gas entrainment in radio jets.

Theoretical apologias contrived to salvage the CDM scenario may have the good fortune of being right; they fit and naturally explain all the available data, and I am not aware of any evidence which contradicts this picture. Actually, these new developments derive naturally from many earlier models of dissipative galaxy formation, which employed gradual dissipative merging of gas-rich proto- or pre- or simply galactic fragments into the bigger and better galaxies of today (e.g., Toomre 1977; Tinsley & Larson 1979; Silk & Norman 1981; Silk 1986, 1987a, 1987b; etc.). But in the synthesis of new data and new theoretical wrinkles, we may have a unified (standard?) model of galaxy formation emerging.

If this picture of galaxy formation by sputtering rather than a splash is correct, planning of PG searches may need some revision. The redshift range of interest is 2 ± 1, which is well suited for ground-based CCD observations and IR imaging may be useful only if most PG's are very dusty, but all candidates so far appear to be very clean. As proposed above, PG's may go through "windows of observability", corresponding to the particularly strong merging and starbursting events, when the ambient dust may be destroyed and may be hard to form. Another hint is that looking for *low surface brightness* objects with apparent sizes ~ 10 arcsec may be the right approach. Yet another is that it still makes sense to look at and around distant AGN's (radio sources or quasars): their nonthermal emissions may be pathological, but the overall dissipative processes going on around them may be quite normal for their times. (Finding a radio-quiet lookalike of 3C326.1 or some such would be most reassuring, though). If the formation process peaks at $z < 2$, observations with the Hubble Space Telescope may be very fruitful.

Whereas the formation of spheroids (ellipticals, bulges) is probably largely over by $z \sim 1$ (we would have seen more interesting things if there were any at that redshift), the formation of disks and diffuse dwarfs may continue to the present epoch. The disks form stars continuously (see the illuminating discussions by Sandage 1985 and 1986), but some gas-rich, tenuous objects like Malin–1 (Bothun *et al.* 1987) or VII Zw 31 (Sage & Solomon 1987) may be just starting now. Low-metallicity dwarfs like I Zw 18 (Kunth & Sargent 1986) or Mrk 36 (Huchra *et al.* 1982) may be just undergoing their first bursts of star formation. These may be genuine examples of $z \simeq 0$ PG's — galaxy formation is certainly not over yet.

This leads to the question of age-dating of stellar systems. Old stellar populations at $z \simeq 0$ are often assumed to have the same age, but the truth is that population synthesis models are unable to distinguish very well whether the bulk of the stars formed at $z \sim 1$ or $z \sim 10$ (cf. Wyse 1985, or Djorgovski 1988). The amplitude of the 4000 Å break, a good indicator of an effective mean age of a stellar population (Spinrad 1980), is found to decrease systematically at larger redshifts (Dressler 1987). Hamilton (1985) found no significant effect for the reddest field galaxies, suggesting that there is a range of ages among the early-type galaxies at moderately high redshifts (up to $z \sim 0.6 - 0.8$).

We may loosely but adequately define galaxy formation as the process and time in which a substantial fraction of the initial gas is converted into stars of the final product. If a gradual dissipative merging of gas-rich fragments is the primary way of building galaxies, where does one draw a distinction between galaxy formation, and active galaxy evolution? The luminous radio galaxies observed at $z > 1$ are apparently undergoing massive starbursts, probably largely stimulated by gas-rich mergers (Djorgovski *et al.* 1987a; Djorgovski 1988). The estimated star formation rates in these systems reach several hundred $M_\odot/\mathrm{yr}$ — more than the most luminous *IRAS* sources at low redshifts (e.g., Soifer *et al.* 1987, and references therein), and two orders of magnitude higher than the classical starburst in M82 (Rieke *et al.* 1980). These galaxies are already fairly well formed, but they are certainly growing rapidly. Sporadic events of this kind are still happening today (Joseph & Wright 1985; Rieke *et al.* 1985; Schweizer 1986; Sanders *et al.* 1987; Harwit *et al.* 1987; etc.).

Galaxy formation may be viewed as an intrinsically stochastic process in which an open system interacts with its environment, and the environments and thus the formation histories may differ. Global or fundamental properties of finished-product galaxies today and the correlations between them are largely driven by some basic physics, e.g., the virial theorem, but may also reflect their formation histories. There is a real possibility that important correlations, like the Tully–Fisher relation for spirals, or the $R - \sigma - \mu$ relation for ellipticals and bulges, may vary slightly in ther slopes and/or intercepts in different large-scale environments. This would make their use as distance indicators subject to difficult-to-detect systematic errors, which could masquerade as large peculiar velocities. Studies of the universality of galaxian properties or their correlations at low red-shifts are necessary before we can reliably study the dynamics of the Local Supercluster and its environs, and may tell us something interesting about galaxy formation itself.

Our view of galaxy formation thus changes, from a mythical cataclysmic event in a distant past, to an evolutionary process or series of processes continuing through the present era. Philosophically, this has the right ring of maturity; it would be wonderful if it really is true. If so, we may be entering the extragalactic El Dorado. But this alluring picture may still be wrong, and the Lyα galaxies at large redshifts may be fool's gold: certainly, much more work needs to be done.

This paper is based almost entirely on the data, work, and ideas shared with my collaborators, in particular Hy Spinrad, Patrick McCarthy, Mark Dickinson, Michael Strauss, and others. They get the credit for the data and whatever turns out well and true from all this, but I get all the blame for the wild speculations, generalizations, and flakey conclusions. We are grateful to the staffs of Kitt Peak, Lick, MMT, VLA, CFHT, and Cerro Tololo observatories for their able and professional help during numerous observing runs. I was inspired by papers by, and discussions with, Joe Silk, Richard Larson, and many others. Finally, I would like to acknowledge the pleasant and productive atmosphere of the Erice Workshop, due largely to our Italian hosts, and to thank to the A.A.S. for a travel grant.

References:

Baron, E., and White, S. D. M. 1987, *Ap. J. Lett.* in press.
Barthel, P. 1987, proceedings of the STScI Conference on the QSO Absorption Line Systems, in press.
Bothun, G. D., Impey, C., Malin D. F., and Mould, J. R. 1987, *A. J.* **94**, 23.
Bruzual, G. 1983, *Ap. J.* **273**, 105.
Chambers, K., Miley, G., and van Breugel, W. 1987, *Nature*, in press.
Davis, M., Efstathiou, G., Frenk, C., and White, S. 1985, *Ap. J.* **292**, 371.
Deharveng, J., Joubert, M., and Kunth, D. 1985, in *Star–Forming Dwarf Galaxies*, eds. D. Kunth *et al.*, p.431. Paris: Editions Frontieres.
Djorgovski, S., Spinrad, H., and Marr, J. 1985a, in *New Aspects of Galaxy Photometry*, ed. J.-L. Nieto, *Lecture Notes in Physics* **232**, p. 193. Berlin: Springer Verlag.
Djorgovski, S., Spinrad, H., McCarthy, P., and Strauss, M. 1985b, *Ap.J.Lett.***299**, L1.

Djorgovski, S. 1987, in *Nearly Normal Galaxies*, proceedings of the Santa Cruz Astrophysics Workshop, ed. S. Faber, p. 290. New York: Springer.

Djorgovski, S., Spinrad, H., Pedelty, J., Rudnick, L., and Stockton, A. 1987a, *Astron. J.* **93**, 1307.

Djorgovski, S., Strauss, M., Perley, R., Spinrad, H., and McCarthy, P. 1987b, *Astron. J.* **93**, 1318.

Djorgovski, S. 1988, in *Starbursts and Galaxy Evolution*, proceedings of the Moriond Astrophysics Workshop, ed. Th. Montmerle, in press. Paris: Editions Frontieres.

Djorgovski, S., Spinrad, H., and Dickinson, M. 1988, in preparation.

Dressler, A. 1987, in *Nearly Normal Galaxies*, proceedings of the Santa Cruz Astrophysics Workshop, ed. S. Faber, p. 276. New York: Springer.

Eggen, O., Lynden-Bell, D., and Sandage, A. 1987, *Ap. J.* **136**, 748.

Eisenhardt, P., and Lebofsky, M. 1987, *Ap. J.* **316**, 70.

Ellis, R. 1987, in *Observational Cosmology*, IAU Symposium 124, ed. A. Hewitt *et al.*, p. 367. Dordrecht: Reidel.

Fabian, A., Nulsen, P., and Arnaud, K. 1984, *M.N.R.A.S.* **208**, 179.

Filippenko, A. 1985, *Astron. J.* **90**, 1172.

Fish, R. 1964, *Ap. J.* **139**, 284.

Frenk, C., White, S., Efstathiou, G., and Davis, M. 1985, *Nature* **317**, 595.

Hamilton, D. 1985, *Ap. J.* **297**, 371.

Hartmann, L., *et al.* 1987, *Ap. J.* in press.

Hartmann, L., Huchra, J., and Geller, M. 1984, *Ap. J.* **287**, 487.

Harwit, M., Houck, J., Soifer, B., and Palumbo, G. 1987, *Ap. J.* **315**, 28.

Heckman, T., *et al.* 1986, *Ap.J.* **311**, 526.

Hu, E., and Cowie, L. 1987, *Ap. J. Lett.* **317**, L7.

Huchra, J., Geller, M., Hunter, D., and Gallagher, J. 1982, in *Advances in Ultraviolet Astronomy*, Y. Kondo (ed.), p. 151. Washington: NASA.

Joseph, R., and Wright, G. 1985, *M.N.R.A.S.* **214**, 87.

Koo, D. 1986, in *The Spectral Evolution of Galaxies*, C. Chiosi and A. Renzini (eds.), p. 419. Dordrecht: D. Reidel.

Koo, D., and Kron, R. 1988, *Ap. J.* in press.

Kunth, D., and Sargent, W. L. W. 1986, *Ap. J.* **300**, 496.

Lebofsky, M., and Eisenhardt, P. 1986, *Ap. J.* **300**, 151.

Lilly, S., and Longair, M. 1982, *M.N.R.A.S.* **199**, 1053.

Lilly, S., and Longair, M. 1984, *M.N.R.A.S.* **211**, 833.

McCarthy, P., Spinrad, H., Djorgovski, S., Strauss, M., van Breugel, W., and Liebert, J. 1987a, *Ap. J. Lett.* **319**, L39.

McCarthy, P., Van Breugel, W., Spinrad, H., and Djorgovski, S. 1987b, *Ap. J. Lett.* **321**, L29.

McCarthy, P. J. 1988, Ph. D. Thesis, University of California, Berkeley.

Meier, D., and Terlevich, R., 1981, *Ap. J. Lett.* **246**, L109.

Miley, G. 1987, in *Observational Cosmology*, IAU Symposium 124, eds. A. Hewitt *et al.*, p. 267. Dordrecht: Reidel.

Pritchet, C., and Hartwick, F. 1987, *Ap. J.* **320**, 464.

Rieke, G., *et al.* 1980, *Ap. J.* **238**, 24.

Rieke, G., *et al.* 1985, *Ap. J.* **290**, 116.

Robinson, A., *et al.* 1987, *M.N.R.A.S.* **227**, 97.

Sage, L., and Solomon, P. 1987, *Ap. J. Lett.* in press.

Sandage, A. 1985, in *Star–Forming Dwarf Galaxies*, eds. D. Kunth *et al.*, p. 31. Paris: Editions Frontieres.

274

Sandage, A. 1986, *Astron. Ap.* **161**, 89.
Sanders, D., *et al.* 1987, *Ap. J.* in press.
Schmidt, M. 1987, in *Observational Cosmology*, IAU Symposium 124, ed. A. Hewitt *et al.*, p. 619. Dordrecht: Reidel.
Schneider, D., Gunn, J., Turner, E., Lawrence, C., Hewitt, J., Schmidt, M., and Burke, B. 1986, *Astron. J.* **91**, 991.
Schneider, D., Gunn, J., Turner, E., Lawrence, C., Schmidt, M., and Burke, B. 1987, *Astron. J.* **94**, 12.
Schweizer, F. 1986, *Science* **231**, 227.
Silk, J. 1977, *Ap. J.* **211**, 638.
Silk, J., and Shull, M. 1979, *Ap. J.* **234**, 427.
Silk, J., and Norman, C. 1981, *Ap. J.* **247**, 59.
Silk, J. 1986, in *Spectral Evolution of Galaxies*, p. 15, ed. C. Chiosi and A. Renzini. Dordrecht: Reidel.
Silk, J., and Wyse, R. 1986, in *Structure and Evolution of Active Galactic Nuclei*, p. 173, ed. G. Giuricin *et al.*. Dordrecht: Reidel.
Silk, J. 1987a, in *Dark Matter in the Universe*, IAU Symposium No. 117, p. 335, eds. J. Kormendy and J. Knap. Dordrecht: Reidel.
Silk, J. 1987b, in *Observational Cosmology*, IAU Symposium 124, eds. A. Hewitt *et al.*, p. 391. Dordrecht: Reidel.
Silk, J., and Szalay, A. 1987, preprint.
Smith, H., Cohen, R., and Bradley, S. 1986, *Ap. J.* **310**, 583.
Soifer, B. T., Houck, J. R., and Neugebauer, G. 1987, *Ann. Rev. Astr. Aph.* **25**, 187.
Spinrad, H. 1980, in *Objects of High Redshift*, proceedings of IAU Symposium 92, eds. G. Abell and J. Peebles, p. 39. Dordrecht: Reidel.
Spinrad, H., Filippenko, A. V., Wyckoff, S., Stocke, J., Wagner, M., and Lawrie, D. 1985, *Ap. J. Lett.* **299**, L7.
Spinrad, H. 1986, *Publ. Astron. Soc. Pacific* **98**, 269.
Spinrad, H., and Djorgovski, S. 1987, in *Observational Cosmology*, proceedings of IAU Symposium 124, eds. A. Hewitt *et al.*, p. 129. Dordrecht: Reidel.
Spinrad, H. 1988, in proc. of the IAP workshop *High-Redshift and Primeval Galaxies*, J. Bergeron and B. Rocca–Volmerange (eds.), in press.
Stasinska, G. 1982, *Astr. Ap. Suppl.* **48**, 299.
Strauss, M. *et al.* 1988, in proc. of the IAP Workshop *Primeval and High–Redshift Galaxies*, J. Bergeron and B. Rocca–Volmerange (eds.), in press.
Sunyaev, R., Tinsley, B., and Meier, D. 1978, *Comm. Astrophys.* **7**, 183.
Tinsley, B., and Larson, R. 1979, *M.N.R.A.S.* **186**, 503.
Toomre, A. 1977, in *The Evolution of Galaxies and Stellar Populations*, B. Tinsley and R. Larson (eds.), p. 420. New Haven: Yale University Observatory.
Tyson, J. A. 1986, *Astron. J.* **92**, 691.
Tytler, D. 1987, *Ap. J.* **321**, 49.
Wolfe, A. 1986, *Phil. Trans. Roy. Soc.* **320**, 503.
Wolfe, A., *et al.* 1987, *Ap. J. Suppl. Ser.* **61**, 249.
Wyse, R. 1985, *Ap. J.* **299**, 593.
Yates, M., Miller, L., and Peacock, J. 1986, *M.N.R.A.S.* **221**, 311.

COMMENTS ON THE REALITY OF THE BUTCHER-OEMLER EFFECT

David C. Koo
Space Telescope Science Institute
3700 San Martin Drive
Baltimore, MD 21218   U.S.A.

ABSTRACT. We propose that some of the evidence, especially recent spectroscopic data, for evolution in the properties of galaxies in the central cores of high redshift clusters remain questionable. Problems include biased selection of the clusters, contamination by extraneous galaxies, and non-inclusion of several corrections including the expansion of the universe. In particular, we suggest that clusters viewed approximately in the plane of flattened superclusters will preferentially be selected, thus magnifying the contamination by other clusters and "field" galaxies. The ratio of the accepted volume to that of the cluster core may easily reach beyond 400 for redshifts z greater than 0.3. Reasonable assumptions on the density structure of clusters and superclusters predict 10% or more contamination as observed.

1. INTRODUCTION

For nearly a decade since the classic paper by Butcher and Oemler (1978a), their claim for more blue galaxies in the cores of rich compact clusters at higher redshifts has fired the imagination of observers and theorists. Recently, attention has shifted from deep photometric towards spectroscopic surveys. Exciting results from such surveys include not only confirmation that many of the blue galaxies apparently in the cluster center have redshifts close to that of the cluster itself, but also evidence for unexpectedly large numbers of galaxies with unusual spectra, including "E+A", strong emission lines, or Seyfert 1 or 2 type features (see conference articles by Gunn, Ellis, Mellier, Henry, etc.). Before we accept these results as evidence for cluster evolution, however, we need to establish the extent to which the effects are unique to clusters rather than to the general field population (or to the outer parts of clusters) and the extent of possible contamination problems.

Taking perhaps an extreme viewpoint, I would like to propose that many of the effects attributed to the cluster-core galaxies represent a more general phenomenon occurring in less dense environ-

R. G. Kron and A. Renzini (eds.), Towards Understanding Galaxies at Large Redshift, 275–284.
© *1988 by Kluwer Academic Publishers.*

ments and that much of the apparent evolution is the result of how clusters were selected, of contamination of the core area by galaxies outside, and of not including corrections such as the expansion of the universe. Indeed, Butcher and Oemler (1984b) themselves suggested the possibility that the evolution seen in clusters may be largely independent of environment, based upon their spectroscopic survey of a non-compact cluster at z = 0.38. Using multicolors as crude estimators of redshift, I find that even field galaxies appear to have larger fractions (~70% at z ~ 0.4 versus ~45% today) of intrinsically very blue galaxies (Koo 1986), but this conclusion relies on models of galaxy evolution and needs confirmation from a spectroscopic survey. Although analysis of the deep field-galaxy redshift surveys have yet to be completed for such a confirmation, preliminary results suggest that larger fractions of field galaxies at moderate redshifts z > 0.3 did undergo enhanced star formation, as seen in the distribution of either emission-line strengths (Ellis 1987) or bluer colors (see talk by Koo and Kron). At present, however, the fraction of "E+A" type field galaxies still appears low but a quantitative comparison has yet to be made, so that perhaps the frequency of such galaxies seen in cluster cores is giving direct conclusive evidence for cluster evolution (see talk by Gunn or Dressler 1987). In the meantime, let me continue with my concerns.

2. SICILIAN PIZZA MODEL

The picture I have is shown in Figure 1. Based upon strong hints of flattened extended distributions of galaxies (like a Sicilian pizza) seen in the redshift surveys of Haynes and Giovanelli (1986), or De Lapparent et al. (1986), I simply imagine that the types of galaxies seen within an apparent cluster core (1 Mpc diameter where Ho = 50 km s^{-1} Mpc^{-1} has been used throughout this paper) depend on the vantage point. Viewed face-on (Fig. 1-I), a sausage clump (cluster) may NOT be contaminated much by bits of pepper (blue galaxies) or anchovies ("E+A"); viewed in the plane of the pizza (supercluster), one may see increased contamination along the line of sight (Fig. 1-III); and in the extreme case, one may see two or more sausage clumps superimposed (Fig. 1-II) along with substantial contamination. The following sections describe a number of what I call anti-Butcher-Oemler effects, all of which may contribute to the illusion of high-redshift cluster cores containing more unusual galaxies, if such galaxies are common outside of cluster cores.

3. ANTI-BUTCHER-OEMLER EFFECTS

3.1. Moving-Target Effect

Before describing the anti-BO effects, a bit of history shows how slippery the target can be. The original BO effect was the claim that the central cores of two rich compact clusters at high redshift,

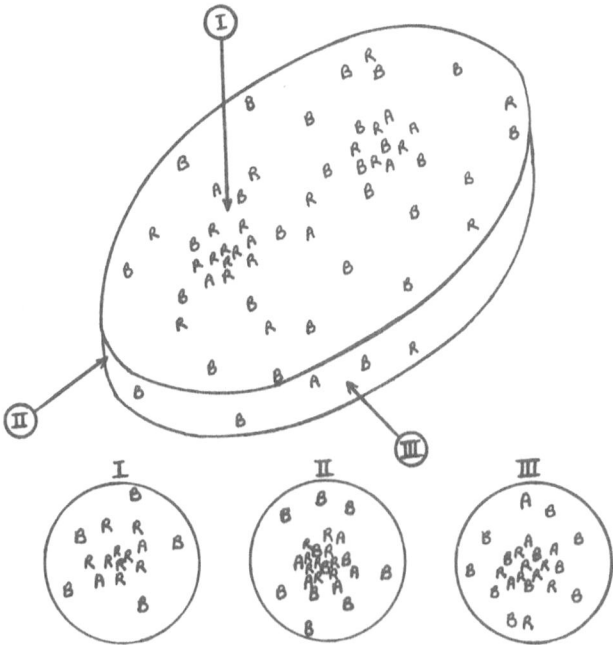

Figure 1.   Several views of a flattened supercluster of galaxies are
shown within the circular frames, where R's (sausage bits) represent
red early-type galaxies usually found in high density environments,
A's (anchovies) represent unusual "E+A" type galaxies, and B's (pepper bits) represent blue galaxies.

3C 295 (z = 0.46) and 0024+16 (z = 0.39), exhibited a large fraction
(50%-60%) of blue galaxies that were presumed to be spirals (Butcher
and Oemler 1978a), whereas comparable clusters like Coma or others do
not (Butcher and Oemler 1978b).   Soon after, DeGioia and Grasdalen
(1980) criticized the work by pointing out that the effect could be
explained by the lack of corrections for the known correlation
between intrinsically fainter non-spiral (i.e. gE or S0) galaxies
being bluer (color-mag effect) and the possibility of poor background
subtraction.   Mathieu and Spinrad (1981) gave further support to the
latter suggestion when their independent photometry of 3C295 did not
show an excess of blue galaxies.   In addition, the lack of any blue
excess in an even higher redshift cluster 0016+16 at z = 0.54 (Koo
1981) demonstrated that the BO effect could not be universal.
Moreover, Kron (1982) argued that the apparent excess of blue
galaxies in a well-studied, nearby, rich compact cluster, A2199,
would mean either more problems at larger redshifts if not real, or
that some nearby clusters also have blue excesses.   Wirth and
Gallagher (1980) found that spiral fractions in two nearby clusters
are much greater than previously measured, a result that may also
diminish differences between the low and high redshift clusters.

Couch and Newell (1984) later claimed confirmation of the classic BO effect by expanding the sample of clusters with deep photographic color photometry, including corrections for the color-mag effect and improvements in the background field subtraction.

Then Butcher and Oemler (1984a) changed their definition of blue to include only **very** blue (rest frame B-V < 0.7) galaxies within a central area containing 30% of the cluster galaxies (typically 1-2 Mpc diameter) to quite faint absolute magnitudes (M_v < -20). They strengthened their conclusion of rapid cluster evolution by showing that nearby clusters had very-blue-galaxy fractions of $\lesssim$ 5%, field had 44%, and distant clusters (z ~ 0.5) had 25%. Cluster 0016+16, however, with 2% and still at the largest redshift in their sample, remained an anomaly in their picture. Furthermore, the presence of both high and low fractions is a phenomenon that appears to extend below redshifts of z ~ 0.3, thus placing further doubt to the homogeneity of the cluster sample and to the suggested evolution with redshift.

Spectroscopy had already begun in full force by this time, with Dressler and Gunn (1982, 1983) leading the way and claiming for 3C295 extreme evolution in the number of active-galactic-nuclei. They also discovered the presence of unusual "E+A" type galaxies which were relatively red but best explained as a combination of a normal gE type spectrum with an "A" type spectrum (strong Balmer lines) as seen from a stellar population 1 Gyr after a burst of star formation, but with no strong emission lines, unlike equally blue spirals today. Moreover, since many of the blue galaxies were not cluster members, the original BO effect could be questioned as resulting from poor background subtraction. A new effect was thus born. A similar study of 0024+16 (Dressler et al. 1985), however, showed a genuine excess of spiral-like blue galaxies centered at the cluster redshift, thus supporting the original BO effect.

As previously mentioned, Butcher and Oemler (1984b) showed that unusual "E+A" and active galaxies could be found even in non-compact distant clusters, putting into some question the extent to which the effects were unique to dense environments. More recently, Dressler and Gunn (Dressler 1987, talk by Gunn) found that the anomalously red cluster 0016+16 contains a high number of "active" galaxies, where active is now defined as non-passive, i.e., not like that of normal early-type galaxies. They did confirm, however, the small number of very blue galaxies as did Ellis et al. (1985), who restudied the cluster using multiple intermediate band CCD photometry. Thus 0016+16, which remains a cluster that has no excess of blue or very blue galaxies, is now cited as showing strong evolution on the basis of an excess of "E+A" spectra. To complicate further the issue of what exactly is meant by the BO effect and cluster evolution, Dressler and Gunn report the detection of a significant shift of the distribution of 4000Å breaks to smaller amplitudes among otherwise RED galaxies in clusters at redshifts z > 0.5. This is an important result that provides direct evidence for evolution, but its relationship to the BO effect, if any, remains unclear.

Other spectroscopic surveys of distant clusters can be found among the proceedings of this conference with a general consensus that cluster evolution has been detected. Note that the criteria for evolution is no longer confined to the presence of very blue galaxies but also includes not only the findings of Dressler and Gunn above but also slight ultraviolet excesses among the reddest galaxies as described by MacLaren, Ellis, and Couch (1987) or larger numbers of close, probably interacting, pairs of galaxies as suggested by Henry. The diversity by which distant clusters appear to be signaling evolution is itself quite compelling and opens exciting doors of opportunity to understand better the evolution of galaxies. But before attributing all these phenomenon to be clues to the evolution of cluster galaxies, I believe closer attention to problems of contamination is needed to separate how much of each evidence belongs to the dense cluster core, to the less dense outer parts of clusters, to perhaps overlapping but less compact clusters in the same supercluster, or to galaxies belonging to the general field or small groups, all residing in a supercluster around the rich cluster.

3.2. Tail-Does-Not-Wag-the-Dog Effect

Tails of distributions, i.e., extreme examples of any class, are expected to be peculiar. Among the four compact clusters studied by Butcher and Oemler (1984a) with redshifts $z > 0.3$, A370 at $z = 0.37$ has the highest known redshift among Abell clusters, 0024+16 at $z = 0.39$ has the highest known redshift among Zwicky clusters, 3C295 at $z = 0.46$ stood as the highest known redshift among galaxies for over a decade (and is a strong emission line radio galaxy of course), and 0016+16 at $z = 0.54$ was noticed by Kron and observed by Spinrad because its appearance was so extreme in richness, redness, and faintness among clusters visible over an area covered by about a dozen deep 4m plates. As emphasized in this conference by Chincarini, Ellis, and Gunn, a sample of distant clusters is needed whose selection is well understood. The distant clusters of Gunn and colleagues are certainly an improvement and the best available, but until some simulations, as suggested by Chincarini, are made using realistic spatial distributions and colors of galaxies, groups, clusters, and superclusters, an accurate assessment of the selection biases and contamination problems of the sample would remain difficult.

3.3. Rich-Get-Richer Effect

Very rich clusters are known to be strongly correlated with other clusters, groups, and galaxies (Bahcall and Soneira 1983). This correlation will certainly aggravate any contamination problems, especially in spectroscopic surveys where the background is generally not defined by control samples in nearby fields but rather by redshifts alone. Such large-scale aggregations may even preclude accurate "background" galaxy-surface-density estimates in photometric surveys, which tend to use controls within at most a few tens of Mpc

280

away. This strong tendency for other rich structures to be nearby greatly increases the probability that the selection of the clusters at high redshift are often due to or at least affected by chance projections of non-rich-cluster galaxies. In the case of the BO clusters with z > 0.3, I would surmise that the two very blue clusters, A370 and 0024+16, are the result of two overlapping clusters (Fig. 1-II), one that is compact, rich, and quite red with a looser one quite blue, or the result of an elongated structure viewed pole-on; cluster 0016+16, on the other hand, is perhaps being viewed as in Fig. 1-I, and hence shows mainly red galaxies (but the 9 "E+A" galaxies out of 33 cluster members observed by Dressler and Gunn remains unusual); finally 3C 295 is uncertain but could be a cluster viewed projected into the plane of a flattened supercluster and hence has an apparent excess of non-early-type galaxies (Dressler and Gunn 1983). Though more time consuming, some spectroscopic control samples should be included in future work. Such control fields might be taken adjacent to the cluster core region if the effects within the core are to be distinguished from non-core regions, or further away if the effects of the supercluster are to be discriminated against that of the cluster itself. In the end, without any known method for accurate spatial positioning along the radial direction, only statistical analysis of many clusters combined with realistic simulations of their selection will provide solid evidence for evolution within the cluster core (unless a large fraction of the cluster core galaxies undergo similar changes). The field galaxy samples will also provide crucial clues of what effects are unique to clusters.

3.4. Sad-Giant Effect

As was emphasized by Osterbrock (1984), the bluest (i.e., saddest) and brightest (i.e., most giant) galaxies are most likely to be Seyfert 1 or 2 active galaxies. Thus if a large volume of space is included as being part of the cluster, and if the bluest and brightest objects in such a volume are selected for spectroscopic followup, there will be a bias towards finding relatively more such objects than in comparable volumes in nearby surveys that reach fainter luminosities. As for strong but narrow emission lines in distant galaxies, it is worth noting that, with the exception of the work on 32 galaxies near clusters by Dressler et al. (1985) and a few others many years ago (e.g. by D. Wells as compiled by Pence 1976), virtually no large-aperture spectroscopy is available for a wide variety of galaxies. Thus considerable care must be taken to intercompare the emission line strengths seen from a small aperture enclosing the red bulge of an otherwise very blue disk galaxy in nearby surveys to that from apertures that cover such disks at large redshifts. An extension of the Dressler et al survey would certainly be a worthy thesis topic.

3.5. Tuck-in-Your-Tummy Effect

As mentioned above, two relatively nearby clusters in projection would pose serious contamination effects. Yet these two clusters may not even be discriminated by spectroscopy, e.g. as two partially overlapping Gaussians in redshift space. In fact they may appear to have the same redshift if they are infalling fast enough towards each other in an expanding universe. The amplitude of such an effect is not negligible, perhaps 500 km sec^{-1} or more (i.e., many Mpcs worth of separation) between clusters, as suggested recently from the work on large-scale motions of nearby galaxies (see talk by Burstein). Indeed, coherent motions in general aggravate the problem of ambiguity between spatial positions and redshifts. Kaiser (1987) estimates that rich clusters themselves have infall-turnaround radii near 1500 km sec^{-1}; Bingelli et al (1987) suggest infall motions by late-type galaxies around Virgo. A fundamental assumption in the studies of distant rich compact clusters is that they are fully virialized and can be compared to clusters of similar structure nearby. But if even the best example of a relaxed cluster, Coma, has sub-structure today (Fitchett and Webster 1987), choosing the equivalent progenitors at high redshifts would indeed be difficult. Moreover, even the use of a metric diameter (typically 1 or 2 Mpc) for studying distant clusters may be a questionable practice. In fairness to those claiming detection of spectral changes in cluster galaxies with time, corrections for evolution in the structure of clusters are likely to exaggerate the observed effects (Gunn 1987).

3.6. FOG Effect

One of the most serious contamination problems occurs as a result of the so called "Finger-of-God" effect. This effect produces an apparent elongation of a cluster in redshift space due to the large cluster velocity dispersion and obscures (fogs) what is or is not a cluster-core member by a very large factor. For example, for a cluster core of 1 Mpc diameter (i.e., 50 km sec^{-1} worth), the FOG effect, in the case of say ±3 sigma, where sigma is a velocity dispersion of 1500 km sec^{-1}, encompasses a total volume (assuming a cylinder to first order) about 9000/50 or 180 times larger or nearly 140 Mpc^3! This is the volume which the spectroscopists would include as being part of the cluster core itself, and to my knowledge, has NOT been corrected for in published discussions of blue or active galaxies in distant clusters. Since 100 Mpc^3 would usually contain less than one field galaxy on average, field contamination appears to be negligible. Projected onto a rich cluster, however, this volume may not only include the outer parts of the cluster itself but also a volume with a density much higher than the typical field, depending upon the viewing angle and selection biases, as discussed above. In support of this probable contamination, the distributions of bluer galaxies of the "cluster" are less compact, not only in the plane of the sky (Butcher and Oemler 1984a), but also in redshift space (Dressler 1987).

3.7. Stuff-the-Sausage Effect

To compound problems from the FOG effect, the expansion of the universe multiplies the sausage-like contaminating volume by $(1+z)^{1.5}$ with $q_o = 0.5$ or $(1+z)^2$ with $q_o = 0$, where I have taken account of the change in velocity dispersion with redshift, but not any dynamics such as infall. This effect results from the adoption of a single metric diameter within which the contents of clusters at different redshifts are compared, since the cores of compact, rich clusters are presumed to be virialized and thus stable in size as the universe expands. Some galaxies in the very outer parts of clusters or in the field, but otherwise projected onto the same metric diameter in a large-redshift cluster, would partake in the expansion and lie beyond the metric diameter at lower redshifts. Together with the FOG effect, the ratio of contaminating volume to cluster core volume, using the above values, is about 500 by $z = 0.5$.

4. SUMMARY

To put the variety of anti-BO effects into perspective, a very rough estimate of the amplitude of various effects can be derived by adopting the recent results of the CFA redshift slice (as reported by Huchra 1988) where flattened structures of 10 Mpc in thickness, about 50 Mpc in size, and volume densities five times average were measured. To first order then, if a rich cluster were imbedded within such structures, there is about 10% chance that it would be viewed close to the plane of the supercluster. If so, we can estimate the contamination by the supercluster as well as by cluster members outside the core. For the cluster, we adopt spherical symmetry, a constant density to a radius of 0.25 Mpc, and a r^{-3} density drop beyond to the presumed edge of 3 Mpc. Furthermore, we assume that the spectroscopists include a ±3 sigma volume within a cylinder of 1 Mpc metric diameter, independent of z, containing about 50 cluster galaxies (44% of the total cluster), where the sigma of the cluster's intrinsic velocity dispersion is 1500 km sec^{-1}. We find that about 25% of these visible cluster members lie outside the core of 1 Mpc, thus resulting in substantial contamination to begin with, assuming that our interest is in cluster core galaxies alone. If 40% of these are E+A galaxies, then the observed fraction of 10% by $z \sim 0.5$ (Dressler 1987) would be fully accounted for, though admittedly, evolution has occurred but not in the core of clusters. To this possible source of contamination should be added the supercluster galaxies occupying the cylinder of 1 Mpc diameter, where we assume a uniform filling of the entire FOG sausage by the enhanced super-cluster volume density, (about 0.015 Mpc^{-3} to reach $M_v < -20$) and that the stuff-the-sausage effect is applicable to superclusters. The results are presented in Table 1, where the contaminating volumes can be compared to the cluster's spherical core-volume of 0.52 Mpc^3.

Table 1
Supercluster Contamination
of Distant Clusters (% of total = 50)

Redshift	qo = 0			qo = 0.5		
	Volume (Mpc³)	No. Gal.	(%)	Volume (Mpc³)	No. Gal.	(%)
0.015	140	2.0	(4)	140	2.0	(4)
0.30	240	3.4	(7)	210	2.9	(6)
0.50	315	4.4	(9)	260	3.7	(7)
0.75	425	6.0	(12)	330	4.6	(9)
1.0	560	7.9	(16)	400	5.6	(11)

These numbers (or fractions) alone do not quite account for the average observed fractions of emission-line galaxies in distant clusters (15% at z ~ 0.5 according Dressler 1987) by about a factor of two, but could easily be underestimated by such factors if these clusters are biased by having another cluster or moderately rich group be in projection and within the FOG cylinder.

In summary, we find that if distant clusters are frequently projections of rich clusters and superclusters, the resulting contamination by non-cluster-core galaxies in spectroscopic surveys may be significant at the 10% level or more. To explain all the observations, especially the large number of E+A galaxies, with this simple idea appears far-fetched at the moment, especially without more detailed knowledge of the true distribution of galaxies and better simulations, but contamination must be a problem at some level. Since only a few, if any, of the anti-BO effects have been accounted for, the burden of proof lies on the shoulders of those in favor of cluster-core evolution to demonstrate (or more precisely define! or redefine!!) their claims.

5. ACKNOWLEGEMENTS

The chairman, R. Kron, is to be blamed for encouraging my provocative stance, but I take responsibility and offer my apologies for any, though unintended, disrespect or harshness in tone expressed. I am sincerely impressed, fascinated, and challenged by the high-quality, exciting results of my colleagues that represent years of difficult observations, and have purposely taken an extreme viewpoint to highlight areas which need more attention. A. Dressler, R. Ellis, J. Gunn, and R. Kron are all thanked for helpful discussions.

6. REFERENCES

Bahcall, N. and Soneira, R.   1983 *Ap. J.*, **270**, 20.
Bingelli, B., Tammann, G. A., and Sandage, A.   1987 *ESO preprint no. 498*.
Butcher, H. and Oemler, A.   1978a *Ap. J.*, **219**, 18.
_____1978b *Ap. J.*, **226**, 559.
_____1984a *Ap. J.*, **285**, 426.
_____1984b *Nature*, **310**, 31.
Couch, W. J. and Newell, E. B.   1984 *Ap. J. Suppl.*, **56**, 143.
DeGioia, K. and Grasdalen, G. L.   1980 *Ap. J. (Lett)*, **239**, L1.
De Lapparent, V., Geller, M. J., and Huchra, J.   1986 *Ap. J. (Lett)*, **302**, L1.
Dressler, A.   1987 in *Nearly Normal Galaxies: From the Planck Time to the Present*, ed. S. M. Faber (Springer-Verlag, New York) p. 276.
Dressler, A. and Gunn, J. E.   1982 *Ap. J.*, **263**, 533.
_____1983 *Ap. J.*, **270**, 7.
Dressler, A., Gunn, J. E., and Schneider, D. P.   1985 *Ap. J.*, **294**, 70.
Ellis, R. S.   1987 in *IAU Symp. No. 124, Observational Cosmology*, ed. A. Hewitt, G. Burbidge, and L. Z. Fang (Reidel, Dordrecht) p. 367.
Ellis, R. S., Couch, W. J., MacLaren, I. and Koo, D. C. 1985 *MNRAS*, **217**, 239.
Fitchett, M. and Webster, R.   1987 *Ap. J.*, **317**, 653.
Gunn, J. E.   1987 in *Nearly Normal Galaxies: From the Planck Time to the Present*, ed. S. M. Faber (New York:Springer-Verlag) p. 455.
Haynes, M. P. and Giovanelli, R.   1986 *Ap. J. (Lett)*, **306**, L55.
Huchra, J. 1988 in *IAU Symp. No. 130, Evolution of Large Scale Structures in the Universe*, in press.
Kaiser, N. 1987 *MNRAS*, **227**, 1.
Koo, D. C.   1981 *Ap. J. (Lett)*, **251**, L75.
_____ 1986 *Ap. J.*, **311**, 651.
Kron, R. G.   1982 *Vistas in Astronomy*, **26**, 37.
MacLaren, I., Ellis, R. S., and Couch, W. J. 1987 *MNRAS*, in press.
Mathieu, R. and Spinrad, H.   1981 *Ap. J.*, **251**, 485.
Osterbrock, D. E.   1984 *Ap. J. (Lett)*, **280**, L43.
Pence, W.   1976 *Ap. J.*, **203**, 39.
Wirth, A. and Gallagher, J. S. 1980 *Ap. J.*, **242**, 469.

THE RISE AND FALL OF THE ACTIVE GALACTIC NUCLEI

A. Cavaliere
II Università di Roma, Italy
P. Padovani
Università di Padova, Italy; STScI, Baltimore, U.S.A.
F. Vagnetti
II Università di Roma, Italy

ABSTRACT: A review of the fall and of the rise stage in the QSO evolution, that stresses the emerging connnections with the galactic histories.

1. INTRODUCTION

Morphological and spectroscopic studies of the environment of QSOs, whenever possible at *low or moderate z*, confirm the long surmised link of these objects with *nuclei of galaxies* in a stage of activity or hyper-activity (for a review, cf. Fried 1986). In fact, the QSOs, like the local AGNs, appear to be located at the centers of their host galactic systems.

The bodies of these hosts often look disturbed or interacting (cf. Hutchings et al. 1984, Smith et al. 1986, Yee this Workshop), and are suspected of being anomalously bright by simultaneous starburst activity. The next question is: to what extent are the *high-z* QSOs associated with the "formation" era of their host protogalaxies, when massive gravitational wells first took shape in an even more hectic environment.

Two areas of overlap are hence emerging between the births and evolutions of QSOs and of galaxies, both of intriguing complexity. This Workshop has stressed complexity from the galactic end: galaxy formation appears to be a long, laborious process including growth and mergings of protogalactic clumps, disk collapse, spiral patterns looming out, with star formation going on all the way since the clump era to be often rekindled by star bursting after interactions. I am to stress complexity from the QSO end.

2. FALL

Since about 20 years (Schmidt 1968) we have growing evidence that the QSO population underwent "evolution": the comoving luminosity function $N(L,t)dL$ depends very effectively on the cosmic epoch t, and in fact – at given L – it *falls*

$$\frac{\partial N}{\partial t} < 0 \qquad (2.1)$$

with t increasing along the light cone to us, with the short time scale $\tau_{eff} \simeq 1$ Gyr.

R. G. Kron and A. Renzini (eds.), Towards Understanding Galaxies at Large Redshift, 285–297.
© *1988 by Kluwer Academic Publishers.*

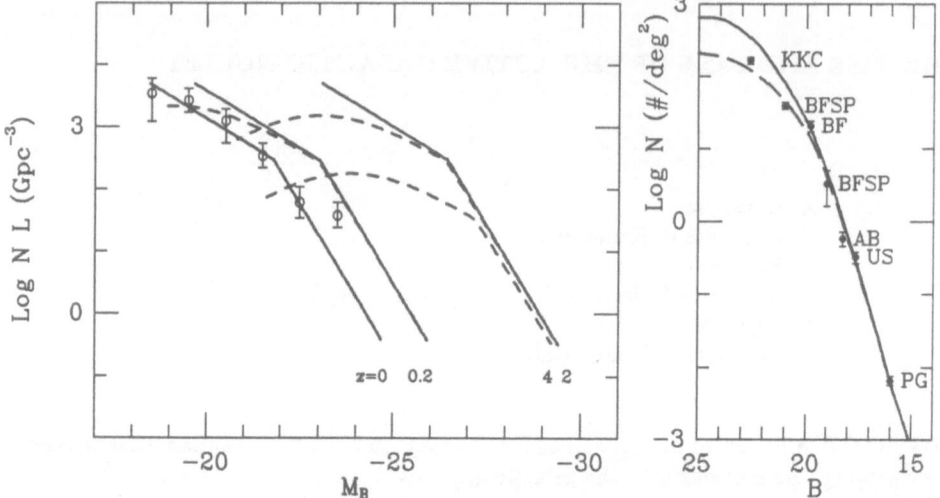

Fig. 1. *Left*: the evolution of the luminosity function $N(L, z)$ for two luminosity evolution models: uniform with $\tau_L \simeq 3$ Gyr (continuous lines); and differential, after Cavaliere et al. 1985 (dashed lines). From $z \sim 5$ to 2, the QSOs are born with $\tau_S = 1$ Gyr. Note how number increase and luminosity decrease produce near invariance of the population with $M_B \lesssim -27$ as $\tau_S \simeq \tau_L/(\gamma - 1)$. Local data by Cheng et al. 1985. *Right*: the integral counts for the same models. Data points: KKC: Koo et al. 1986; BFSP: Boyle et al. 1987 and references therein; BF, AB: Marshall et al. 1984; US: Mitchell et al. 1984; PG: Schmidt and Green 1983. Note that the differential model minimizes also the XRB (Cavaliere et al. 1985), in keeping with the arguments by Cavaliere et al. 1981, De Zotti et al. 1982.

The evidence once relied upon the steeply rising number counts $N(< B) \propto \text{dex}(0.9\,B)$ out to magnitudes $B \simeq 19$, but now is increasingly based upon direct observation of the resolved $N(L, z)$ (Boyle et al. 1987 using multi-objects spectroscopy resolve $\Delta z \simeq 0.5$ out to $z \simeq 2.2$). The data are consistent with a "luminosity evolution" (Mathez 1976, Braccesi et al. 1980): *as if* the distribution per magnitude interval $\propto N(L, z)L$ shifted bodily to lower L with z decreasing, by a factor $\sim 10^{-2}$ from $z \simeq 2.2$ to $z \simeq 0$ (cf. fig. 1).

The simplest implication would have each individual distant QSO dimming down to become a Seyfert 1, 1.5 in our local environment: as pointed out by Cavaliere, Giallongo and Vagnetti 1985 and by Weedman 1986, such a continuity is consistent with the data (to within their uncertainties, $\pm 50\%$) not only as for the total numbers at $z \simeq 2$ vs. $z \simeq 0$, but even as for the double-power-law shape of the luminosity functions (or rather the sections that are actually observed, cf. Cheng et al. 1985). The individual time scale required for such a dimming ($\dot{L} < 0$) is $\tau_L \equiv L/|\dot{L}| \simeq 3$ Gyr at high luminosity, corresponding to a scale $\tau_{eff} = \tau_L/(\gamma - 1) \simeq 1$ Gyr for a population with the observed luminosity distribution in the form of a steep power-law at the bright end, $N(L) \propto L^{-\gamma}, \gamma \simeq 3.5$ (cf. fig. 1).

But thinking of it, an individual time scale of order 3 Gyr is rather curious, being $\ll t_o$ and yet $\gg$ the natural time scale associated with an accreting black hole. In fact, given accretion onto a massive black hole as the most appealing prime mover to energize QSOs on account of stability, inevitability and efficiency $\eta \sim 10^{-1}$ (cf. Rees 1984), the basic time scale is $\eta t_E = \eta c \sigma_T / 4\pi G m_p \sim 4 \; 10^{-2}$ Gyr. This actually governs the simplest accretion flow: self-limited at the Eddington luminosity, from an unlimiting supply, to yield $L \propto exp(t/\eta t_E)$.

To force in an intermediate scale $r_{\dot{L}} \simeq 3$ Gyr it takes a "Great Coordinator": a limited mass stockpile on the way of exhaustion, or decreasing social interactions of the host galaxies with companions failing to replenish the supply, or suitable initial conditions ordering the sequence of black holes formation, have been envisaged to drive the activity patterns of the *individual* objects, and in fact underly the three major classes of models for the *population* evolution.

The diversity of these scenarios may seem surprising, but in fact it is permitted by the structure of the continuity equation

$$\frac{\partial N}{\partial t} = S(L,t) - \frac{\partial(\dot{L}N)}{\partial L} \qquad (2.2)$$

(Cavaliere, Morrison and Wood 1971) governing the population evolution $\partial N/\partial t < 0$. The latter is driven by a combination of the source function S describing destruction (or production) of objects, with the average shifts in luminosity (dimming or brightening, $\dot{L} <$ or > 0) of the objects described by the 2nd term on r.h.s. The literature existing (Cavaliere et al. 1983, Blandford 1986, Cavaliere and Vagnetti 1987) shows that reasonable but diverse combinations of the 1st and 2nd term on r.h.s. can produce very close forms and behaviours of $N(L,t)$, consistent with the data.

3. CONSTRAINTS TO THE ACTIVITY PATTERN

This situation stimulates interest in constraining the individual activity patterns to undo the tangle. In fact, the mass accreted will be constrained with the coming of age of nuclear mass determinations: these may be carried out through the relationship $M \propto \alpha v^2 d/G$ ($\alpha \simeq 1$) both in *active* and in *inactive* galaxies with the attendant and complementary uncertainties (Woltjer 1959, Sargent et al. 1978, Dibai 1984, Joly et al 1986, Wandel and Mushotzky 1986, Kormendy 1987). The resulting constraints have been worked out in some detail by Cavaliere and Padovani (in preparation) comparing two simple relationships for the specific luminosity L/M in the AGNs, conveniently normalized in the form of the Eddington ratio L/L_E of the bolometric to the Eddington luminosity $L_E = Mc^2/t_E = 1.3 \; 10^{46} M_8$ erg/s.

In the Appendix we shall expand somewhat on these relationships to convey a feeling for their robustness, confirmed by detailed numerical work on the basis of eq. 2.2. But their basic content is intuitive: given that the average L decreases with the cosmic t increasing while $M \propto \int dt L(t)$ builds up mainly at the beginning, the longer the activity lasts in each nucleus, the smaller L/M will be locally. From the other end, the shorter is the activity, the more numerous will be the generations of AGNs required to reproduce the apparent statistical invariance of the objects at all $z \lesssim 2$; easily all galaxies reasonably bright may be involved as hosts.

The first relationship

$$\frac{L}{L_E} = \frac{\eta t_E L}{\int dt L(t)} \quad , \qquad (3.1)$$

self-evidently holds in general for an individual object deriving its power from accretion. It yields the following results (see Appendix for details) for the three basic activity patterns that span the domain of interest:

C (for *continuous*). Starvation caused by a limited and not renewable mass stockpile characterizes the extreme case of literal luminosity evolution, with each object dimming *continuously* (after a short formation phase at $z > 2$). Here $2\,10^{-4}\eta_{-1} \lesssim L/L_E \lesssim 5\,10^{-3}\eta_{-1}/T_1$, the upper limit being provided by the extreme case of differential evolution, with no change at the faint end since the turning-on epoch (look-back time $T \equiv T_1\,10$ Gyr).

R (for *recurrent*). An intermediate class is provided by discontinuous activity of a larger set of galaxies, with episodes lasting $t_i < r_{\underline{i}}$ and *recurring* after $t_r > t_i$ on average, for a cumulative effective duration t_e: this behaviour may be driven by accretion rekindled by declining episodes of interactions of the host with companion galaxies (Norman and Silk 1983, Gaskell 1985; for recent evidence cf. Smith et al. 1986, Yee and Green 1987), and can emulate case C on average as numerical work proves. Here $L/L_E \sim 5\,10^{-2}\eta_{-1}/t_{-2,e}$.

E (for *event*). At the other extreme, activity may represent a *single, short event* in the host galaxy lifetime, equivalent to see mainly the black hole formation phase (Blandford 1986, Cavaliere and Vagnetti 1987): if, e.g., this proceeds at the fiducial (but not guaranteed) rate $\dot{M} \sim \eta^{-1}M_E$ for a duration inversely correlated with epoch in many successive generations of different active nuclei, it can emulate a luminosity evolution yet implying $L/L_E \sim 1$ at all z. Suitable initial conditions are required: a tuned production rate of black holes or a tuned sequence of latency times.

Fig. 2 in its lower part illustrates the three kinds of mass distribution expected from the above activity patterns. In its higher part, it visualizes also the corresponding mass distributions predicted for currently inactive nuclei.

To link in with the statistical approach to follow, note that eq. 3.1 has also an approximate statistical meaning (to within factors $1/2 \div 2$) when computed for the representative objects near the "break" of the luminosity functions.

On the fully statistical side, one can take up (see Appendix) the approach of Soltan (1982) and of Phinney (1983), to relate the average Eddington ratio with the observed density of light (dominated by $B \simeq 20 \pm 1$, at $z \simeq 1.5$), or with the mass density of AGNs ρ_S associated to the former under the accretion scenario; the other variable is the total number N_a/δ of nuclei ever been active, that then can be evaluated:

$$\langle \frac{L}{L_E} \rangle = \frac{t_E N_a \langle L \rangle}{c^2 \delta \rho_S} \quad . \tag{3.2}$$

The relationship applies also to $z \neq 0$ as long as the objects involved contribute substantially to ρ_S, and is independent of the value – local, or at a fiducial z – of the bolometric corrections that appear both in L and in ρ_S.

Numerically, we find $\langle L/L_E \rangle \simeq 10^{-4}\eta_{-1}/\delta$ locally, upon using the following data: the optically-selected QSO counts, as given and reviewed by Koo et al. (1986), yielding $\rho_S \simeq 7\,10^{13}\,\kappa_1/\eta_{-1}M_\odot$ Gpc^{-3}, with account duly taken of the "big bump" redshifted into the B band; the luminosity function of the Seyfert 1 and 1.5 galaxies (Cheng et al. 1985), to derive $\langle L \rangle \simeq 2\,10^{44}\kappa_1$ erg/s and $N_a \simeq 6\,10^3$ Gpc^{-3}, a few % of the bright galaxies.

When $\delta \simeq 1$, then eq. 3.2 yields – to within a factor $\simeq 2$ not worse than the data uncertainties – the same result as eq. 3.1 for patterns of class C, a check for the consistency

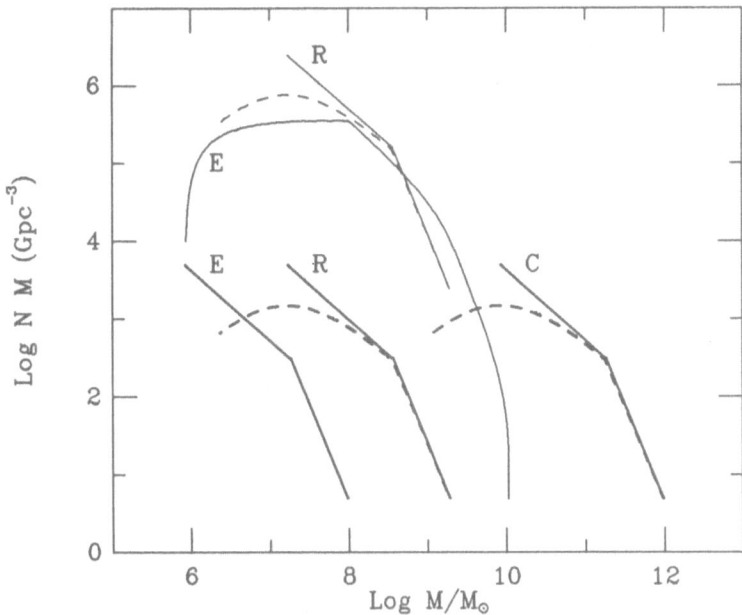

Fig. 2. The mass distributions for Active (lower part) and Inactive (upper part, thin lines) Galactic Nuclei predicted by the three activity patterns discussed in the text: C characterized by a local Eddington ratio $L/L_E \simeq 10^{-4}$, R by $L/L_E \simeq 5\ 10^{-2}$ and E by $L/L_E \simeq 1$ always. The basic luminosity functions are those of fig. 1.

of the two approaches. For patterns of class R, it yields for the total numbers of nuclei entertaining activity $N_a/\delta \sim 3\ 10^6$ Gpc^{-3}. But such extrapolations to local and weak emitters do not hold when the corresponding objects never contributed materially to ρ_S, as is the case with pattern of class E. For this case, eq. 3.2 gives instead $N_a/\delta \sim 7\ 10^5$ Gpc^{-3} already for $z \gtrsim 1.5$, the range providing about half of the contributions to ρ_S. All numbers scale $\propto H_o$ and are given for $H_o = 50$ km/s Mpc.

4. DISCUSSION

The main point is that $\langle L/L_E \rangle > 10^{-2}$ locally will constitute a threshold ruling out literal continuity, however well simulated by the evolutionary data. Recent data reductions and collations – in X-rays by Wandel and Mushotzky (1986) (but see Lawrence et al. 1987), and in the IR-optical-UV by Padovani and Rafanelli (1987) – seem to be settling into the range $10^{-2} \div 10^{-1}$. If this will hold out, then the range from the intermediate patterns R to the extreme case E will be selected. Telling signatures of these two alternatives follow.

Pattern E – that ironically tends to obliterate two features long held typical of black hole activity, stability and recurrence – will be enforced upon us in its neatest form by finding: values $L/L_E \sim const \sim 1$ holding for nuclei active at any z; and masses of local nuclei that should result systematically larger in currently *inactive* nuclei. This finding would hold at any z, mirroring the observed increase of average luminosities with increasing

z: locally, the overall mass distribution of the inactive nuclei should extend up to values $\sim 10^2$ larger than those of the local AGNs (cf. fig. 2).

Patterns of class R are consistent with the morphological evidence (cf. Smith et al. 1986, Yee and Green 1987) of frequent associations of nuclear activity at intermediate powers and redshifts with *distortions* of the galactic bodies: presumable signs of "social interactions" of the host with companion galaxies in groups. It will require an evolutionary increase of L/L_E with z from a local value $\sim 5 \ 10^{-2}$ up to ~ 1 or more out to $z \simeq 2$. It also implies that holes more massive than $10^6 \div 10^7 M_\odot$ will be found in all galaxies reasonably bright, with $L_B > 0.1 L_*$.

Patterns of types R and E are constrained from the other end by observations such as those of Smith et al. (1986) that tend to associate *bright* AGNs with *bright* galaxies (being yet unclear whether the body's enhanced luminosities indicate larger masses or instead enhanced starburst activities). In fact, as stressed by Phinney (1983), the total number of nuclei ever active is ultimately limited by the flat shape of the galaxy luminosity function (cf. Felten 1986). Already when $\langle L/L_E \rangle \gtrsim 10^{-2}$ holds, eq. 3.2 predicts that galaxies with $L_B < 0.6 \ L_*$ be involved; similarly, $L_B \lesssim 10^{-1} L_*$ when $\langle L/L_E \rangle \gtrsim 5 \ 10^{-2}$ holds.

But patterns R easily agree with galaxies bright by star bursting, as they hinge for the nuclear fueling upon the same interactions conducive to trigger extensive rejuvenation of the star populations: growing evidence indicates strongly a similar connection for the radiogalaxies (cf. Windhorst et al. 1985, Heckman et al. 1986, Danese et al. 1987). As an addition particularly suited to this Workshop, Cavaliere, Giallongo and Vagnetti (in preparation) consider the effects (random variances plus systematic trends) introduced by the emission lines and by different kinds of photometric uncertainties, and renormalize upwards to 4 Gyr the fall timescale. So the latter is approaching the timescale for the evolution of starburst galaxies found necessary by Danese et al. 1987 to explain the counts of mJy radiosources and the IRAS counts.

Pattern E, instead, is less directly related to interactions, whilst it involves at least as many galaxies ever gone through an active phase: in fact, from eq. 3.2 some 10^2 generations are required down to $z \simeq 1.5$, and at least some additional $2 \ 10^2$ from there to now.

A final caveat on telling apart patterns R from E on the basis of local values of L/L_E only: E is not necessarily associated with $L/L_E = 1$. For example, Lightman, Zdziarski and Rees (1987) find $L/L_E \sim 0.5$ even for $\dot{M} \sim 10 \dot{M}_E$ in spherical sources powerful at $h\nu > 1/2$ MeV. Then L/L_E in case R should not exceed $5 \ 10^{-3}$ *and* evolve much with z.

In any case, as the evidence shifts towards values $L/L_E \gtrsim 10^{-2}$, less and less room is left to any sizeable population – as sometimes envisaged – of active nuclei that may be selected against by unfavourable orientations either of beamed emission or of accretion disk inclinations.

5. RISE

All scenarios must face a beginning, that first range Δt when the early QSOs began to populate the $L - z$ plane:

$$\frac{\partial N}{\partial t} > 0 \qquad (5.1)$$

at some $z \gtrsim 2.5$. Looking back to higher z, this *rise* in cosmic time will appear as a decline, over a z-range corresponding to a time scale $\tau_{\dot{N}}$, say.

The relevant data are changing on a scale of $\sim 1/2$ yr. A paucity of high-z objects especially of medium and low luminosities was long inferred from preliminary data including Osmer (1982); it was confirmed at magnitudes $B \gtrsim 20$ from deep surveys based on multicolor selection (Koo, Kron and Cudworth 1986), and independently from slitless spectroscopy searches at $M_B \simeq -25$ and (over larger areas) at $M_B \simeq -27$ (Schmidt, Schneider and Gunn 1986 a,b). Face values from a synoptic compilation of many slitless surveys (Véron 1986) again confirmed a decline by $\sim 1/5 - 1/10$ of the QSO surface density around $z \simeq 2.5$ (sharper in the data from large telescopes including weaker objects), followed by a slower decrease out to $z \simeq 3.5$. On the other hand, the hunt for high-z QSOs was rekindled by slitless searches over large areas by Hazard and McMahon (1985), and has been taken up at an accelerating pace with various techniques by Hazard, McMahon and Sargent (1986), Anderson and Margon (1987), Schmidt, Schneider and Gunn (1987a, b), Warren et al. (1987) .... , to show that high-z QSOs do exist with spectra similar to their closer counterparts. These data are far from being assessed as for their statistical weight and their selection bias, and the detailed structure of the decline will remain for some time under scrutiny. Meanwhile, two alternative descriptions are emerging.

Koo and Kron (1987) describe the smooth behaviour of the data in terms of a $N(L, z)$ that at high-z has a rather flat, extended power-law shape, but at its bright end bends over to a steeper slope at progressively lower L with z decreasing: a sharply differential evolution with the powerful sources burning out fast while the weaker ones are very longeve (cf. Cavaliere et al. 1983). The apparent decline is here interpreted in terms of the observable flux *sliding down*, as it were, along $N(L)$ when z increases.

The other alternative, closer to a scenario of short-lived objects and concerned with the requirement of limiting the AGN contributions to the X-Ray Background to under 50%, envisages the *amplitude* of the luminosity functions declining for $z \gtrsim 2$, while the average L still increase (see fig. 1). If so, the rise time indicated for the apparent turn-on in cosmic time of the bulk of the high-z QSOs – i.e. $\tau_S \sim 1$ Gyr, intriguingly longer than ηt_E yet shorter than $H^{-1}(z \simeq 2.5)$ – stimulates questions concerning any connection with the process of galaxy "formation". By a long-standing surmise one envisages a high rate of deep collapses and rapid accretions during the era of protogalaxy collapse and settling; more specifically, when definite and sufficiently massive gravitational centers first took shape in a highly non-steady environment. In this vein, one qualitative point is that any such QSO-protogalaxy connection must be rooted at $z > 5$: not only this is obvious for the QSOs now being found at $z > 4$, but it must hold also for those at $z \sim 2 - 3$ *if* the bulk of the decline is really sharp.

6. CONSTRAINTS TO THE RISE

Quantitatively, the source $S(t)$ may be connected (Cavaliere and Szalay 1986) with the rate of galaxy formation dN_g/dt as schematically visualized by:

$$S(t) = \varepsilon \frac{dN_g}{dt}(t' = t - t_q) \qquad \varepsilon < 1 \tag{6.1}$$

The effective rise time τ_S is set by the balance of three effects.

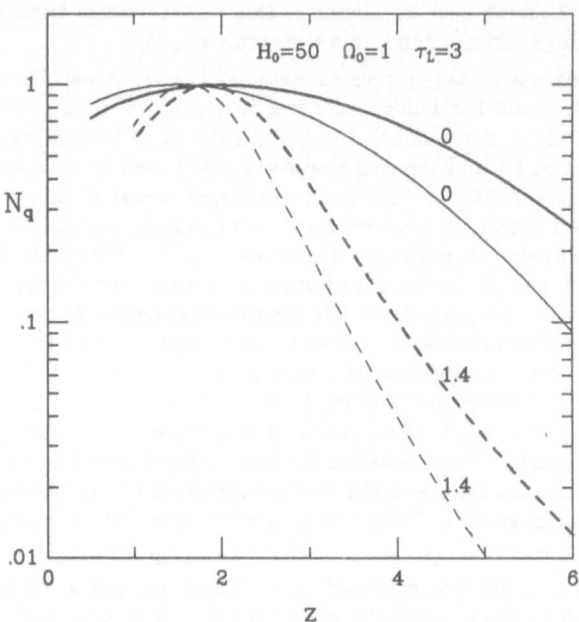

Fig. 3. The high-z behaviour of QSO number as given by the birth time scale discussed in § 6. Continuous lines: $N_g(z)$, the normalized PS number of galaxies with M_g ending collapse at $z = 2$ (thick line) or 1.5 (thin); for $\langle t_q \rangle$, $\sigma = 0$, $N_q(z) \propto N_g(z)$. Dashed lines: resulting QSO numbers, with the lags indicated in Gyr and $\sigma = 0.5$ Gyr. All curves normalized to their maximum. $\Omega_o = 1$ (for $\Omega_o < 1$ all behaviours are flatter).

First stands the unavoidable lag t_q of QSO turn-on relative to the host protogalaxy formation, corresponding to the time to build up a massive black hole, with a possible addition for output rise to detectability: an average value $\langle t_q \rangle \sim 1$ Gyr at $z \simeq 2.5$ tends to steepen the QSO turn-on as the formations of the parent objects are mostly confined to considerably earlier cosmic times when all time scales were shorter.

Second, this lag will be unavoidably subjected to a statistical distribution with dispersion $\sigma = \langle (t - \langle t_q \rangle)^2 \rangle^{1/2}$, reflecting variances of the initial conditions, and alternative routes to collapse (from prompt dynamical infall, to Eddington-limited accretion, to slow dissipation of angular momentum j or of random kinetic energy $\langle v^2 \rangle$, before the runaway final collapse, cf. Rees 1984): this instead tends to smooth out the rise time.

Third, an intrinsic smoothing effect arises even if we restrict the range of the host masses to a narrow range around a given M_g (with a value $\sim 10^{12}$ or $\sim 10^9 M_\odot$ depending on the range of z where the black hole formation begins, see below); even so, the z range for triggering a QSO is unavoidably widened by the actual distribution of collapse times for the nucleation of a given M_g to occur. This is built in the expressions for $N_g(M_g, t)$ to be used in eq. 6.1, that are derived from any model of galaxy formation based on the

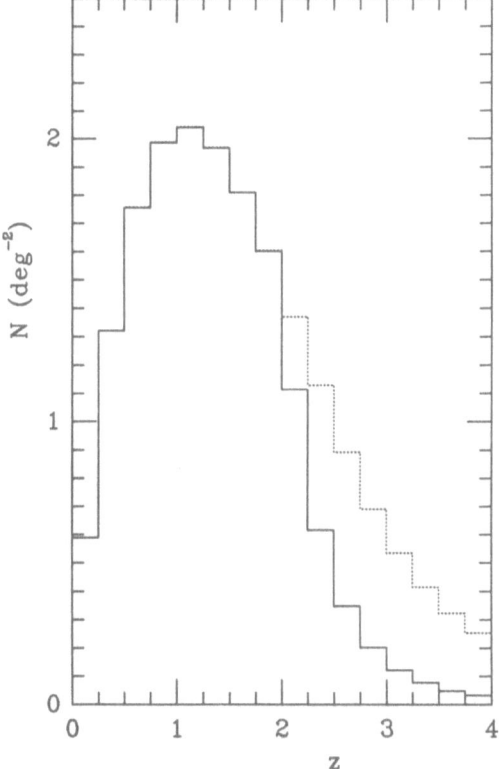

Fig. 4. The z-distribution of the QSOs with $B \leq 20$ as predicted by the differential evolution model in fig. 1 (continuous line) compared with a model otherwise similar except for $S \equiv 0$ (dotted line).

gravitational instability of a random field of initial perturbations $\delta\rho/\rho$ (see Bardeen et al. 1986).

In the framework of the Hierarchical Clustering scenario at each epoch a typical mass $M_c(t)$ preferentially collapsed, but a decreasing number of larger M_g also collapsed being associated with larger initial perturbations, while a number of smaller M_g still survived reshuffling into larger units: these effects shape $N_g(M_g, t)$. In the Appendix, it is recalled and updated from Cavaliere and Szalay 1986 the simplest yet representative model for N_g and dN_g/dt.

On this basis, fig. 3 visualizes how simultaneous formation of the bulk of the QSOs and of fully grown galaxies ($M_g \sim 10^{12} M_\odot$ *ending* formation at $z = 2$ or $z = 1.5$, lag $\langle t_q \rangle = 0$) even with dispersion $\sigma = 0$ would set a scale for the QSO number $N_q(z)$ too slow compared with that suggested by the present data. Short rise times require $\langle t_q \rangle \gtrsim 1$ Gyr for $\Omega_o = 1$ (most proto-hosts then would form at $z_g > 5$), or with $\langle t_q \rangle \simeq 3$ Gyr for $\Omega_o = 0.2$ $(z_g \simeq 10)$. In addition, the dispersion must be limited: $2\sigma \lesssim \langle t_q \rangle$. Such a relatively uniform formation process may suggest for the bulk of the holes Eddington-limited accretion with its duration $\propto \ln M$. Bright objects forerun if, plausibly, large masses collapse more directly or as stockpiles provide larger accretion rates.

If one accepts values $z \gtrsim 5$ for the onset of an extensive hole formation, then in keeping with the thrust of the Hiererchical Clustering scenario (Silk and Wyse 1986, Baron and White 1987) embodied in eqs. A.2, A.3 one expects the formation process to begin within $10^9 - 10^{10} M_\odot$ clouds or clumps that are gradually coalescing (with much dissipation) to build up a protogalaxy: the host masses involved will scale down from the values $\sim 10^{12} M_\odot$ at $z \simeq 2$ following $M_g \propto t^{2/3a} \propto (1+z)^{-6/(n+3)}$, where $n \simeq -2$ is suggested by the Cold Dark Matter scenario (cf. Blumenthal et al. 1984).

To account for the effects from the shape of the luminosity function, like the "sliding down" that always exists, one may solve eq. 2.2 complete with a source $S(L,t) > 0$ as above with a rise time $\tau_S \simeq 1$ Gyr, and an effective, average $\dot{L} < 0$ with time scale $\tau_{\dot{L}} \simeq 3$ Gyr (whether due to a class R or to a class E pattern). The resulting luminosity functions and counts are illustrated in fig. 1; fig. 4 shows the expected z-distribution. Note in fig. 1 how the ultrabright objects, once turned on, tend to compensate the true number rise from $z \sim 4$ to 2 with the apparent fall due to luminosity evolution; the balance is good under the condition $\tau_S \simeq \tau_{\dot{L}}/(\gamma - 1)$, numerically satisfied by the values set independently for τ_S, $\tau_{\dot{L}}$ and $\gamma \simeq 3.5$.

Summing up, the *closest* epoch for the triggering events is that yielding – against the dispersions associated to $N_g(M_g, t)$ and to hole formation – a value for τ_S matching the intrinsic $\tau_{\dot{N}}$ set by the data. At high z the limiting factor was the small mass of the clumps then prevailing: we envisage an *upper* limit $z \lesssim 7$ $(\Omega_o = 1)$ for the bright objects formed or forming in clumps with total mass $\sim 5\ 10^9 M_\odot$.

7. DISCUSSION AND CONCLUSIONS

The above is to stress the interest attached to the behaviour of the high-z QSOs, likely to constitute one of the few probes visible into those early epochs (cf. Tyson, Koo, this Workshop, as for the brightness of primeval galaxies).

A look at alternative explanations of the apparent decline. Obscuration, preferentially by absorbtion in dusty intervening galaxies (cf. Bechtold et al. 1984, Ostriker and Heisler 1984, Netzer 1985), would be smeared out in z by statistical spreads in dust production and variance in galaxy cross-section along the line of sight; this would imply much precursory reddening not evident in the spectra of high-z QSOs (Cavaliere and Szalay 1986, Wright 1986). On the other hand, it would leave unexplained the similar decline in the radio band, especially of flat-spectrum sources (Peacock 1985), among which the radio-loud quasars are well represented. Finally, many medium and low L objects beyond $z \simeq 2$, even though optically obscured, still would overproduce the XRB (cf. Cavaliere and Vagnetti 1987).

The latter constraint applies also to explanations invoking a systematic change (of which there is little evidence yet) in the line spectrum only, as might be caused by heavy intrinsic extinction or filtering of a canonical primary emission, such as to elude conventional identification criteria.

So one is led back to consider extensive turning on of QSOs largely originated at pre-galactic epochs and in protogalactic clumps, probably too weak and too faint then to be observed by their star light only. If this is so, one may entertain two speculations: several active "nuclei" may form in some protogalaxies, implying spectral signatures (cf. Gaskell 1983) of binary, or even multiple, objects among high-z QSOs; some isolated objects may be expected after all, straggling from their original protogalaxies.

Acknowledgements. We thank G. De Zotti, E. Giallongo, R. Kron and C. Norman for many useful discussions.

APPENDIX

Values of L/L_E.

Pattern C. In the simplest case of a uniform time scale $\tau_{\dot{L}}$ for all objects, the result is $L/L_E = \eta t_E L/\tau_{\dot{L}} L_{max}$ (the integral in eq. 3.1 is evaluated as $L_{max}\tau_{\dot{L}}$ for all $\Delta t > \tau_{\dot{L}}$ from $z \simeq 2.2$, because the dimming is so rapid); locally $L/L_{max} \simeq 10^{-2}$ holds, so that $L/L_E \simeq 2\ 10^{-4}\eta_{-1}$ is to be expected. Little changes obtain in the Eddington ratios (within a factor of 2) for a non-uniform evolution, e.g., if $\tau_{\dot{L}}$ were itself to change $\propto t$ corresponding to $L(z) \propto (1 + z)^k$, even with $k \simeq 3.5$ as favored by Boyle et al. 1987. A more sizeable increase of L/L_E for the weak objects obtains with a differential luminosity evolution: in the extreme case of no change at the faint end since the formation epoch (look-back time $T \equiv T_1\ 10$ Gyr), an upper limit $L/L_E \leq \eta t_E/T \simeq 5\ 10^{-3}\eta_{-1}/T_1$ obtains.

Pattern R. Here $L/L_E = \eta t_E L/t_e L_e$ holds: if the (average) dimming is fast, then the factor L/L_e may be down to $\sim 10^{-2}$, but by the same token the weighted $t_e \to t_i$ (the mass is contributed mainly by the first few episodes, including the initial hole formation); the result is $L/L_E \sim 5\ 10^{-2}\eta_{-1}/t_{-2,e}$.

Pattern E. $L/L_E \sim$ const ~ 1 at all z by construction in the simplest case of accretion at the Eddington limit.

To derive eq. 3.2, assume $M \propto L$ to begin with; then $\langle L/L_E \rangle \propto \rho(L)/\rho(M)$, where $\rho(L)$ and $\rho(M)$ are the densities of observed bolometric luminosity and of the inferred mass density, for active nuclei. Under the accretion scenario, express the latter quantity as $\rho(M) = \delta\rho_S$, where $\rho_S = (4\pi\kappa/\eta c^3) \int dz\ (1 + z) \int dS\ S\ N(S,z)$ is the *total* mass density accumulated in AGNs, computed from the bolometric fluxes κS weighted with the differential number counts $N(S)$ (the dominant contribution comes from the range $B = 20\pm1$); the "duty cycle" δ is obviously bound to be ≤ 1, with the upper limit corresponding to literal continuity. Finally, express the luminosity density as $\int dLN(L)L = N_a\langle L \rangle$ where $\langle L \rangle$ is the average luminosity of AGNs and N_a is their comoving number density. Summing up, we find

$$\langle \frac{L}{L_E} \rangle = \frac{t_E N_a \langle L \rangle}{c^2 \delta \rho_S} \quad . \tag{A.1}$$

The result is invariant to within a factor 1 to 2 if $M \propto L^\alpha$ with $0 \lesssim \alpha \lesssim 1$ were assumed instead of the linearity supported by the observations of Padovani and Rafanelli (1987).

A simple model for $S(t)$ in eq. 6.1, representative for the framework of hierarchical clustering, is based on the mass distribution (Press and Schechter 1974)

$$N_g(M_g, t) \propto [M_g/M_c(t)]^a e^{-[M_g/M_c(t)]^{2a}/2}. \tag{A.2}$$

Here $a = 1/2 + n/6$ in terms of the effective index n for the power spectrum of the initial linear perturbations, and the typical protogalactic mass settling at t scales like

$$M_c(t) = M_*(t/t_*)^{2/3a} \left[\frac{1+A}{1+A(t/t_*)^{2/3}} \right]^{1/a} \qquad (A.3)$$

where the term $A = (1 - \Omega_o)(H_o t_o)^{2/3}/(\pi \Omega_o)^{2/3}$ describes the slowing down or effective halt of collapses in an open universe (Cavaliere et al. 1977). For use in eq. 6.1, in both eqs. A.2 and A.3 one has to replace t (the *retarded* time when QSOs turn on) with $t' = t - t_q$ (the time when the actual formation started), and $N_g(M_g, t)$ is to be integrated over the relevant mass range. In addition, a convolution with the probability distribution of t_q is implied, see Cavaliere and Szalay 1986 for details.

REFERENCES

Anderson, S.F., and Margon, B. 1987, *Nature* **327**, 125.

Bardeen, J.M., Bond, J.R., Kaiser, N. and Szalay, A.S. 1986, *Ap. J.* **304**, 15.

Baron, E., and White, S.D.M. 1987, preprint.

Bechtold, J., Green, R.F., Weymann, R.J., Schmidt, M., Estabrok, F.B., Sherman, R.D., Wahlquist, H.D. and Heckman, T.M. 1984, *Ap. J.* **281**, 76.

Blandford, R. D. 1986, in *Quasars*, Proc. I.A.U. Symp. Nr. **119**, Bangalore, India, p. 295.

Blumenthal, G., Faber, S., Primack, J. and Rees, M.J., 1984, *Nature* **311**, 517.

Boyle, B. J., et al. 1987, *M.N.R.A.S.* **227**, 717.

Braccesi, A., Zitelli, V., Bonoli, F., and Formiggini, L. 1980, *Astr. Ap.*, **85**, 80.

Cavaliere, A., Danese, L. and De Zotti, G. 1977, *Ap. J.* **217**, 6.

Cavaliere, A., Danese, L., De Zotti, G. and Franceschini, A. 1981, *Space Sci. Rev.* **30**, 101.

Cavaliere, A., Giallongo, E., Messina A., and Vagnetti, F. 1983, *Ap. J.*, **269**, 57.

Cavaliere, A., Giallongo, E., and Vagnetti, F. 1985, *Ap. J.*, **296**, 402.

Cavaliere, A., Morrison, P. and Wood, K. 1971, *Ap. J.* **170**, 223.

Cavaliere, A. and Szalay, A.S. 1986, *Ap. J.* **311**, 589.

Cavaliere, A., and Vagnetti, F. 1987, *Workshop "Supermassive Black Holes"*, Fairfax, Virginia, to appear in the Proceedings.

Cheng, F. Z., Danese, L., De Zotti, G., and Franceschini, A. 1985, *M.N.R.A.S.*, **212**, 857.

Danese, L., De Zotti, G., Fasano, G., and Franceschini, A. 1986, *Astr. Ap.*, **161**, 1.

Danese, L., De Zotti, G., Franceschini. A., and Toffolatti, L. 1987, *Ap.J. (Letters)*, **318**, L15.

De Zotti, G., Boldt, E. A., Cavaliere, A., Danese, L., Franceschini, A., Marshall, F. E., Swank, J. H., and Szymkowiak, A. E. 1982, *Ap. J.*, **253**, 47.

Dibai, E. A. 1984, *Soviet Astr.*, **28**, 245.

Felten, J. E. 1986, preprint.

Fried, J. 1986, in *Structure and Evolution of Active Galactic Nuclei*, p.309, Giuricin, Mardirossian, Mezzetti and Ramella eds., Reidel, Dordrecht.

Gaskell, C.M. 1985, *Nature* **315**, 386.

Hazard, C. and McMahon, R. 1985, *Nature* **314**, 238.

Hazard, C., McMahon, R. and Sargent, W.L.W. 1986.

Heckman, T.M., et al. 1986, *Ap.J.* **311**, 526.

Hutchings, J., Crampton, D., Campbell, B. 1984, *Ap. J.* **280**, 41.

Joly, M., Collin-Souffrin, S., Masnou, J. L., and Nottale, L. 1985, *Astr. Ap.*, **152**, 282.

Koo, D.C. and Kron, R.G., 1987, preprint.

Koo, D.C., Kron, R.G., and Cudworth, K.M. 1986, *P.A.S.P.*, **98**, 285.

Kormendy, J. 1987, *Workshop "Supermassive Black Holes"*, Fairfax, Virginia, to appear in the Proceedings.

Lawrence, A., Watson, M. G., Pounds, K. A., and Elvis, M. 1987, *Nature*, **325**, 694.

Lightman, A.P., Zdziarsky, A.A., and Rees, M.J. 1987, *Ap. J. (Letters)* **315**, *L113*.

Marshall, H.L., Avni, Y., Braccesi, A., Huchra, J.P., Tananbaum, H., Zamorani, G. and Zitelli, V. 1984, *Ap. J.* **283**, 50.

Mathez, G. 1976, *Astr. Ap.*, **53**, 15.

Mitchell, K.J., Warnock III, A. and Usher, P.D. 1984, *Ap. J. (Letters)* **287**, L3.

Netzer, H. 1985, *Ap. J.* **289**, 451.

Norman, C., and Silk, J. 1983, *Ap. J.*, **266**, 502.

Osmer, P.S. 1982, *Ap. J.* **253**, 28.

Ostriker, J.P. and Heisler, J. 1984, *Ap. J.* **278**, 1.

Padovani, P., and Rafanelli P. 1987, *Ap. J.*, submitted.

Peacock, J.A. 1985, *M.N.R.A.S.* **217**, 601.

Phinney, E. S. 1983, Ph. D. thesis, Univ. Cambridge, England.

Press, W.H. and Schechter, P. 1974, *Ap. J.* **187**, 425.

Rees, M. J. 1984, *Ann. Rev. Astr. Ap.*, **22**, 471.

Sargent, W. L. W., Young, P. J., Boksenberg, A., Shortridge, K., Lynds, C. R., and Hartwick, F. D. A. 1978, *Ap. J.*, **221**, 731.

Schmidt, M. 1968, *Ap. J.* **151**, 393.

Schmidt, M. and Green, R.F. 1983, *Ap. J.* **269**, 352.

Schmidt, M., Schneider, D.P. and Gunn, J.E. 1986a, *Ap. J.* **306**, 411.

——————— 1986b, *Ap. J.* **310**, 518.

——————— 1987a, *Ap.J. (Letters)* **316**, L1.

——————— 1987b, preprint.

Silk, S.J., and Wyse W.R.F.G. 1986, in *Structure and Evolution of Active Galactic Nuclei*, p.173, Giuricin, Mardirossian, Mezzetti and Ramella eds., Reidel, Dordrecht.

Smith, E. P., Heckman, T. M., Bothun, G. D., Romanishin, W., and Balick, B. 1986, *Ap. J.*, **306**, 64.

Soltan, A. 1982, *M.N.R.A.S.*, **200**, 115.

Véron, P. 1986, *Astr. Ap.* **170**, 37.

Wandel, A., and Mushotzky, R. F. 1986, *Ap. J. (Letters)*, **306**, L61.

Warren, S.J. et al. 1987, *Nature* **325**, 131.

Weedman, D.W. 1986, in *Structure and Evolution of Active Galactic Nuclei*, p.215, Giuricin, Mardirossian, Mezzetti and Ramella eds., Reidel, Dordrecht.

Windhorst, R.A., Miley, G.K., Owen F.N., Kron, R.G., and Koo, D.C. 1985, *Ap. J.* **289**, 494.

Woltjer, L. 1959, *Ap. J.*, **130**, 38.

Wright, E.L. 1986, *Ap. J.* **311**, 156.

Yee, H.K.C. and Green, R.F. 1987, preprint.

PROPERTIES OF GALAXIES NEAR QUASARS

H. K. C. Yee
Département de Physique
Université de Montréal
C.P. 6128, Succursale A
Montréal, PQ H3C 3J7, CANADA

ABSTRACT. Several results from extensive imaging surveys of fields
around quasars are summarized. These include the frequency of finding
close companions to low-redshift radio-quiet quasars, the quasar-galaxy
covariance amplitudes of radio-loud quasars from z=0.15 to z=0.65, and
the preliminary analysis of three-colour photometry from two clusters
of galaxies associated with quasars.

Quasars at redshifts < 1.0 provide excellent markers for discover-
ing galaxies and galaxy clusters at moderately high redshifts. A comprehensive program of imaging and spectroscopic surveys of galaxies in
quasar fields has been carried out over the past few years by myself
and collaborators. In this paper, recent results are summarized and
some new results are presented. Unless specified otherwise, H_0=50 km
sec^{-1} and q_0=0 are used throughout this paper.

1. LOW-REDSHIFT RADIO-QUIET QUASARS

Using imaging data from 37 PG quasars having 0.05<z<0.30, the net frequency of occurrence of close companions to quasars is derived (Yee,
1987). After correcting for chance projections, ~40% of the quasars
are found to have at least one associated companion brighter than ~-19
mag and within a projected radius of 100 kpc. Using the binary galaxy
sample of Turner (1976) as comparison, the frequency of finding a companion to a PG quasar is ~6 times greater than that for field galaxies,
but is similar to that of the lower luminosity Seyferts galaxies
(Dahari, 1984). The luminosity distribution of the companions is consistent with that drawn randomly from the luminosity function (LF) of
local normal galaxies. This indicates that the proximity of a bright
quasar produces relatively small effects on the properties of the companion. Spectacular examples of interaction with prominent tidal tails
and accompanying star formation episodes are probably rare or have
lifetime considerably shorter than that of the induced quasar activity.
No correlations between the companions and the quasars are found, sug-

R. G. Kron and A. Renzini (eds.), Towards Understanding Galaxies at Large Redshift, 299–303.
© 1988 by Kluwer Academic Publishers.

gesting that, while a companion may trigger the quasar, it does not determine the level of activity of the quasar.

2. THE ENVIRONMENTS OF QUASARS AT $z \sim 0.6$

Quantitative measures of the richness of galaxy environment of quasars can be estimated by deriving the quasar-galaxy covariance amplitude (Yee and Green 1987, hereafter, YG). Average galaxy counts in the sky obtained from control fields observed with the same procedures are used to correct for background galaxy contamination in the quasar fields. In order to compare the environments of quasars with those of galaxies and clusters of galaxies, we need to make the assumption that the galaxies associated with quasars have the same average LF as that of other galaxies. The background correction allows one to derive the LF of the associated galaxies at the redshifts of the quasars and the shape of the quasar-galaxy spatial covariance function. Data from the fields of 31 radio-loud quasars from Yee, Green and Stockman (1986, hereafter YGS) and Green and Yee (1984) are used to derive crude LFs of galaxies at three redshift bins with <z>=0.24, 0.42 and 0.61. The LFs are derived by averaging magnitude distributions of excess galaxies in the fields of quasars within the appropriate redshift range. These LFs are then assumed to be similar to the average LFs of galaxies at these redshifts. The normalization of the LFs is achieved by modelling the control field counts, using as inputs the measured LFs and various q_0's. The count data are best fitted, with a large uncertainty, by world models with q_0's between 0.0 and 0.5 an evolution in M^* of -0.9 ± 0.5 mag at $z \sim 0.6$.

It is found that the average environment of radio-loud quasars at $z \sim 0.6$ is about three times richer in galaxies than that of their counterparts at $z \sim 0.4$. A significant fraction, possibly as high as 35% of bright radio-loud quasars at $z \gtrsim 0.5$ are found in clusters as rich or richer than Abell class 1 clusters. Since there is a paucity of quasars found in rich clusters at low redshifts, this result suggests that at past epochs, the conditions in some rich clusters were conducive to the formation of quasars. Furthermore, there has been a rapid evolution of the physical conditions in some rich clusters in a period as short as 2 G-yrs causing the quasars harboured by rich clusters to dim by as much as 3 to 4 magnitudes.

3. MULTI-COLOUR IMAGING OF GALAXY CLUSTERS ASSOCIATED WITH QUASARS

Besides permitting the study of the environments of quasars, CCD direct imaging of fields around quasars has proven to be an efficient method of identifying galaxy groups and clusters at high redshifts (z<1.0). Imaging surveys by myself and collaborators have found over 20 moderately rich galaxy clusters with redshifts between 0.2 and 0.8 which we can study individually. As examples, preliminary results from three-colour imaging of the clusters associated with the quasars Pks 0405-12 (z=0.574) and Pks 0812+02 (z=0.402) are presented. Direct images in Gunn i (8200Å), r (6500Å), and g (4950Å) were obtained from the CTIO 4 m and 1.5 m telescope, respectively. Both clusters have been confirmed by spectroscopic observations to be associated with the quasars (Marr

and Spinrad [1986] for Pks 0405-12 and our own unpublished multi-slit
spectroscopy for Pks0812+02). Completeness limits in r are estimated
to be 23.3 and 22.7 mag for the two fields, respectively. However, for
Pks0405-13, only galaxies brighter than 22.9 are used in the analysis
so that background count corrections from YGS can be applied. These
limits correspond to absolute magnitudes -20.4 and -19.6 in the observed r band for Pks0405-13 and Pks0812+02, respectively.

Using the LF models of Yee and Green (1987), the quasar-galaxy
covariance amplitudes are 1260 ± 260 and 550 ± 180 Mpc$^{1.77}$ for Pks0405-12
and Pks0812+02, respectively. Thus, Pks0405-12 is situated in a cluster of richness class greater than 1, and Pks812+02 in a cluster of
richness between that of class 0 and 1. The cluster associated with
Pks0405-12 is rich enough that an LF of the excess galaxies can be
estimated from this field alone. The measured LF has an M^*_r(obs) of
-21.5 mag, and is consistent with the M* value of -21.7 mag derived by
YG from excess galaxies from 9 different quasar fields in the same
redshift range. Azimuthally averaged profiles for the two clusters,
centered around the quasars, are also derived. They fit the King profile with core radii of 200 and 210 kpc for Pks0405-12 and Pks0812+02,
respectively. The computed core radii are larger than the average core
radius of the groups and poor clusters associated with low-redshift
quasars (Yee and Green, 1984) but smaller than the radius of 250 to 400
kpc expected for rich Abell clusters. Thus, both clusters can be characterized as compact and centrally concentrated.

To study the colours of the galaxies in the fields, model colours
of galaxies with no evolution are constructed using spectral energy
distributions of galaxies of 4 morphological types: E+S0 (from Yee and
Oke, 1978), Sbc, Scd, and Im (from Coleman, Wu and Weedman 1982). The
loci of the colours in the colour-colour diagram are plotted as functions of redshift in Figure 1. Data from galaxies in the Pks0405-12
field are also plotted. The majority of the galaxies within a radius
of 250 kpc from the quasars have colours consistent with being early
type galaxies at the redshift of the quasar. Note that within this
small area, only 3.6 background galaxies are expected. At 250 to 500
kpc, however, the number of galaxies with colours consistent with being
foreground galaxies or galaxies with morphological types later than Sbc
at z=0.574 dominates.

In Figure 2, the histograms of g-r colours are plotted for the two
fields. The shaded regions of the histogram represent galaxies with
colours bluer than $(g-r)_c$ which is defined as 0.1 mag redder than an
Sbc galaxy at the redshift of the quasar. The definition of $(g-r)_c$,
which is about 0.2 to 0.3 mag bluer than an elliptical galaxy, is similar to the colour criterion used by Butcher and Oemler (1985, hereafter
BO) for defining their blue galaxies. To correct for background contamination, crude colour distribution corrections derived from galaxies
at the edge of the Pks0405-12 CCD frame are used. For Pks0405-12, it
is estimated that 85% of the background galaxies are bluer than
$(g-r)_c$=1.25 mag; and for Pks0812-02, it is assumed that 75% of background galaxies have g-r<1.0 mag. The total background counts are
taken from YGS. Using these corrections, it is found that the fraction
of red galaxies is similar in both clusters. At r 250 kpc, ~53% of the
galaxies around each quasar are redder than expected Sbc colours. In

the region between 250 and 500 kpc from the quasars, however, only ~ 33%
of galaxies in both fields are redder than Sbc. Thus, the two clusters
are very similar in every aspect with the exception of a factor of two
difference in richness.

The g-r distributions contain two interesting preliminary results.
One is that the blue galaxy fraction increases as a function of radius
from the quasar (assumed to be the cluster center) which is consistent
with the Gunn and Dressler (Gunn, 1988, in these proceedings) conclusion that the BO blue galaxies preferentially avoid the center of the
clusters. More interesting is the possibility that the blue fractions
of the clusters containing quasars may be considerably larger than that
of the ordinary clusters studied by BO. At the redshifts of the quasars, the blue galaxy fraction expected for BO clusters is ~0.2 to 0.3.
While the BO clusters are richer on the average, and thus, there may be
some unknown effects at play, it should be noted that both clusters
have a BO concentration parameter of C>0.5, rivalling the most concentrated BO clusters. Thus, it is of great interest both to increase the
sample of photometric study, and to obtain spectroscopic confirmation
of the blue galaxy fraction of clusters associated with quasars.

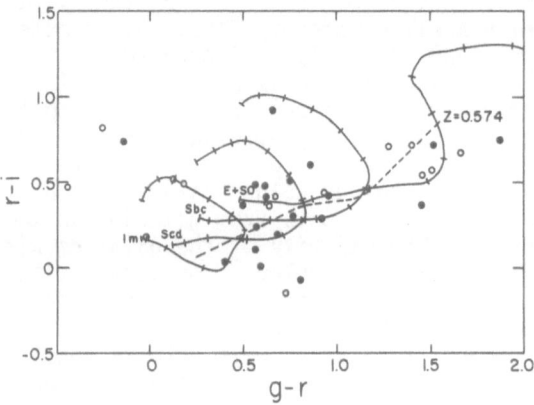

Fig. 1. Colour-Colour plot of galaxies in the field of Pks0405-12.
Solid curves represent colour loci of non-evolving galaxies shifted as
functions of redshifts. Each tick mark represents a redshift increment
of 0.1 with the z=0 ends marked by the morphological type. Solid dots
(●) represent galaxies within 250 kpc projected distance of the quasar
and opendots (○), galaxies between 250 and 500 kpc from the quasar.
The dashed line traces the expected loci of galaxies at z=0.574. Typical error bars are between 0.1 and 0.15 mag. The 3 points at the extreme left are affected by saturated lines and bright stars in the CCD
images.

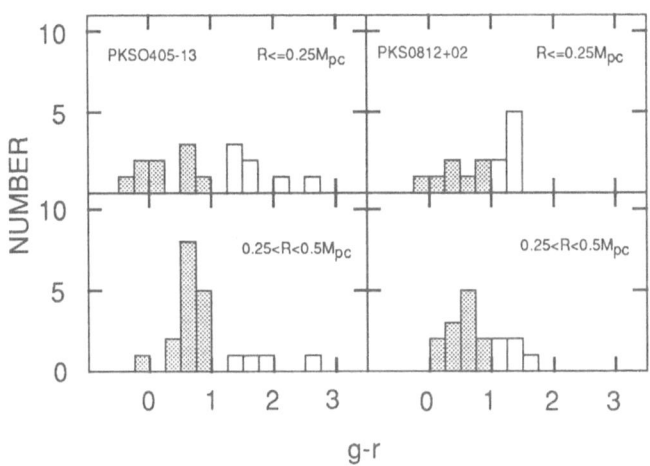

Figure 2. Distributions of g-r colours of galaxies around the quasars as a function of radius from the quasars. Shaded regions represent galaxies bluer than $(g-r)_c = (g-r)_{Sbc} + 0.1$.

REFERENCES

Butcher, H., and Oemler, A. 1984, Ap.J., **285**, 426 (BO).
Coleman, G.D., Wu, C.C., and Weedman, D.W. 1980, Ap. J. Suppl., **43**, 393.
Dahari, O., 1984, A. J., **89**, 966.
Green, R.F., and Yee, H.K.C. 1984, Ap. J. Suppl., **54**, 495.
Gunn, J.E. 1988, in these proceedings.
Marr, J. and Spinrad, H. 1985, P.A.S.P., **97**, 684.
Turner, E.L. 1976, Ap. J., **208**, 20.
Yee, H.K.C. 1987, A. J., in press.
Yee, H.K.C. and Green, R.F. 1984, Ap. J., **280**, 79.
Yee, H.K.C. and Green, R.F. 1987, Ap. J., **319**, 28 (YG).
Yee, H.K.C., Green, R.F., and Stockman, H.S. 1986, Ap. J. Suppl., **62**, 681 (YGS).
Yee, H.K.C. and Oke, J.B. 1978, Ap. J., **226**, 753.

QUASAR ABSORPTION LINES AND HIGH REDSHIFT GALAXIES

Arlin P. S. Crotts
McDonald Observatory
University of Texas
Austin, TX 78712 U.S.A.

ABSTRACT: If metal-line systems occur only where the line of sight to a QSO intercepts a galaxy, we already have information on many hundreds of galaxies at redshifts greater than unity. Here we review research on the connection between galaxies and absorption line systems seen in QSO spectra. Most information about absorption due to galaxies is confined to low-ionization species because strong high-ionization transitions are limited to the vacuum UV. Conversely, the bulk of information about QSO metal-line systems comes from redshifts where galaxies are difficult to observe and may be, due to evolution, very different from present day galaxies. Studies underway to bridge this gap include comparisons of the velocity structure of individual absorbers and galaxies and their clustering, both spatial and in velocity, as well as the galaxian environment of QSO absorbers, and the composition and ionization state of the gas in QSO absorption systems versus galactic coronal clouds. These tend to indicate differences between the properties of high-z QSO absorbers and low-z galaxies, suggesting recent evolution in the population causing the absorption.

1.   DETECTION SENSITIVITY AS A FUNCTION OF ABSORBER REDSHIFT

Until a large telescope is placed above the Earth's atmosphere, study of absorption at low redshift is limited to a small number of transitions of relatively rare species (except in the case of a few bright objects which can be studied with the International Ultraviolet Explorer). Confined to the optical, the sensitivity to detection of absorption in QSO spectra is a strong function of redshift:

TABLE 1: Strongest transitions in different redshift ranges

ATOMIC SPECIES	OPTICAL z RANGE	LINE STRENGTHS
Ca II; and Na I	$0 < z \leq 1$	large f values, but rare ion species
Mg I and II; Fe II; and Mn I	$0.2 \leq z \leq 2$	large f values, more common ions, particularly Mg II
H I; C II, IV; Si II, III, IV; N V; O I, VI; Fe II; Al II; and more	$1 \leq z \leq 6$	large f values and common species, particularly C IV and Lyman alpha

(c.f. Morton and Smith 1973, and Aaronson et al. 1975)

305

R. G. Kron and A. Renzini (eds.), Towards Understanding Galaxies at Large Redshift, 305–311.
© 1988 by Kluwer Academic Publishers.

Note that only low ionization states are accessible in the optical and near UV.

Investigators tend to separate absorbers into classes described by the line widths and strengths of these species, rather than the form and origin of the bodies causing the absorption, or their location with respect to other objects. These classes include: 1) hydrogen (Lyman series) only systems, 2) narrow ($\Delta v \leq 1000$ km s^{-1}) heavy-element or "metal-line" systems, and 3) broad ($\Delta v \gg 1000$ km s^{-1}) metal-line systems. Little is known about the hydrogen-only systems below $z \approx 1.6$, whereas the broad metal-line systems seem to be caused by material intrinsic to the QSO (Burbidge 1971). Smaller classes of objects have been proposed, e.g. "damped Lyman α systems" (Wolfe et al. 1986), which may fall between classes 1 and 2. Also, Tytler (1986) proposes that classes 1 and 2 are simply opposite ends of a continuum of absorber properties.

2.. THE NUMBERS OF GALAXIES AND ABSORPTION SYSTEMS AT LOW REDSHIFT

Several cases are known where an absorption system appears at the same redshift as a local galaxy a small angular distance away from the QSO (c.f. Blades et al. 1981, Boksenberg and Sargent 1978), so evidently some low-redshift metal-line systems are caused by galaxies. Are all low redshift metal-line systems then due to galaxies? The reason for assuming so is compelling. After all, the absorption is caused by non-primordial elements, therefore nuclear processes, and assumably a luminous object such as a galaxy. To date, all cases of low redshift (Ca II or Na I) QSO absorption systems have been found to sit at the same distance as a galaxy sitting near the sightline to the QSO. The model is often presented, therefore, that galaxies are surrounded by gaseous coronae of material that they have expelled that causes absorption.

Can one then assign a well-defined cross-section to single galaxies to describe how often a QSO sightline passing them at a certain distance will intercept a column density of gas sufficiently large to cause detectable absorption? The cases of low redshift Ca II absorption correspond to a cross-sectional radius for a Schechter [1976] L_* galaxy (assuming the absorption cross-section is proportional to the isophotal area) of $R_{Ca\ II} = \pi^{-1} \sigma_{abs}^{-1/2} \approx 18$ h^{-1} kpc for $W_{rest} \geq 0.6$Å (Morton et al. 1986). This is close to the corresponding value for low redshift ($\langle z \rangle = 0.50$) Mg II: $R_{MgII} = 25$ h^{-1} kpc assuming $q_0 = 0$ (Tytler et al. 1987 - corrected for spherically shaped coronae). The Mg II line is 14 times stronger than the Ca II transition for gas of solar composition, so this implies a sharp truncation of the density profile of low ionization gas (note added after presentation: recent work shows significant increase in the Mg II cross-section at higher z: $R_{MgII} = 42$ h^{-1} kpc for $\langle z \rangle = 1.67$, again for $W_{rest} \geq 0.6$Å [Lanzetta et al. 1987]). Compare this to the profiles of galaxies at 21 cm at the level of $n_{H\ I} = 1.8 \times 10^{20} cm^{-2}$, where the median effective radius is $R_{H\ I} \approx 15$ h^{-1} kpc (Bosma 1978). Since a column density of hydrogen of only about $3 \times 10^{18} cm^{-2}$ may contain enough metals to be detected as Mg II absorption, one cannot easily extrapolate this cross-section to small enough column densities, but sees that a sharp cutoff is indicated.

We can compare the cross-sections of low-redshift, low-ionization species to those of high redshift C IV, where we find $R_{C\ IV} = 46$ h^{-1} kpc for $\langle z \rangle = 1.71$ (Young et al. 1982 - corrected for spherical coronae and $W_{rest} \geq 0.6$Å). This implies either that galaxies are surrounded by more loosely associated high-ionization coronae, or galactic coronae have undergone drastic evolution in size and perhaps ionization between $z = 1.7$ and now. Searches for UV absorption around low z galaxies, with sensitivities of $W_{rest} = 0.5$ to 1Å and sightlines passing within 65 h^{-1} kpc of large galaxies, yield no definite C IV absorption detection in nine out of nine cases (York 1987), so either the high ionization clouds are almost completely unassociated with galaxies or the low versus high-ionization radius differences are due to evolution (the results from Lanzetta et al. 1987 argue for the latter).

Different methods might be exploited to provide better information on this issue. A number of investigators have proposed observations using large, space-based telescopes to look for the stronger UV metal absorption lines in the spectra of QSOs in the direction of low redshift galaxies. One might in this way reach more sensitive W_{rest} limits for higher ionization species and circumvent the unfortunate lack of knowledge about C IV at the redshifts where galaxies are visible. I propose here that one might also conduct a high spatial density survey in the optical of both galaxies to $z \approx 0.5$ and background QSOs. The correspondence of Mg II absorption in the QSOs could be mapped relative to the incidence of nearby galaxies. The frequency of Mg II absorption is such that approximately 7% of the QSOs would show a system at $z < 0.5$. In this way the cases of Mg II absorption caused by galaxies could be mapped in the manner of Ca II, and perhaps some leverage could be obtained on how this changes for increasing redshifts. Of course, such a study would benefit greatly from multi-object spectroscopy.

3. GALAXY CLUSTERING VERSUS THE CLUSTERING OF METAL-LINE ABSORBERS

Since the ease in studying the clustering properties of a population is a strong function of the density of objects detected, the only detailed information about the clustering properties of metal-line absorbers comes from redshifts high enough for strong UV transitions to be seen. Metal-line absorption systems with $1.5 \leq z \leq 2.5$ show velocity clustering for splittings up to 500 $km \ s^{-1}$, with a value of the two-point function $\xi \approx 6$ (Sargent 1987). This value was obtained by looking at velocity splittings between metal-line absorbers along single lines of sight. If each absorber has internal structure on this scale, at least part of this signal could be due to these internal velocities, and not true clustering. Indeed, some investigators have argued that the similarity in the velocity dispersion of Galactic corona clouds and the QSO metal-line widths argue for a single cause for both phenomena (Sargent 1977). In order to look at the pairing of metal-line absorbers that are truly clustered in space, one can look at pairs of QSOs close to each other on the sky and determine if the presence of an absorber in one sightline enhances the probability of finding an absorber in the other sightline at nearly the same redshift. Scanning the entire published literature, one can find a marginal signal for positive clustering power on scales less than about 2 h^{-1} comoving Mpc (Crotts 1985). More data on close pairs is needed to make a definite statement about the small scale spatial ξ, and these data are quickly being obtained. In comparison the hydrogen-only absorbers have been shown to have much weaker clustering over the same line-of-sight velocity scales, with $\xi \approx 0.5$ (Webb 1987), and even weaker clustering on spatial scales of less than 2 h^{-1} Mpc (Crotts 1987).

4. ABSORBER COMPOSITION AND IONIZATION STATE

Most absorbers break up into several components at about 20 $km \ s^{-1}$ resolution, and even more at 10 $km \ s^{-1}$ (as a lesson, note the increasing complexity with greater resolution of the Ca II lines in the direction of SN 1987a in the LMC:

Resolution:	200 $km \ s^{-1}$	20 $km \ s^{-1}$	3 $km \ s^{-1}$
number of components:	2	6	24

References: Andreani and Vidal-Madjar [1987]). Components in metal-line systems have velocity widths of $b = 2$ to 20 $km \ s^{-1}$, so the possibility remains that a low b, high-column density line can be hidden within the structure of lower column density profiles. This implies that abundances at even high resolution for QSO absorbers may be insufficient to eliminate possible uncertainties in composition up to factors of 10 to 100 (Boksenberg et al. 1979). Resolution on the order of 5 $km \ s^{-1}$ may be needed to circumvent this ambiguity.

One can still study the ionization state of the absorbers by comparing the strengths of lines due to differently charged ions of the same atom. Wolfe (1983) did this for both high redshift QSO absorption systems and clouds in the corona of our Galaxy using the UV transitions of C II and C IV. These show that even unbiased samples of the two populations are inconsistent with being drawn from the same parent distribution at the 99% confidence level. They are inconsistent in the sense that C IV is much more prevalent at higher redshift.

308

5. VELOCITY STRUCTURE OF GALACTIC CORONAE AND QSO ABSORPTION LINES

We are presently in the ironic position of having much better information about the velocity structure of many high redshift QSO absorption systems than we do about the absorption caused by any low redshift galaxies except perhaps our own. Absorbers typically have line structures with components spread over a few hundred kilometers per second, but sometimes reaching up to about 1000 $km\ s^{-1}$ (York et al. 1984). Since these velocity widths are within the same range as the dispersion of velocities of clouds in the Galactic corona, we can investigate the hypothesis that the two phenomena are identical.

I would like to discuss two cases studied recently that compare in detail the structure of QSO absorbers and galaxies, taking advantage of accidents of geometry found in the sky. The double QSO 0957+561 is gravitationally lensed in such a way that two bright images of the same QSO reach the earth along two very different paths (Walsh et al. 1979). Both paths intersect each of two different absorbers such that the path separations are on scales interesting in terms of the size of a galaxy (Young et al. 1981). I present here recent data obtained at the KPNO 4-meter telescope using the Cassegrain echelle spectrograph at 22 $km\ s^{-1}$ resolution. The absorber at $z = 1.125$, where the path separation is approximately 4 h^{-1} kpc, has very similar structure in the two beams (see Figure 1). Both have a strong C IV component centered near $z = 1.12502$. The "B" image has another strong component at $z = 1.12455$ that is blended with a 3σ detection of a narrow component at $z = 1.12435$. This is very close to the redshift of the narrow component seen in the "A" image at $z = 1.12429$ (with typical one-sigma redshift errors of about 0.00005). All C IV lines have a total width of about 200 $km\ s^{-1}$.

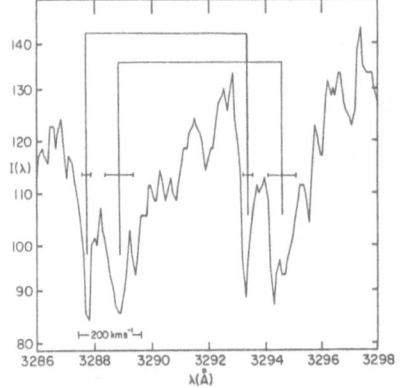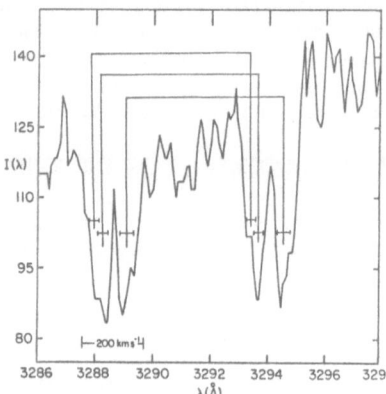

Figure 1: $z = 1.125$ C IV absorption lines of Q0957+561, a) is the A image of the QSO, and b) the B image. Redshift components shown are simple gaussian fits, spectra shown are in air wavelengths and not sky-subtracted.

The $z = 1.391$ absorbers in both beams (separation = 0.3 h^{-1} kpc) have C IV structures that are even more similar. In this system we can also observe the structure of many other lines, particularly lower-ionization states, which are more comparable to transitions seen in the optical and near UV. These too show nearly identical structure between the two beams. For example, the Al II ($\lambda1670.81$) line shows a 300 $km\ s^{-1}$ trough in both images with a deep component at $z = 1.39078$ in one beam and 1.39091 in the other (see Figure 2).

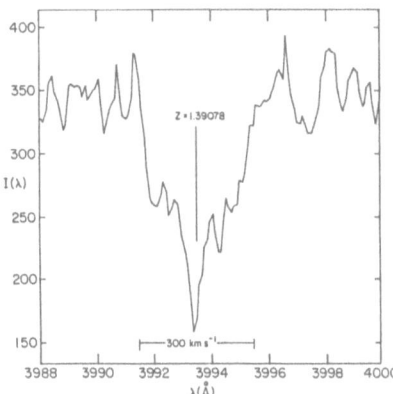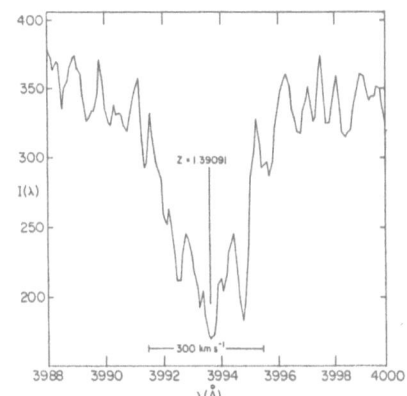

Figure 2: as in Figure 1, except for the Al II (λ1670.81) line of the $z = 1.391$ system.

These data are almost identical in resolution and S/N to spectra taken by IUE of sources in the Magellanic clouds (Savage and deBoer 1979, 1981, deBoer and Savage 1980), and the separation of the gravitationally lensed paths fall in the range of separations of these stars in the Magellanic clouds. With these we can compare the coherence in velocity structure between different lines of sight for two QSO metal-line systems to the velocity coherence in the Galactic corona and clouds in the Magellanic stream. I have used a cross-product over the line-width of absorption depth in two beams such that a value of unity corresponds to identical distributions, and a value of 0.5 corresponds to two different random distributions. Figure 3a shows the result for the Q0957+561 and IUE data (where line profiles within 50 $km\ s^{-1}$ of heliocentric rest have been excluded since these are probably due to the Galactic disk). It is apparent that the QSO lines show more velocity coherence than Galactic corona sightlines *even including structure within the Magellanic clouds*. The difference between the two populations is even greater if structure within 50 $km\ s^{-1}$ of the source is excluded (see Figure 3b), presumably leaving only Galactic coronal clouds (this comparison is some-what weak, since the S/N of the remaining line profiles is considerably lower than those in Q0957+561). There seems to be little evidence, however, for structures in the corona that extend over 0.3 or 4 kpc in a manner similar to those in the QSO absorber. This would imply that either the space between the Earth and the Magellanic Clouds is not a fair representation of the structure of the larger Galactic corona, or QSO absorption systems have structure markedly different from the outer parts of our Galaxy (perhaps they are composed of extended sheets formed by shocks or tidal forces).

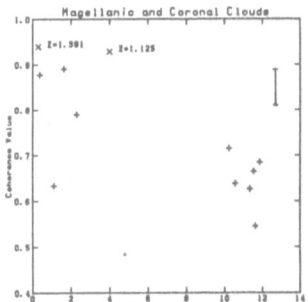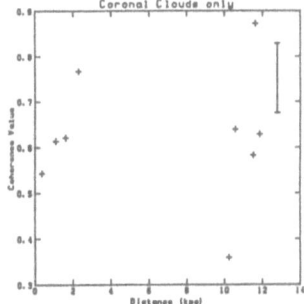

Figure 3: the degree of similarity of the velocity structure along lines of sight to the Magellanic Clouds and to Q0957+561 (indicated by the absorption redshift) as a function of sightline separation: a) is for the QSO and gas including the Magellanic Clouds, b) for the Magellanic Clouds, excluding gas within 50 $km\ s^{-1}$ of the sources. Typical error bars are also shown in the upper right.

At least one configuration in the sky allows us to observe a bright AGN in the background of a galaxy at high enough redshift so that strong UV transitions are moved into the visible. The BL Lac object AO 0235+164, at a redshift of 0.94 (Cohen *et al.* 1986), sits two arcseconds away from a bright galaxy (Spinrad and Smith 1975) at $z = 0.524$ (Smith *et al.* 1977). It shows absorption at the redshift of the galaxy, as well as $z = 0$ and 0.851 (Burbidge *et al.* 1976). High resolution spectroscopy also obtained at KPNO by the author shows that the Mg II lines of the $z = 0.524$ system are composed of at least five narrow velocity components spread over 350 $km\ s^{-1}$, in a manner typical of QSO metal-line systems (incidently, this system was studied in 21 cm at extremely high resolution by Wolfe *et al.* 1982, so we should be able to compute an accurate Ca II / Mg II / H I ratio for this cloud). In this case we have an absorption system at relatively high redshift that we know to be associated with a galaxy, but has structure similar to that of many metal-line systems that are not known to be. This suggests that if our Galaxy's corona has a velocity structure different from metal-line systems in QSOs, it may be because the nature of absorbers or galactic coronae have changed since the Universe was about one-half its present age.

6. LUMINOSITY ASSOCIATED WITH METAL-LINE ABSORBERS

Two studies have recently begun to determine what luminous objects are associated with QSO metal-line systems with $z \leq 1$. Bergeron reported (at the Space Telescope Science Institute workshop on QSO absorption on May 20) on observations that she has made at CFHT of the spectra of objects within small distances (less than a few tens of arcseconds) of QSOs that show low redshift absorption. I will not mention these in detail because not all of this work is published (c.f. Bergeron 1986, also Miller *et al.* 1987), but she does report that out of 13 fields surveyed, eight showed an object at the absorption redshift with a detectable 4000Å break, and three had strong narrow lines, particularly [O II] 3727Å.

York and Yanny (and recently, myself, in cooperation with them) have been imaging large fields around QSOs showing absorption at $z \leq 1$ in narrow bands centered on 3727Å $\times (1 + z_{abs})$, as well as medium-width bands on either side of the 4000Å and Balmer breaks in order to characterize the environment of QSO absorbers. These results, some of which are also unpublished, show that in four fields there are two positive and two null detections for strong 3727Å emission, and at least one of these detections is confirmed by follow-up spectroscopy (Yanny *et al.* 1987). Other 3727Å and 4000Å break data are not completely reduced. Both of these studies indicate a large incidence of strong 3727Å emitters in the immediate vicinity of QSO metal-line absorbers, suggesting a link with recent star formation. At least in the case of our study, these sites of star formation appear to be very small (nearly unresolved) and close to the line of sight to the QSO.

7. CONCLUSIONS

1) The number densities of absorption lines due to various species indicate that all low-ionization, low-redshift systems may be caused by galaxies, but that significant evolution has taken place in their numbers since $z \approx 1.7$ such that either the typical cross-section or the actual number of absorbers has drastically decreased with time.

2) The ionization state of high redshift carbon absorption is significantly hotter than measurements for clouds in the corona of the Galaxy.

3) In cases where high redshift metal-lines associated with galaxies can be studied at high resolution, they show detailed velocity structure similar to that found in most absorbers at too high a redshift for most galaxies to be seen.

4) The coherence in velocity structure over distances from 0.3 to 4 kpc in QSO C IV absorption lines is much greater than for Galactic coronal gas between the Earth and Magellanic Clouds. This indicates the presence of larger structures in QSO absorbers, such as sheets or shells of gas.

5) Absorbers with $0.2 < z \leq 1$ are often associated with galaxies, particularly small galaxies with active star formation.

These indicate that the gaseous coronae of galaxies at $z = 1$ to 2 were much larger, and composed of structures larger than those seen in the corona of our Galaxy. This may be correlated with the presence of enhanced star formation at earlier epochs. These conclusions are tempered by the caveats that the population of clouds we have studied in our own Galaxy may not represent low-redshift coronal clouds in general, and possibly some high redshift absorbers may not be closely associated with galaxies at all. Further data to address these uncertainties, and strengthen, refute or add to the above points would be very useful.

REFERENCES

Aaronson, M., McKee, C. F., and Weisheit, J. C. 1975 *Ap. J.*, **198**, 13.

Andreani, P., and Vidal-Madjar, A. 1987 I. A. U. Circulars Nos. 4320 and 4323.

Bergeron, J. 1986 *Astron. Astrop.*, **155**, L8.

Blades, J. C., Hunstead, R. W., and Murdoch, H. S. 1981 *Mon. Not. Royal Astron. Soc.*, **194**, 669.

Boksenberg, A., Carswell, R. F., and Sargent, W. L. W. 1979 *Ap. J.*, **227**, 370.

Boksenberg, A., and Sargent, W. L. W. 1978 *Ap. J.*, **220**, 42.

Bosma, A. 1978 Ph.D. thesis (Groningen).

Burbidge, E. M. 1971 *Pontif. Accad. Sci. Scripta Varia.*, **35**, 121.

Burbidge, E. M., Caldwell, R. D., Smith, H. E., Liebert, J. and Spinrad, H. 1976 *Ap. J.*, **205**, L117.

Cohen, R. D., Smith, H. E., Junkkarinen, V. T., and Burbidge, E. M. 1986 *preprint* (UCSD).

Crotts, A. P. S. 1985 *Ap. J.*, **298**, 732 (Erratum: **305**, 581).

_____. 1987 *in preparation*.

de Boer, K. S., and Savage, B. D. 1980 *Ap. J.*, **238**, 86.

Lanzetta, K. M., Turnshek, D. A., and Wolfe, A. M. 1987 *preprint* (STScI).

Miller, J. S., Goodrich, R. W., and Stephens, S. A. 1987 *A. J.*, **94**, 633.

Morton, D. C., and Smith, W. H. 1973 *Ap. J. Suppl.*, **26**, 333.

Morton, D. C., York, D. G., and Jenkins, E. B. 1986 *Ap. J.*, **302**, 272.

Sargent, W. L. W. 1977 Proc. of *The Evolution of Galaxies and Stellar Populations*, eds. B. M. Tinsley and R. B. Larson (Yale Observatory: New Haven), p. 427.

Sargent, W. L. W. 1987 Proc. of STScI Workshop *QSO Absorption Lines: Probing the Universe*, eds. C. Blades, C. Norman and D. Turnshek (University Press: Cambridge), *in press*.

Savage, B. D., and de Boer, K. S. 1979 *Ap. J.*, **230**, L77.

_____. 1981 *Ap. J.*, **243**, 460.

Schechter, P. 1976 *Ap. J.*, **203**, 297.

Smith, H. E., Burbidge, E. M., and Junkkarinen, V. T. 1977 *Ap. J.*, **218**, 611.

Spinrad, H., and Smith, H. E. 1975 *Ap. J.*, **201**, 275.

Tytler, D. 1986 *preprint* (Columbia Univ.).

Tytler, D., Boksenberg, A., Sargent, W. L. W., Young, P., and Kunth, D. 1987 *Ap. J. Suppl.*, **64**, 667.

Walsh, D., Carswell, R. F., and Weymann, R. J. 1979 *Nature*, **279**, 381.

Webb, J. K. 1987 Proc. I. A. U. Symp. No. 124 *Observational Cosmology*, eds. A. Hewitt, G. Burbidge and L.-Z. Fang (Dordrecht: Reidel), p. 803.

Wolfe, A. M. 1983 *Ap. J.*, **268**, L1 (Erratum: **271**, L43).

Wolfe, A. M., Davis, M. M., and Briggs, F. H. 1982 *Ap. J.*, **259**, 495.

Wolfe, A. M., Turnshek, D. A., Smith, H. E., and Cohen, R. D. 1986 *Ap. J. Suppl.*, **61**, 249.

Young, P., Sargent, W. L. W., Boksenberg, A. 1982 *Ap. J. Suppl.*, **48**, 455.

Young, P., Sargent, W. L. W., Boksenberg, A., and Oke, J. B. 1981 *Ap. J.*, **244**, 415.

York, D. 1987 *private communication*.

York, D. G., Green, R. F., Bechtold, J., and Chaffee, Jr., F. H. 1984 *Ap. J.*, **280**, L1.

Yanny, B., Hamilton, D., Schommer, R., Williams, T., and York, D. G. 1987 *Ap J.*, *in press*.

THE GREGORIAN UPGRADE AT ARECIBO AND ITS IMPACT ON SOME ASPECTS OF EXTRAGALACTIC RESEARCH

Riccardo Giovanelli[1]
NAIC[2], Arecibo Observatory
USA

ABSTRACT. The performance of the 305m telescope at Arecibo is flawed with a number of limitations that are related with the current optical design. These shortcomings are correctible with the replacement of line feeds with a Gregorian subreflector system and the construction of a ground shield around the edge of the main reflector. The upgrade will substantially increase the sensitivity of the telescope and allow optimal implementation of advances in electronic technology. Its potential impact on selected topics of extragalactic spectroscopy is discussed in some detail.

1. INTRODUCTION

With its large collecting area, the 305m telescope is the instrument of choice for a variety of high sensitivity experiments, including many in areas of cm-wave spectroscopy. The large size of the mirror was made possible by its virtue of being anchored to the rim of a natural depression; its spherical shape and movable focal feed system allow observations of objects within 20° from the zenith. However, this arrangement is marred by spherical aberration: the focal point of rays reflected near the rim of the dish is 30 m lower than that of rays reflected near its center. The problem has been solved with the construction of "line feeds", cumbersome contraptions which collect radiation reflected from various sectors of the reflector and correct for spherical aberration. They have well served the Observatory in the past. Their limitations, however, are becoming increasingly severe obstacles to the demands of astronomical research and, especially, to the optimal utilization of the main resources of the telescope. A program of radical upgrade has been designed, that will revolutionize the optical setup of the instrument, making possible huge increases in sensitivity, spectral coverage, frequency agility and interference suppression. This program is currently under consideration for funding by the National Science Foundation.

In the following section, we discuss the limitations of the current system. In Section 3, some details are given on the proposed upgrade and in Section 4 the potential impact on some aspects of extragalactic research is discussed, with particular emphasis on 21cm spectroscopy.

2. LIMITATIONS OF THE CURRENT SYSTEM

(a) *Vignetting and Ground Radiation.* The typical line feed used for 21cm observations illuminates only half the area of the reflector, and the illuminated region moves on the reflector as a source is

---

[1] Arecibo Observatory, P.O. Box 995, Arecibo P.R. 00613, USA

[2] The National Astronomy and Ionosphere Center is operated by Cornell University under cooperative agreement with the National Science Foundation.

313

R. G. Kron and A. Renzini (eds.), Towards Understanding Galaxies at Large Redshift, 313–318.
© *1988 by Kluwer Academic Publishers.*

tracked in hour angle. At zenith angle (ZA) higher than 10°, the illumination patter spills over the edge of the reflector. Loss of gain results, accompanied by a steep increase of ground (at 300°K) radiation pickup. At $ZA = 20°$, the gain has decreased by one-third, and the ground radiation contributes about 50°K to the system temperature (T_{sys}), as illustrated by the top curve in Figure 1. Such increase more than doubles the T_{sys} obtained at low ZA. In order to match the quality of an observation below $ZA = 10°$, more than 10 times the integration time is necessary at $ZA = 20°$.

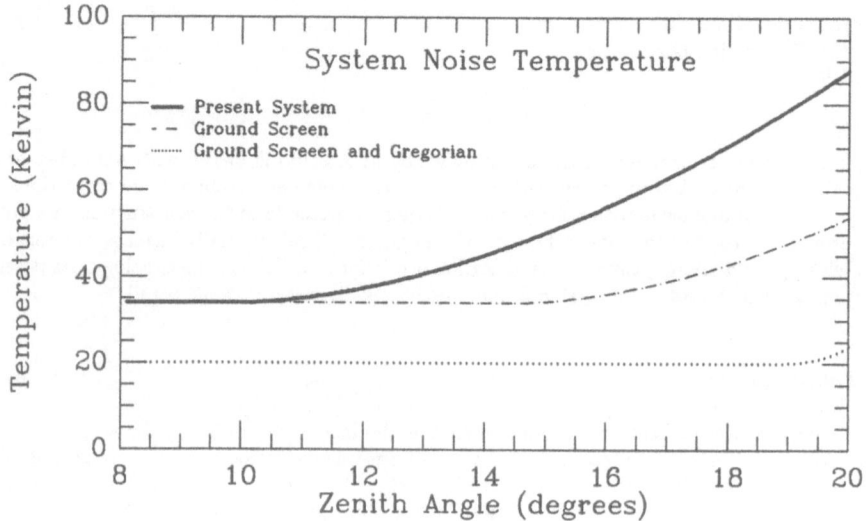

Figure 1. $T_{sys}(ZA)$ for a current state-of-the-art receiver at Arecibo. The heavy solid line illustrates the present situation. T_{sys} remains fairly constant for $ZA < 10°$. The dotted line illustrates the expected behavior after the construction of the ground screen around the relector's rim (see Section 3), but still maintaining line feeds. Finally, the bottom solid line describes the expected T_{sys} after the construction of the ground screen and the replacement of line feeds with a Gregorian subreflector system.

(b) *Line Feeds are Noisy Devices.* Figure 2 illustrates the changes in receiver noise temperature made possible at 21cm by advances in engineering technology over the last quarter century. Current HEMT devices yield receiver noise temperatures of 2 or 3°K. Microwave background radiation, galactic synchrotron at high latitudes and atmospheric contributions add another 7°K or so. The remainder of the system noise is due to losses in feeds, junctions, transmission lines, etc. Current T_{sys} attainable at 21cm at Arecibo are in the upper 30's. Replacement of current transistors with HEMT's may help bring it close to 30°K. About 2/3 of that is due to losses, in great part associated with line feeds, which are mostly responsible for the widening gap between real (upper dotted line in Fig. 2) and optimal performance. Overall T_{sys} below 20°K can be obtained with current receiver technology on standard paraboloids. With line feeds, Arecibo is forced to values nearly twice as high, as well as to a growing inability to take advantage of other progress in electronic technology.

(c) *Line Feeds are Narrow Band Devices.* Line feeds' gain depends on frequency in roughly gaussian fashion, $G(\nu)$, centered on a frequency ν_p for which the feed is optimally designed. The half-power width of $G(\nu)$ depends exclusively on the length of the feed: the longer the feed, the narrower its frequency response. Typically, a 12m feed has a 45 MHz half-power bandwidth. Longitudinal

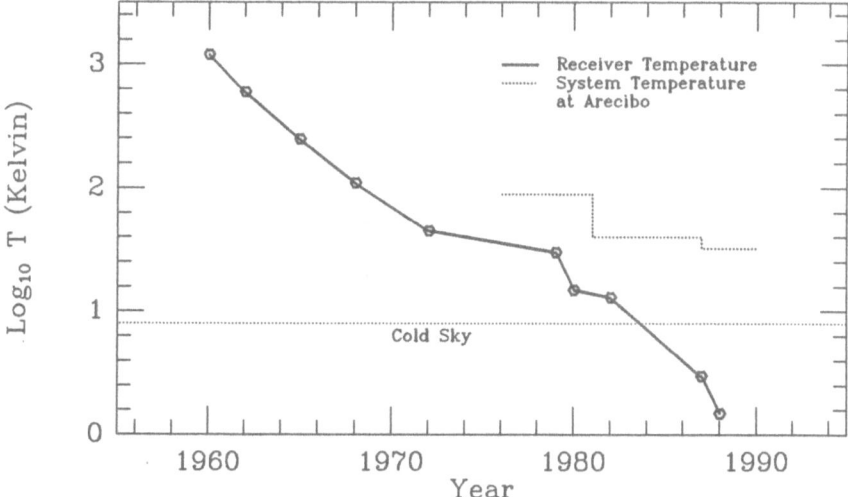

Figure 2. Improvements in receiver noise temperature at 21 cm since 1962. The 305m telescope started operating at 21 cm after the 1974-75 reflector surface upgrade. A HEMT receiver currently under construction is expected to obtain a T_{sys} near 30°K. The "cold sky" line includes the contributions of the microwave background, galactic synchrotron and the atmosphere.

displacement of the feed by mechanical means partially eases this limitation, as ν_p can be shifted, albeit at the cost of a reduction in peak gain. This extends the usable frequency range of a feed to about 80-100 MHz, with little change of the instantaneous bandwidth from 45 MHz. Thus observations can only be done in those narrow frequency ranges for which feeds are available or can be constructed, and those programs that demand frequency agility, such as redshift surveys, require a repeated number of separate integrations on the same source.

In addition, the construction of high frequency feeds is a very delicate operation, one that can be reliably carried out only up to frequencies near 3 GHz. While the recently refaired 305m reflecting surface has an rms tolerance of 2.2 mm, allowing useful observations up to 8 GHz, those cannot be pursued due to the unavailability of good quality feeds. We are thus deprived of the use of the frequency domain where the telescope resolution would be the finest.

(d) *Line Feeds are vulnerable to Interference.* Because line feeds are suspended from the focal region platform above the neighboring hills' crestline, they are eminently vulnerable to horizontally propagating man-made signals, and because of their physical size, shielding is impractical. We are thus unable to protect radio astronomical observations from the growth of noise pollution and effectively are blinded in many important frequency bands.

3. THE GROUND SCREEN AND THE GREGORIAN UPGRADE

The attack to the problems described in the preceding Section consists mainly of two steps. First, the construction of a *Ground Screen*, i.e. an 18 m high fence around the rim of the reflector. The screen will reflect "cold sky" into the feed when, at high ZA, its illumination pattern spills over the reflector's edge. This screen will largely reduce the amount of ground radiation pickup, as seen in Fig. 2, but will not affect the reduction of gain with increasing ZA. The second main step, and by far the most important, consists in replacing line feeds with a subreflector system consisting of two mirrors, a 20m secondary and a 9m tertiary, enclosed within a rigid radome, 24m in diameter, a *Gregorian* optical arrangement. The radome will be a metallic enclosure, truncated in the bottom, which will provide support for the subreflectors and protection from horizontally-propagating interference. Movable along a track under the azimuth arm, it will replace one of the two carriage houses in the suspended platform 150 m above the main reflector. The surfaces of the subreflectors are not conic sections; many solutions exist for their shape, providing latitude to achieve, for example, the most desirable illumination patterns. Among those considered, an offset illumination pattern, *i.e.* one that will preferentially illuminate downhill portions of the main reflector, will eliminate spillover noise pickup, except for the very highest ZA, as illustrated in Fig. 2, limiting at the same time deterioration of gain. The final solution may include a slightly elliptical illumination pattern, which will increase the maximum gain from 8 to 11 or 12 K/Jy. The task of feeding the antenna is finally reduced to that equivalent to feeding a standard paraboloid. Gone are the feed losses and their narrow bandwidth limitations. T_{sys} as low as those attainable on paraboloids will become possible, as well as broad band, frequency-agile observations. Most importantly, perhaps, the frequency horizon will be expanded to the limits set by the main reflector surface.

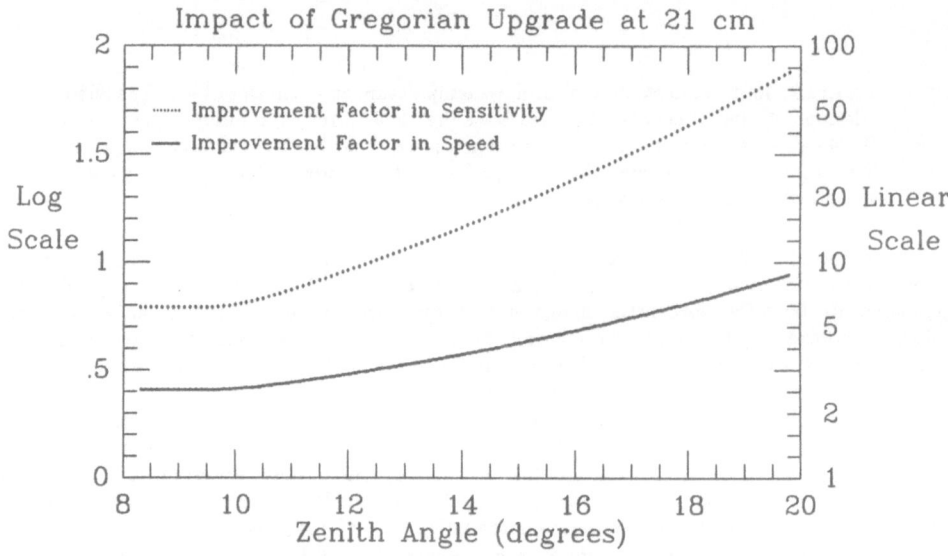

Figure 3. The curves incorporate the compounded effect of the increase in gain and the reduction in T_{sys} deriving from the implementation of the Gregorian subreflector system and the Ground Screen upgrades.

The compounded effect at 21 cm of gain and T_{sys} improvements obtainable from a Gregorian *cum* ground screen upgrade is illustrated in Fig. 3. At low ZA, the performance will be about 2.5 times better in sensitivity (or 6 in speed), while at high ZA's the improvement factor climbs dramatically: at the highest ZA, the system will be nearly 100 times faster. A ZA-weighted average indicates an overall improvement of a factor 4.5 in sensitivity, which may be thought equivalent to quadrupling of the area of the main reflector.These figures do not, however, tell the whole story, as we shall see next.

4. IMPACT OF THE UPGRADE ON RESEARCH

In this Section we will concentrate on illustrating the advantages of the planned upgrade as they apply in a couple of examples of interest in the context of the main topic of the workshop.

(a) 21 cm Redshift surveys. The 305m telescope is a powerful redshift machine. It produces approximately 1000 new, high quality redshifts per year. The depth of currently accessible samples is well matched by that of the UGC (Nilson 1973) and CGCG (Zwicky et al. 1962-68) galaxy catalogues. However, the need for much deeper samples is arising. For example the study of the distribution of galaxies of magellanic luminosity over volumes spanning cosmic voids, taxes current observational facilities to the limits of their performance.

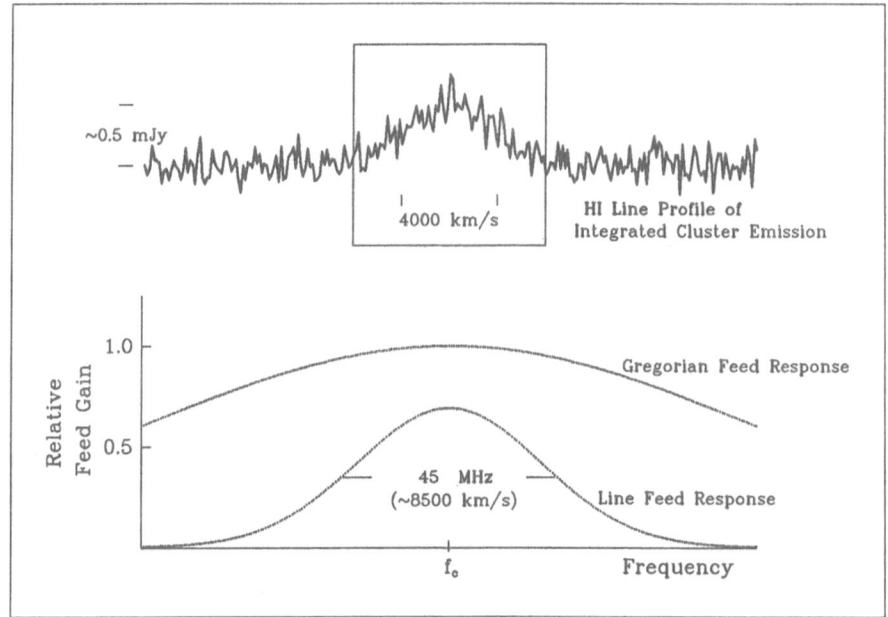

Figure 4. The noisy profile is a simulation of the integrated line emission of a cluster of galaxies at $z = 0.5$. The two dotted curves represent the gain functions $G(\nu)$ of a line feed and of the Gregorian subreflector feed system. The boxed-in section of the "spectrum" outlines the maximum instantaneous bandwidth coverage with current instrumentation. The cluster was assumed arbitrarily to yield a 5σ line.

The increase in speed of the upgraded Arecibo for this type of work goes beyond the numbers illustrated in Fig. 3. For example, one will have the ability to instantaneously explore much broader bandwidths than the 45 MHz of a 12m line feed and the increased speed in data taking will allow more expeditious observing schemes. A relatively conservative estimate of the factor by which the speed of the system will increase for this type of work is on the order of 60, for objects of the type of those observed now. It should be kept in mind, however, that in this mode, detectability depends on integrated flux, rather than surface brightness. Thus, to detect an m=17.5 spiral with the Gregorian will take a time comparable to detecting an m=15.3 spiral with the current sytem. For a galaxy of normal HI content, such an effort may now take 10 minutes. Normal spirals as faint as m=18.0

will be common targets of 21cm redshift surveys with the Gregorian, offering a sample of objects accessible from Arecibo of about 500,000. Gas-rich dwarf and irregular objects as faint as m=20 will also be relatively easy targets. Confusion, a serious problem with continuum surveys with single-dish antennas, will still remain a minor nuisance, occurring in 1% of the galaxies surveyed, prevalently in clusters and tight groups. In very distant clusters, however, the possibility to include most galaxies within the beam provides interesting opportunities.

(b) *Distant Clusters.* Much debate in the course of this workshop has concentrated on the issue of whether the blue galaxy fraction in clusters of galaxies increases with redshift, as first pointed out by Butcher and Oemler (1981). The HI content of spirals in clusters has been shown to be a good indicator of evolutionary phenomena. The integrated HI emission at a redshift of 0.5 would yield a signal on the order of 0.5 mJy. This flux is not prohibitive by present standards. The reason the experiment has not been succesfully carried out yet is that such weak emission would be spread over a velocity range comparable with the velocity dispersion of the whole cluster, a line width of a few thousand km/s. Such a line width translates into 20 MHz or so, which is nearly half the current instantaneous bandwidth of an Arecibo line feed. Assuming that the line feed at the right frequency were to be built - and given their narrow frequency response, several of them would be needed to observe clusters covering a wide range in redshift - the spectral "line" would be half as wide as the feed passband and nearly impossible to separate from instrumental ripple. A measure of the integrated flux of such a feature would not be respectable. In Fig. 4, a 5σ line has been simulated, and plotted on the same frequency scale as the feed $G(\nu)$, on the assumption that the feed were perfectly centered at the frequency of the redshifted cluster's 21cm emission. The boxed-in section of the spectrum corresponds to the frequency interval that could be accessible now with the right feed, the total display to what may be accessible with a Gregorian system. Clearly, in the second case, the signature of the cluster appears as a far more distinct feature in the spectrum, easily detectable in a few hours integration. This is just an example of the type of basic experiments that could be done with an upgraded 305m telescope, now impossible no matter how large the investment of telescope time.

Information and illustrations presented here have been freely borrowed from the NAIC proposals for the implementation of the Ground Screen and for the Gregorian subreflector system submitted to the NSF, the work of many talented individuals at and outside NAIC. This presentation is mostly a brief summary of selected information in those proposals.

5. REFERENCES

Butcher, H. and Oemler Jr., A. 1981, *Ap. J.* **285**,426
Nilson, P. 1973, *Uppsala General Catalog of Galaxies, Uppsala Ann. Obs.* 6 (UGC)
Zwicky, F., Herzog, E., Karpowicz, M., Kowal, C.T. and Wild, P. 1961-68, *Catalogue of Galaxies and of Clusters of Galaxies*, 6 volumes, Pasadena: Cal. Inst. of Tech. Press (CGCG)

HIGH RESOLUTION IMAGING IN THE INFRARED USING ADAPTIVE OPTICS

Jacques M. Beckers
Larry E. Goad
Advanced Development Program
National Optical Astronomy Observatories*
Tucson, AZ   85726-6732

ABSTRACT. Adaptive optics is an especially powerful tool for high resolution imaging in the infrared beyond 2μm, where it can cover all or a
large fraction of the sky and where the complexity and cost is modest as
compared with the telescope itself. We describe the principles behind
the prototype of a so-called polychromatic adaptive optics system now
being assembled at NOAO to be used at the 150 cm McMath and 380 cm
Mayall telescopes on Kitt Peak. In it, the wavefront errors due to
seeing and the telescope itself are being corrected for IR wavelengths,
while the wavefront disturbances are being sensed at visible and near IR
wavelengths.

1. INTRODUCTION

In the broadest sense, astronomical adaptive optics can be defined as
the technique in which the optics in a telescope are adjusted continuously with the aim of improving image quality for both short and long
exposures. Adaptive optics in this broad definition might correct for
telescope aberrations, tracking errors and dome seeing, as well as for
atmospheric seeing outside the telescope dome.

Rapid guiding, aimed at compensating telescope tracking errors and
some atmospheric seeing effects, is the simplest form of adaptive
optics. A number of experiments on Mauna Kea have demonstrated the
substantial improvement in image quality that it can achieve. A more
complex form of adaptive optics consists of rapid guiding on sub-aper-
tures of a telescope. Such an experiment was tried with modest success
on the six subapertures of the MMT, and with substantial success by
Smithson (Smithson 1985) on the Sacramento Peak Vacuum Telescope and
most recently by Lelievre (1987) on the CFHT.

The term "active optics" has now been generally adopted for relatively slow adjustments to the telescope optics, aimed primarily at the

*Operated by the Association of Universities for Research in Astronomy,
Inc., under contract with the National Science Foundation.

R. G. Kron and A. Renzini (eds.), Towards Understanding Galaxies at Large Redshift, 319–338.
© *1988 by Kluwer Academic Publishers.*

correction of telescope aberrations. Normally active optics adjustments
are made on the primary mirror. Examples of the use of active optics
are the Keck Telescope, the ESO-New Technology Telescope (Tarenghi 1986)
and the active support system developed at NOAO for a 1.8 m borosilicate
glass honeycomb mirror (Pearson et al. 1987). Full "adaptive optics"
corrects not just for telescope aberrations and tracking errors but also
for the rapidly changing atmospheric seeing. It therefore has to have
high bandwidth control and is generally quite complex, especially since
it will be aimed at giving diffraction-limited images. Astronomical
adaptive optics systems are under development at the Sacramento Peak
Observatory by R. Smithson and R. Dunn, at ESO by F. Merkle, at Harvard/
Smithsonian by P. Nisenson, and by us at the National Optical Astronomy
Observatories. At the moment, no active or adaptive optics system is
part of an actually functioning astronomical telescope, although it is
expected that both will be part of the astronomy scene well within the
next decade.

In this review, we will describe some of the atmospheric optics
that are relevant to adaptive optics (section 2), the utility of adaptive optics to astronomy (section 3), the predicted image improvements
(section 4), the NOAO prototype adaptive optics system (section 5),
possible astronomical applications (section 6) and finally the use of
artificial stars for atmospheric seeing measurements (section 7).

2. ATMOSPHERIC OPTICS

F. Roddier (1981) describes the effects of atmospheric turbulence on
optical wavefront propagation. Hardy (1982) and Beckers et al. (1986)
summarize the behavior of the optical wavefront at the entrance pupil of
the telescope to the extent that it is of interest to adaptive optics.
This wavefront has both amplitude variations $A(x)$ and phase variations
$\phi(x)$ across the pupil. All astronomical adaptive optics in planning
correct for phase variations only. Roddier and Roddier (1986) showed
that the amplitude variations cause only minor image deterioration.

The phase variations $\phi(x)$ are quite independent of wavelength λ as
long as $\phi(x)$ is expressed in a linear scale (e.g., μm). The same is
true for the wavefront tilts $d\phi/dx$. This is the reason that some adaptive optics programs (Beckers et al. 1986) propose to measure the wavefront tilts at one wavelength (λ_1) to correct the image at another
wavelength (λ_2). Such systems are referred to as "polychromatic adaptive optics." Because of the absence of a reference wavefront, astronomical adaptive optics cannot use measurements of $\phi(x)$ directly.
Instead, one has to rely on the wavefront tilts or wavefront curvatures
(Roddier 1987) to measure the atmospheric optics effects.

When expressed in wavelengths or radians the phase variations
$\phi(x)$ are of course strongly wavelength-dependent. The spatial spectrum
of $\phi(x)$ is generally believed to have a shape corresponding to that of
light propagated through an atmosphere with a Kolmogoroff turbulence
distribution. In it the RMS phase difference between two points
Δx apart equals:

$$\langle[\phi(x) - \phi(x + \Delta x)]^2\rangle^{1/2} = 0.42 \ (\Delta x/r_o)^{5/6} \text{waves} \tag{1}$$

where r_o is the so-called Fried's parameter. It varies with wavelength as $\lambda^{1.2}$ but is often referred to $\lambda = 500$ nm.

Across a circular aperture with a diameter d the RMS phase variation equals:

$$\Delta\phi_{RMS} = 0.16 \ (d/r_o)^{5/6} \text{waves} \tag{2}$$

of which most is due to wavefront tilts or image displacement and only part (36%) is due to other wavefront corrugation or image deterioration.

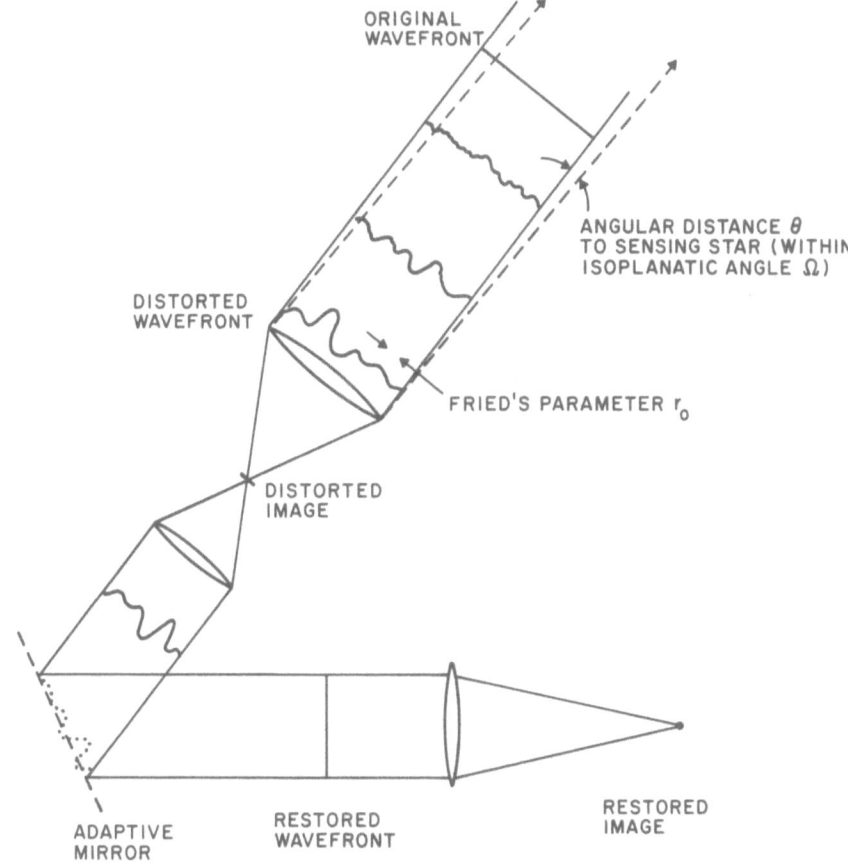

Figure 1: Schematic of adaptive optics concept.

For $d < r_o$ the image of a point source is therefore close to a diffraction-limited Airy disk; for $d > r_o$ the seeing effects deteriorate the image. For large telescopes the image width is approximately equal to λ/r_o, and therefore decreases somewhat with wavelength ($\lambda^{-0.2}$). In this article we use r_o = 20 cm at λ = 500 nm (for 0.5 arcsecond seeing) unless otherwise noted.

Of interest to adaptive optics is the time scale of the wavefront variations. In astronomical applications the wavefront changes are primarily caused by the winds in the layer where the seeing originates (rather than the movement of the astronomical object in the sky). The typical time scale τ, also referred to as "speckle lifetime," equals therefore

$$\tau = r_o / V_{wind} \tag{3}$$

where V_{wind} is the average wind velocity. In this article we use V_{wind} = 10 m/sec as a typical average wind velocity giving τ = 20 msec.

Also of interest to adaptive optics is the so-called isoplanatic patch size. The isoplanatic patch is defined as the area on the sky for which wavefront disturbances are approximately the same. The diameter Ω of this area is:

$$\Omega = 2/3 \; r_o / \bar{h} \tag{4}$$

where $\bar{h}$ is the average height of the seeing (see section 4.4). For $\bar{h}$ = 4000 meters this results in 7 arcseconds at λ = 500 nm.

3. SKY COVERAGE

Figure 1 schematically illustrates the functioning of an adaptive optics system. The atmospheric wavefront tilts are sensed on a sensing star that lies within the isoplanatic patch ($\theta < \Omega$) around the object under study, for the wavelength under study (λ_2). The wavefront tilts are corrected by an adaptive mirror on scales corresponding to r_o for the wavelength under study (λ_2). Note that Ω, r_o and τ can be substantially larger at λ_2 in the infrared (e.g., 2.2 µm) than at the wavefront measurement wavelength λ_1 (e.g., 0.5 µm).

In Figure 2 we show graphically the variation of r_o, τ and Ω with wavelength. The number of photons collected from an optical sensing star increases rapidly with observing wavelength ($\propto \lambda^{3.6}$) allowing the use of fainter and fainter stars for wavefront sensing. In addition the area of the isoplanatic patch increases as $\lambda^{2.4}$ so that the total sky coverage for adaptive optics increases dramatically toward longer wavelengths, as does therefore its astronomical utility. Sky coverage in the K-band (2.2µm) is approximately 3%; in the M-band (5µm) it approaches full sky. Coverage is of course dependent on r_o, V_{wind}, h and direction in the sky. For V_{wind} = 10m/sec and h = 4,000 meters, figure 3 shows how the limiting R magnitude of the sensing star and the sky coverage depend on r_o and λ_2. Details of these calculations are given by Beckers et al. (1986).

In figures 2 and 3 we have assumed that the wavefront sensing would occur in the R-band of the spectrum (0.59-0.81µm). The R-band has two advantages: it is desirable to keep the separation of λ_1 and λ_2 as small as possible to increase the possibility that the IR object under study has an optical counterpart, and the longest wavelength photon-

Wavelength	Pupil Subarea (r_0)	Time Constant (τ)	Limiting Magnitude	No. Of Stars $(b = 0°)$	Size Isoplanatic Patch	Fraction Of Sky Covered
0.5 μm	• 20 cm	⊓ 20 msec	7.7	6×10^4	• 0.1↑	<0.1 %
2.2 μm	○ 120 cm	⊓ 120 msec	12.8	1.3×10^7	○ 0.6↑	3 %
5.0 μm	○ 320 cm	⊓ 320 msec	15.6	1.7×10^8	○ 1.6↑	80 %
10.0 μm	○ 730 cm	⊓ 780 msec	18.0	1.3×10^9	○ 3.6↑	100 %

Figure 2: Variation of Fried's parameter r_0, speckle lifetime τ and isoplanatic patch size Ω with wavelength as well as the resulting increase of sky coverage with wavelength.

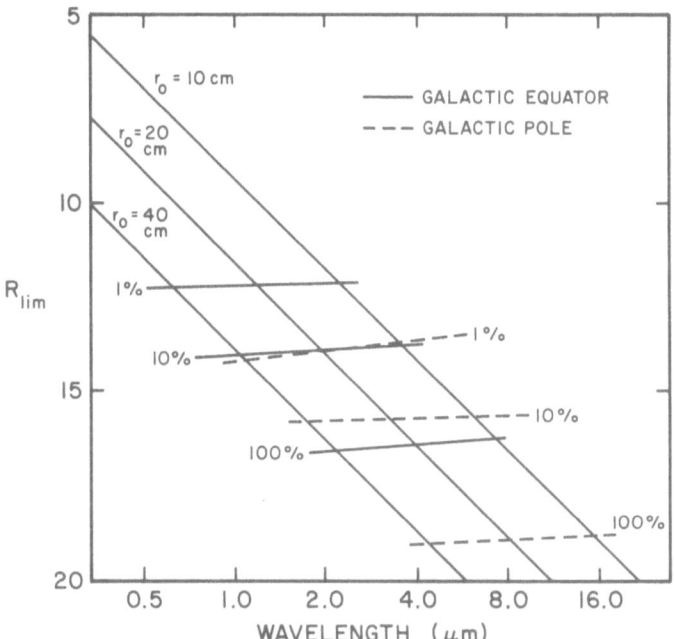

Figure 3: Change of limiting magnitude (R_{lim}) and sky coverage with wavelength.

noise-limited devices that can be used for rapid wavefront sensing use gallium arsenide intensifiers which cut off at 0.90µm. Recently, however, very sensitive array detectors have become available for the near infrared (1-5µm) which may be of eventual use for wavefront detection. The optimum wavelength for wavefront sensing appears to be the K-band (2.0-2.5µm) where the telescope and atmospheric thermal emission is still not a factor and where the detector quantum efficiency is very high (80%). For the integration times involved (<0.1 sec), the noise of the arrays is determined by readout noise. This is the case for both the InSb SBRC array and for the HgCdTe arrays from Rockwell.

Figure 4 uses the measured properties of the 58 x 62 InSb arrays. Because of the larger number of objects available for wavefront sensing, the K arrays compete very favorably with the R-band wavefront sensors. This is especially true in the galactic plane, where the sharp reduction of interstellar absorption causes a rapid increase in sensing stars and hence sky coverage. Since at these integration times the detector arrays are limited by readout noise rather than photon or dark current noise, substantial gains in sky coverage and hence astronomical utility might be expected in the future, as the readout noise is being reduced.

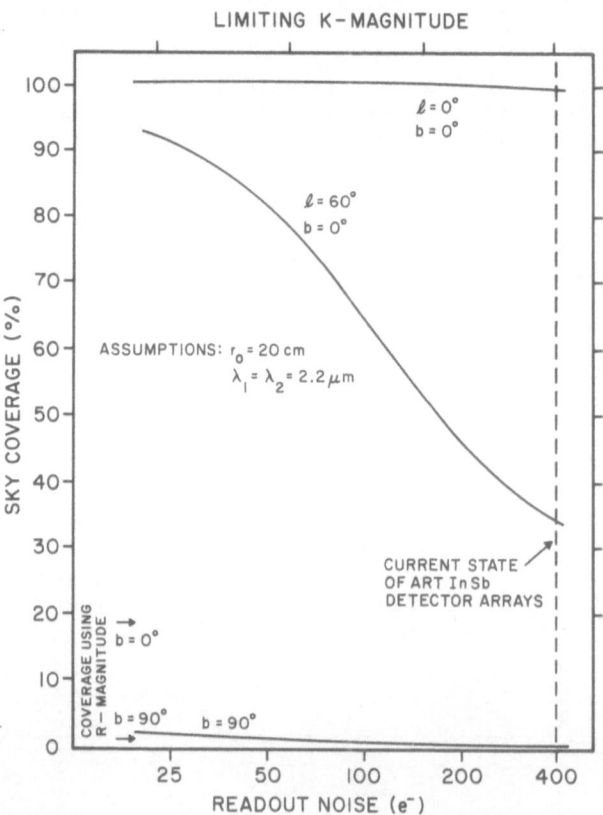

Figure 4: Sky coverage when using a wavefront sensor in the K band.

4. PREDICTED IMAGE IMPROVEMENTS

In a perfect adaptive optics system all the energy of a star image should become concentrated in an Airy disk corresponding to the diffraction of the circular telescope aperture. A number of factors limit the performance of adaptive optics systems. Most of these are listed in Table I and are discussed below.

TABLE I

Imperfections in Adaptive Optics Systems

Item	Resulting Strehl Ratio (in K-Band)
1. Amplitude variations (scintillation)	0.97
2. Chromatic effects ($\lambda_1 \neq \lambda_2$)	0.98
3. Refraction effects away from zenith	1.00
4. Deviations from isoplanicity	1.00–0.45
5. Finite corrector element size	0.80
6. Time delay between wavelength sensing and correction	1.00–0.90
7. Inaccuracies in wavefront sensing (photon noise)	
8. Imperfections in control algorithms	*
TOTAL	0.76–0.31

* to be evaluated

4.1 Amplitude Variations

Roddier and Roddier (1986) calculated the modulation transfer function (MTF) resulting from the uncorrected amplitude variations at different wavelengths. Their results are shown in figure 5. The effects are substantial at visible wavelengths but decrease rapidly with increasing wavelength. The typical MTF shows a decrease from 100% at low spatial frequency (the curve corresponding to the shape of the original seeing disk) to a level value at higher spatial frequency. That means that a sharp Airy disk will be superposed on a residual broad halo-like image with the shape of the seeing disk. We will refer to this as an A & S (Airy and Seeing) profile. Roddier et al. found that this A & S image shape is typical for most adaptive optics imperfections. It is probably due to the fact that the adaptive optics imperfections tend to have small spatial scales which cause the wings of the seeing/halo disk. The large spatial scale variations on the other hand are well corrected by the adaptive optics, causing the central Airy disk-like feature. The A & S image is shown schematically in figure 6. The so-called Strehl ratio (SR) is defined as the maximum intensity in the generally non-perfectly corrected image expressed in units of the maximum for a perfect Airy disk. For a large telescope the SR equals the value of the high-spatial-frequency plateau in the MTF. This SR is listed in Table I.

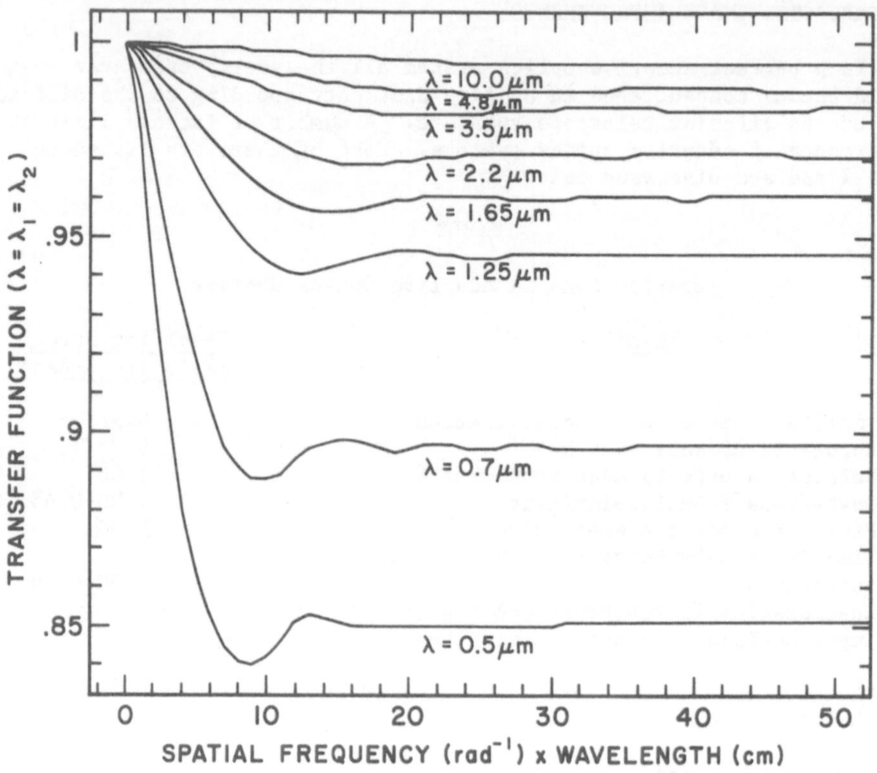

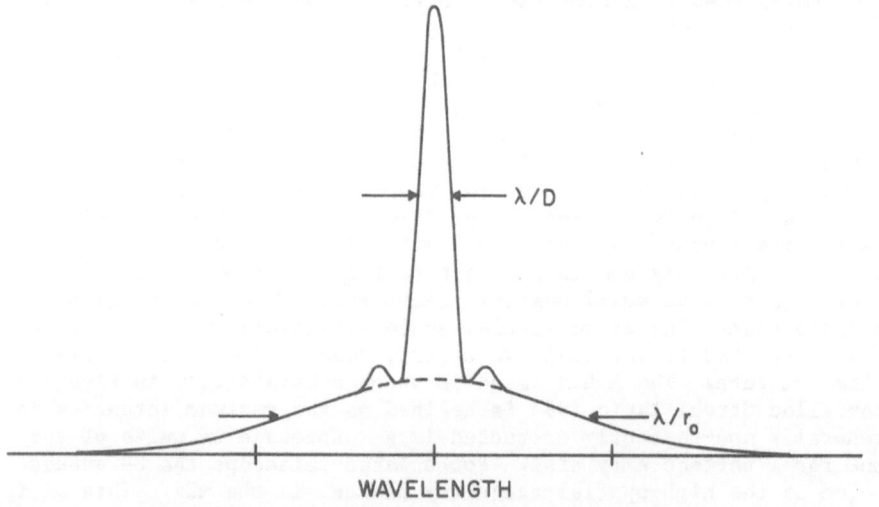

Figure 5: Modulation transfer function (MTF) resulting from uncorrected amplitude variations (scintillation) (from Roddier and Roddier, 1986).

Figure 6: Image shape of partially corrected image.

4.2 Chromatic Effects

Roddier and Roddier (1986) also calculated the MTF resulting from chromatic effects in the wavefront propagation when a different wavelength is used for wavefront measurement (λ_1) and astronomical observations (λ_2). Figure 7 shows their results for $\lambda_1 = 0.5\mu m$ and $\lambda_2 = 2.2\mu m$. It again results in an A & S type profile with a SR of ~ 0.98.

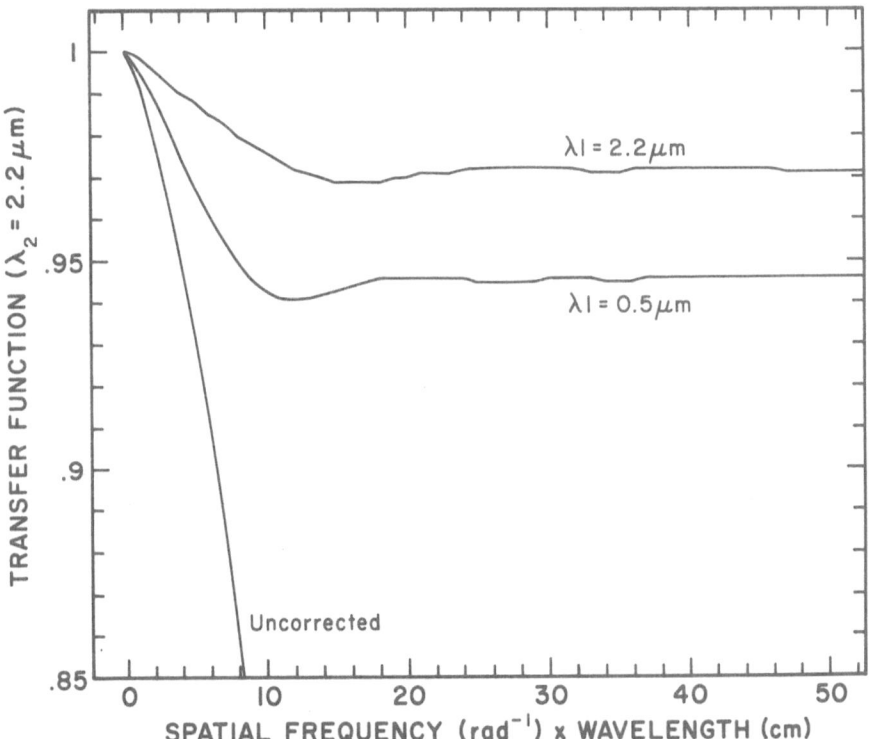

Figure 7: MTF resulting from the use of different wavelenths for wavefront sensing (λ_1) and observation (λ_2) (from Roddier and Roddier, 1986).

4.3 Refraction Effects

Away from zenith the paths through the atmosphere for two different wavelengths deviate slightly. It can easily be shown that this is a minor effect which can be ignored.

4.4 Deviations from Isoplanicity

Roddier and Roddier (1986) calculated the MTF for different angles θ between the sensing star and the object under study. Figure 8 shows the results for $r_0(0.5\mu m) = 10$ cm, $\bar{h} = 4$ km and $\theta = 10"$ and $20"$ at the K-band. The image profile again has close to an A & S shape although not as precise as for the effects discussed above. The central image is also somewhat anisotropic in the direction between object and reference star. The 10" case corresponds to the isoplanatic patch diameter Ω criterion given in equation (4), giving an SR of about 0.45.

At 20" the MTF levels off at 10% (i.e., the Strehl ratio is 10%). This is well outside the isoplanatic patch as defined before. A 10% response can either be viewed as bad or as significant depending on one's perspective and application. For an 8-meter telescope a 10% SR for $r_0 = 10$ cm corresponds to a diffraction-limited central spike with a width of 0.05 arcsecond at 2μm, 450 times brighter than the 1 arcsecond background seeing halo (which contains, however, 90% of the energy).

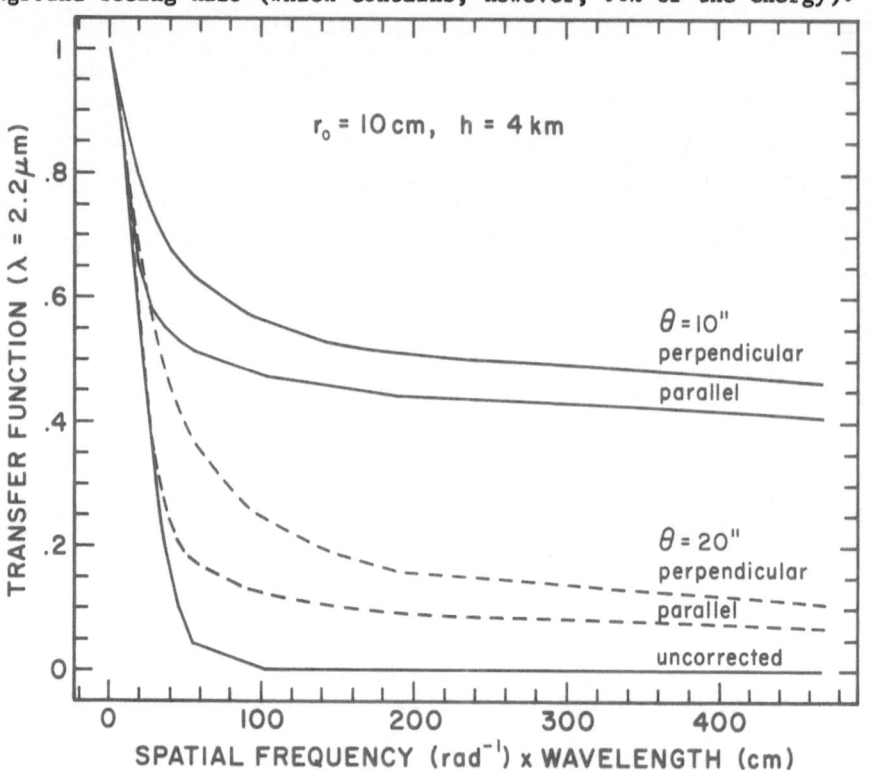

Figure 8: MTF resulting from deviations from isoplanicity. The diameter Ω of the isoplanatic patch at $\lambda_2 = 2.2$ μm equals 20 arcseconds so that $\theta = 10$ arcseconds lies at the boundary of the isoplanatic patch (from Roddier and Roddier, 1986).

4.5 Finite Corrector Element Size

Adaptive mirrors are generally constructed to remove wavefront phase variations, from small spatial frequencies up to frequencies corresponding to scales close to r_0. Higher spatial frequencies are not removed and spatial frequencies near r_0^{-1} are only partially corrected. Hudgin (1977) calculated the remaining wavefront errors for a number of different types of adaptive mirrors. Some of the results are reproduced in figure 9 which shows the mean square residual wavefront error as a function of r_s/r_o, where r_s approximately equals the width of the influence function of each actuated element of the adaptive mirror. Also shown in figure 9 is the Strehl ratio as calculated from the residual wave errors. The mirror that NOAO is using (section 5) has approximately a Gaussian influence function which for $r_s = r_o$ results in an RMS residual wavefront error of 0.08 waves and a Strehl ratio of 80%.

It is interesting to look at some simulations of the wavefront corrections by a segmented adaptive mirror, including both piston and tilt correction (see also figure 9) done by Smithson et al. (1987). Figure 10 shows the results of such a simulation for a mirror segment size of 15 cm under good (~0.5 arcseconds) medium (~1 arcseconds) and poor seeing (~2 arcseconds) conditions. The Strehl ratios for the corrected images are 88%, 60% and 14% respectively. One is struck again by the same behavior of the profile (A & S) as calculated for the previously discussed imperfections.

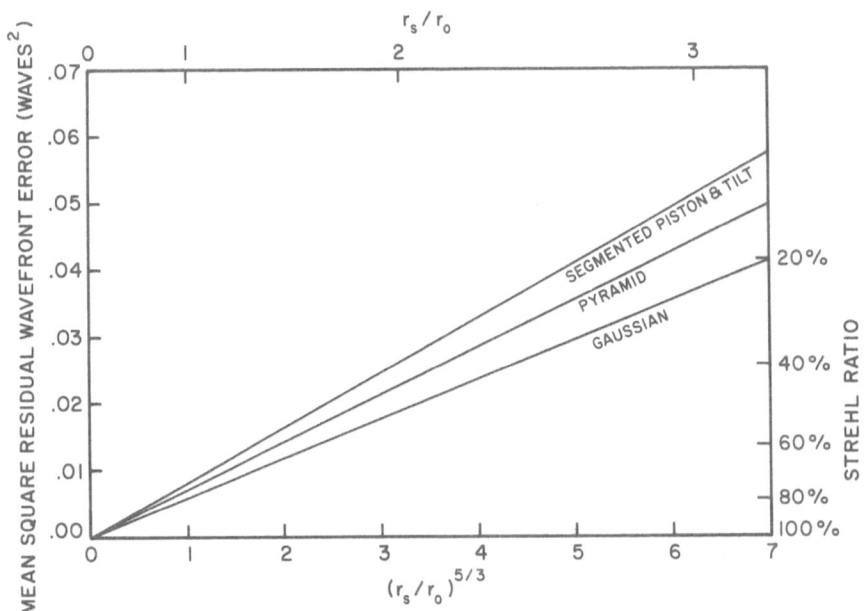

Figure 9: Residual wavefront errors and resulting Strehl ratios for different types of adaptive mirrors. The quantity r_s stands for the distance between the adaptive mirror actuators (from Hudgin 1977 with modification by Smithson 1985).

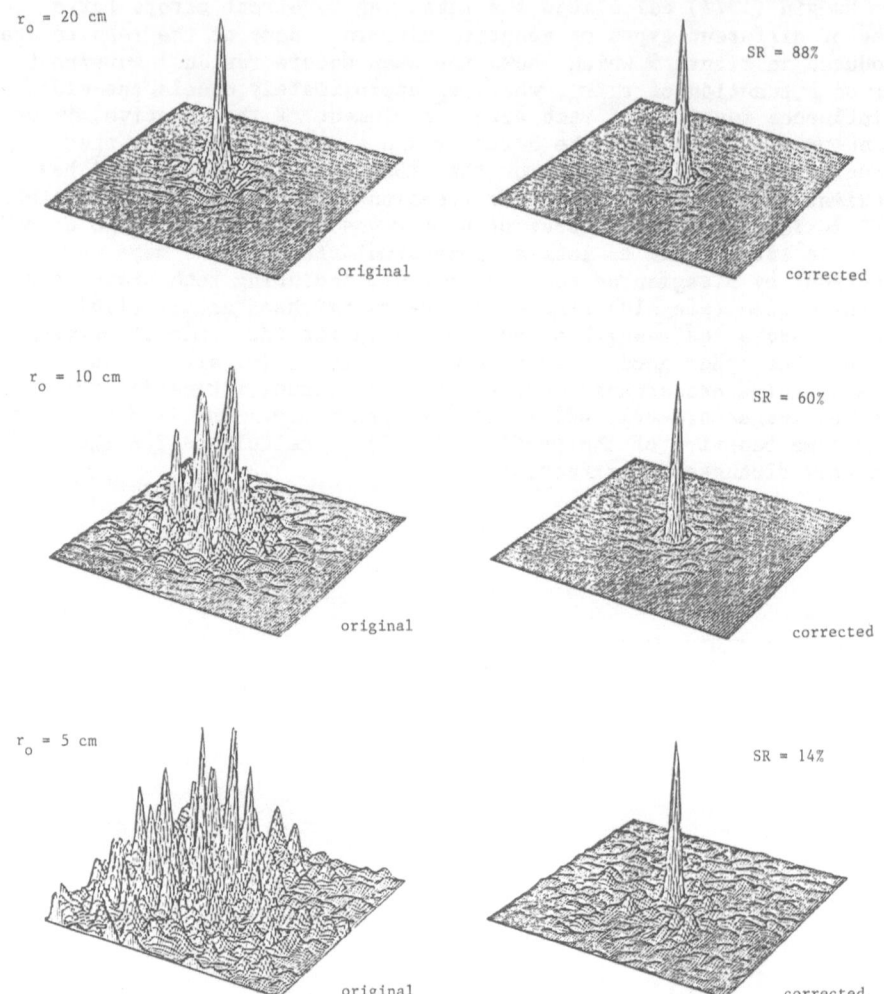

Figure 10: Modeling of image restorations using a segmented adaptive mirror with a segment size of 15 cm for different seeing conditions. The telescope aperture is taken to be 75 cm. (Courtesy R. Smithson).

4.6 Time Delay and Photon Noise Effects

Hardy (1982) discusses the effects of time delay and photon noise on the residual wavefront errors. The size of the errors depends on the wind velocity and on the brightness of the star used to sense the wavefront. Both are actually related, a higher wind velocity requiring shorter integration times and hence a brighter reference star. We refer to Hardy's paper for further information. It suffices to say that the limiting magnitudes used to estimate the sky coverage in section 3 are

bright enough, and the time constants short enough, to cause <0.05 wave
residual wavefront errors according to Hardy's calculations, resulting
in a Strehl ratio of >0.9. Although no simulations were made for the
shape of the corrected point spread function, we again expect it to be
of an A & S nature since both time delay and photon noise affect the
large spatial-scale structures less than the small-scale ones.

4.7   Imperfections in Control Algorithms

We have no information yet on the image deterioration due to limitations
in the control algorithms. One of the goals in our prototype astronomical adaptive optics system (section 5) is to develop optimum algorithms
for astronomical use. In it we will explore the use of different time
constants for large- and small-scale wavefront variations, and the
optimum treatment of photon noise for small-scale corrections, so that
the actual additional deteriorations of the wavefronts that can occur
with noisy signals (see, e.g., Hardy 1982) are avoided.

4.8   Conclusion

In Table I we summarize the estimated changes in the Strehl ratio
resulting from imperfections. These changes are time dependent, position dependent and dependent on the brightness of the star used for
sensing. The resulting A & S point spread function should therefore be
noisy. This makes adaptive optics a poor tool where absolute photometry
is needed. Instead adaptive optics will find its primary use in:

- morphology studies
- spectroscopy
- relative photometry over small areas of the sky
- the detection of faint pointlike sources against
  a bright background

We will return in section 6 to a discussion of the astronomical applications of adaptive optics.

5.   THE NOAO/ADP PROTOTYPE ADAPTIVE OPTICS SYSTEM

At NOAO we have started a pilot program to evaluate the applicability
and utility of polychromatic adaptive optics to astronomy. It is
described in detail by Beckers et al. (1986). This program is ultimately aimed at the application of adaptive optics to future 8-16 meter
telescopes. The pilot program is aimed at correcting the wavefront at
the McMath telescope (aperture 150 cm; 75 cm for the McMath auxiliary)
and the Mayall telescope (aperture 380 cm) on Kitt Peak.

5.1   Adaptive Mirror

Figure 11 gives the properties of the adaptive mirror. It was described
earlier by Hardy (1980). The mirror has a continuous faceplate, which
avoids diffraction effects that would occur at the edges of a segmented

<u>PROPERTIES OF ADAPTIVE MIRROR</u>

<u>MANUFACTURER</u>:	ITEK CORPORATION
<u>NUMBER OF ACTUATORS</u>:	37 + 18 EDGE ACTUATORS
<u>MAXIMUM DEFLECTION</u>:	± 5 MICRONS
<u>ACTUATOR SEPARATION</u>:	2.75 CM
<u>MIRROR DIAMETER</u>:	23 CM
<u>CONTROLLED DIAMETER</u>:	16.5 CM
<u>TYPE OF ACTUATOR</u>:	PZT STACK
<u>SENSITIVITY</u>:	0 005 μm/VOLT
<u>BANDWIDTH</u>:	DC - 150 HZ

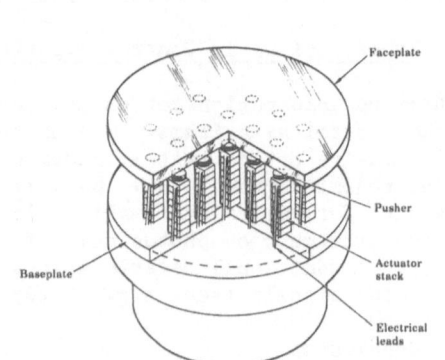

Figure 11: Properties of NOAO Adaptive Mirror.

mirror and which avoids thermal emissions and their variation due to the
segment gaps in the infrared. The mirror has 37 actuated elements with
an additional 18 independently-controlled edge elements used primarily
for edge support. This mirror has been extensively characterized and
calibrated at NOAO. It was found that ambient temperature variations
have a substantial effect on the mirror surface quality, because of
differential expansions between the actuators. These differences can
however be tuned out as long as the ambient temperature changes stay
within a range of ±20C. Although at low temporal frequencies we found
the actuators to exhibit hysteresis and creep, at the frequencies at
which we expect to operate the device the hysteresis effects are
undetectable.

The telescope pupil diameter will be covered by about 6 segments,
corresponding to r_s = 25 or 12.5 cm at the McMath telescope and 63 cm at
the Mayall telescope. For good (1 arcsecond) seeing on Kitt Peak these
r_s values correspond to r_o values at 1.07μm or 0.60μm wavelength at the
McMath telescope and 2.3μm at the Mayall telescope. These are therefore
the wavelengths to which the prototype system is being targeted.

5.2 Wavefront Sensor

A comparison by L. Goad et al. (1986) of different options for sensing
the wavefront led to the selection of a Hartmann-Shack device as the
wavefront tilt sensor. Figure 12 gives the properties of the device. A
lenslet array of 37 hexagonal lenslets 600μm apart, placed at a pupil
image, forms 37 star images on an intensified Reticon array. The intensifier also serves as the device that washes out the speckles in each
star image, reducing the number of detector pixels needed to centroid
the image.

PROPERTIES OF WAVEFRONT SENSOR

MANUFACTURER: ADAPTIVE OPTICS ASSOCIATES

TYPE: HARTMANN-SHACK TYPE

NUMBER OF SUBAPERTURES: 37 (EXPANDABLE TO 64)

SPEED: 600 HZ (350 HZ WITH MODAL CONTROL INCLUDED)

DETECTOR: IMAGE INTENSIFIER COUPLED TO 100 x 100 RETICON ARRAY OR
                       DIRECT INDIUM ANTIMONIDE IR ARRAY

SENSITIVITY: BRIGHT SOURCES: 1/40 OF FWHM OF IMAGE
                       FAINT SOURCES: PHOTON LIMIT

WAVELENGTH USED: FOR STELLAR WORK 0.8 µm OR 2.2 µm, FOR SOLAR WORK 0.5 µm

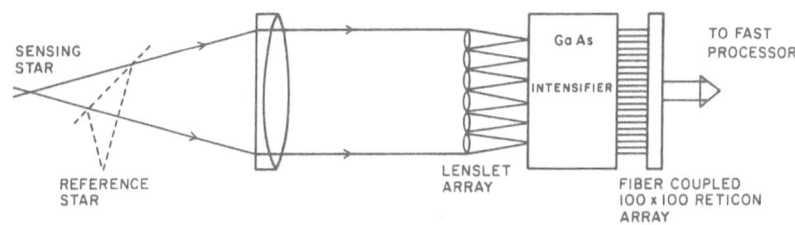

Figure 12:   Properties of NOAO Wavefront Sensor.

5.3 Optical Arrangement

Figure 13 shows the optical configuration used for the NOAO prototype
system. Off-axis Gregorian telescopes are used as the collimator and
the camera. A field lens near the entrance image I_1 images the pupil on
the adaptive mirror. A "tip-tilt mirror" is used both to remove overall
image motion and to fold the beam so that the off-axis aberrations of
the collimator and camera optics compensate. A dichroic beamsplitter
separates the visible from the infrared radiation. The telescope pupil
is imaged on the lenslet array for the visible radiation. Figure 14
shows the relative orientation of the pupil image on the hexagonal
lenslet array and on the PZT actuators of the adaptive mirror.

5.4 Status

The prototype adaptive optics system is now (mid-1987) being commis-
sioned on the McMath telescope. We expect that much of the initial
testing will be done with the McMath 75 cm auxiliary telescope at visible wavelengths. The 150 cm McMath main telescope will be used both for
solar observations in the near infrared, using small sunspots (pores) as
inverse stars to sense the wavefront, and for stellar observations.
Observations on the 380 cm Mayall telescope will emphasize the K-band

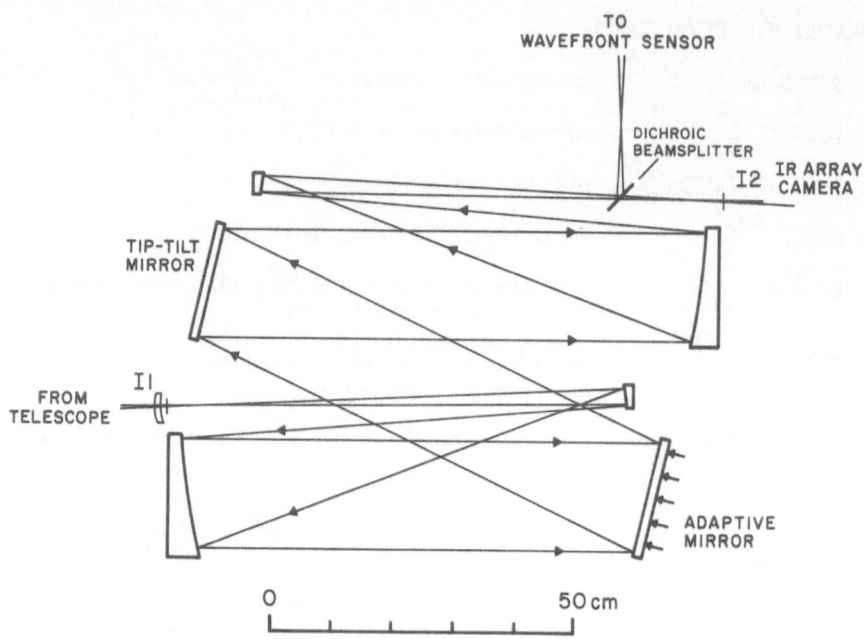

Figure 13: Optical configuration using NOAO adaptive mirror and wavefront sensor.

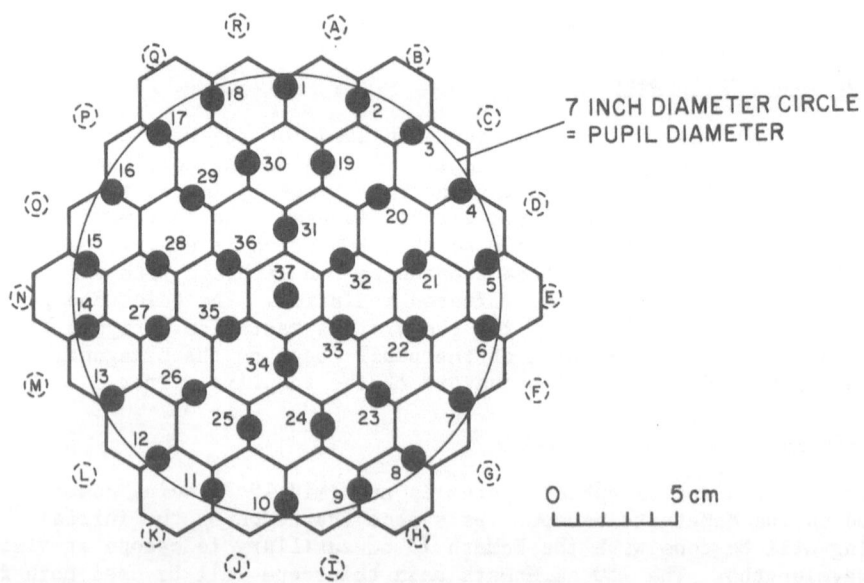

Figure 14: Relative position of telescope pupil on the adaptive mirror and wavefront sensor.

(2.2μm) and longer wavelengths. Initially the prototype system will be used there at the coudé focus, eventually it will be installed in the Cassegrain cage.

6. ASTRONOMICAL APPLICATIONS OF ADAPTIVE OPTICS

Astronomical subjects for study fall broadly into two classes. In the first class are those objects in which there is actually resolvable detail visible at the diffraction limit of the telescope that would be invisible under normal seeing-limited imaging. The second class includes objects that remain pointlike even at the diffraction limit. Adaptive optics helps in their observation by enhancing the contrast with respect to the background (e.g., supernovae in distant galaxies) or by allowing higher resolution spectroscopy.

A large number of astronomical objects fall in the first class. In our solar system this includes the sun itself with its wide variety of phenomena and physical processes occurring at a scale near 100 km (0.15 arcsecond). Observation of solar magnetic and velocity fields at 1.6μm are of special interest, because this is the only wavelength where the solar convection zone itself can be observed and because of the increased sensitivity of spectral lines to Zeeman splitting at the longer wavelengths. Other interesting objects in the solar system include satellites of Jupiter and Saturn, especially Io, the rings of Uranus, and a number of resolvable asteroids. In other, perhaps younger, solar systems like βPic, high angular resolution is of great interest to study early planet formation.

Adaptive optics will play a major role in the study of the properties and evolution of galaxies. In addition to allowing us to study the IR structure of our own Milky Way (see end section 3), adaptive optics will allow us to study the morphology of distant galaxies, and their mass and luminosity distribution. In Table II we calculate the number of distant galaxies within the part of the sky covered in the K- and M-bands, as well as the amount of scattered light expected from the sensing star if care is taken to keep the scattered light in the adaptive optics system to a minimum. A large number of V = 22 (z ~ 0.5) galaxies will be observable with adaptive optics provided that their K brightness is high enough to make them detectable against the K ~ 13 per square arcsecond sky background. Other important galaxy research will include the study of stellar properties and populations in nearby galaxies, the study of starbursts in bright IR galaxies, and the serendipitous imaging of galaxies near bright stars towards the galactic pole, for the systematic extension of galaxy number – magnitude – diameter statistics.

Table II also lists the number of QSO's observable within the fraction of the sky available to adaptive optics. Some AGN's are of course bright enough to serve as their own wavefront measurement source, including 3C273, NGC 1068 and NGC 4151. The same is true for a number of stellar objects in the process of formation (e.g., T Tauri), rapidly evolving AGB stars (e.g. NML Cyg), or compact HII regions.

TABLE II

Application of Adaptive Optics to Distant Galaxies
(for 0.5 arcsecond visual seeing).

	2.2μm (K)	5.0μm (M)
Wavelength		
Size point source (8m telescope)	0.055 arcsec	0.125 arcsec
Limiting R magnitude for sensing star	14.3	17.2
Diameter isoplanatic patch	39"	106"
Sky coverage at Galactic Pole	3%	52%
Number of galaxies between V = 21.5 and 22.5	2.5×10^5	4.4×10^6
Scattered light of sensing star* (magnitudes per square arcsecond)	27	32
Number of QSO's (assuming a total of 10^4)	300	5200
Scattered light of sensing star* (magnitudes per 8 meter diffraction disk)	34	36

*The scattered light was estimated from the data published by King (1971) assuming that the scattering is achromatic. The amount of scattered light most likely decreases with wavelength, so the scattered light values are overestimates. Values listed are at $\theta = \Omega/4$.

This summary of astronomical applications certainly does not exhaust the field. A gain in angular resolution of 10 (from 0.5 to 0.05 arcsecond) might be expected to provide at least as big a gain in observational astronomy as would a change from 10 to 1 arcseconds. Most astronomers these days would not bother observing with 10 arcsecond seeing. Although we don't expect that to be the case soon for 0.5 arcsecond, we do expect that 0.05 arcsecond images will result in a rich harvest.

7. USE OF ARTIFICIAL STARS FOR ADAPTIVE OPTICS

Foy and Laberie (1985) proposed the use of an artificial reference star for atmospheric wavefront sensing. Such an artificial star could be generated by atmospheric scattering of the radiation of a laser, tuned to one of the Na-D lines, off the neutral sodium layer found at an altitude of ~ 100 km above the earth's surface. With a sufficiently powerful laser (~ 10W) such a scattering spot would be bright enough to be usable for wavefront sensing by a Shack-Hartmann detector. The advantage over using real stars is of course that the star can be placed anywhere on the sky, thus increasing the sky coverage to 100% at all wavelengths and in all directions. This technique would also eliminate scattered light since the artificial star is monochromatic. Thompson et al. (1987) have carried out a preliminary experiment of this kind, with promising results. If it can be made to work, this technique will be of great benefit to astronomy. Complexities remain to be worked out, including the differences in ray paths through the atmosphere for stars and for the artificial reference star (which is at a finite distance). In principle that can be done. And then there is the cost of a 6400 element adaptive mirror and wavefront sensor if such a system were used at visible wavelengths on an 8-meter telescope with 1 arcsecond seeing.

8. CONCLUSION

We expect several astronomical adaptive optics systems to be implemented in the near future (ESO, the National Solar Observatory, NOAO/ADP, the Center for Astrophysics). It will be very exciting to follow this rapidly developing field. Most of the present plans emphasize rather conventional techniques compared with what is available in military technology. Unconventional adaptive optics, for example using phase conjugation, are usable only for narrow-band sources and are therefore of little interest to astronomers. Other techniques for astronomy are being explored, however, such as the development of adaptive mirrors and wavefront sensors using wavefront curvature rather than displacement and tilt (Roddier 1987).

J. Grace and J. Goad have been of invaluable help in the preparation of this manuscript. M. Cook and J. DuHamel professionally prepared the figures. Comments from many astronomers helped in the formation of this paper. We especially thank Bob Smithson for allowing us to use his results in this review.

9. REFERENCES

Beckers, J.M., Roddier, F.J., Eisenhardt, P.R., Goad, L.E., and Shu, K.L. 1986, Proceedings SPIE, **628**, 290.

Foy, R. and Labeyrie, A. 1985, Astron. and Astrophys., 152, L29.

Goad, L., Roddier, F., Beckers, J., Eisenhardt, P. 1986, Proceedings SPIE, **628**, 305.

Hardy, J. 1980, in Optical and Infrared Telescopes for the 1990's, 535.

Hardy, J.H. 1982, Proceedings SPIE, **332**, 252.

Hudgin, R. 1977, J. Opt. Soc. America, **67**, 393.

338

King, I. 1971, Publ. Astr. Soc. Pacific, **83**, 199.
Lelievre, G. 1987, Proceedings 9th Santa Cruz Workshop, Springer-Verlag, (in preparation).

Pearson, E. and Stepp, L.M. 1987, Proceedings SPIE, **817**, in preparation.

Roddier, F. 1981, Progress in Optics XIX (Ed. E. Wolff), pages 281-376.

Roddier, F. 1987, Private Communication.

Roddier, F. and Eisenhardt, P. 1986, Proceedings SPIE, **628**, 314.

Roddier, F. and Roddier, C. 1986, Proceedings SPIE, **628**, 298.

Smithson, R.C. 1985, Private Communication.

Smithson, R. and Peri, M. 1987, Private Communication.

Tarenghi, M. 1986, Proceedings SPIE, **628**, 213.

Thompson, L.A. and Gardner, C.S. 1987 (Preprint submitted to Nature).

APPLICATIONS OF THE TULLY-FISHER RELATION AT HIGH REDSHIFT

P.C. van der Kruit
A.J. Pickles
Kapteyn Astronomical Institute
P.O. Box 800
9700 AV Groningen
The Netherlands

ABSTRACT. We discuss the possibilities of using the velocity width – integrated luminosity relation for spiral galaxies at cosmologically interesting redshifts. Our estimates and discussion show that questions of cosmological nature can be usefully addressed by use of optical lines such as Hα, when observed as integrated profiles on a 16-m ground-based telescope such as ESO's VLT.

1.   INTRODUCTION AND AIMS

The velocity width-integrated luminosity relation for spiral galaxies has since its discovery by Tully and Fisher (1977) been accepted as an important tool in deriving distances to galaxies. A large number of studies have been devoted to its calibration on nearby samples, to derivation of distances to nearby groups, the Virgo cluster and of global values for the Hubble constant, and to extension towards redder wavelengths where absorption is less and contribution from young stellar populations minimal. A current highlight is the use of the infrared version of the Tully-Fisher relation (from now on in short TFR) is the study of Aaronson et al. (1986) in which distances to ten galaxy clusters were measured and fundamental motion of the Local Supercluster with respect to the microwave background established.

   Almost all current versions rely on integrated HI profiles for the measurement of the velocity width. Observational convenience clearly favours radio as an application in practice. However this also excludes the use of the TFR for cosmological purposes: Current radio telescopes lack the sensitivity to measure HI profiles at larger distances, while tunability of the receivers and interference in the radio spectrum also limits in practice use to small radial velocities. Only the Arecibo dish with its large collecting area and extensively tunable receivers can reach out to redshifts of order 0.05 with ease and somewhat larger with serious difficulty.

   There are three general area's in which the TFR can be useful at high redshift:

R. G. Kron and A. Renzini (eds.), Towards Understanding Galaxies at Large Redshift, 339–347.
© 1988 by Kluwer Academic Publishers.

- Curvature of the universe. This is the ancient method of using the Hubble Diagram (redshift versus apparent magnitude) to determine q_0 (see e.g. Sandage, 1961, 1972). The major advantage of using the TFR on spirals over first-ranked cluster ellipticals is that the TFR just requires the relation to hold at larger z (or at least be correctable for luminosity evolution) and does not require a particular type of galaxy (e.g. cD's, first-ranked gE's, etc.) to be a standard candle.
- Streaming motion in the Universe. General streamings over large volumes can be studied by e.g. taking a shell of galaxies at the same distance. This can give valuable information on the clustering structure and large-scale mass distribution in the Universe.
- Morphological (and luminosity) evolution of spiral galaxies. As detailed below this is part of the application of the TFR to the problems just mentioned through checks and corrections.

As an example the following will be related specifically to the first problem, i.e. the problem of measuring q_0. From figures presented by e.g. Sandage (1961, 1972) it is obvious that one already needs to go to $z \sim 0.4$ for marginal results and to $z \sim 1$ for more definite conclusions. This can be quantified with the Friedmann equation of luminosity distance

$$D_{lum} = c/q_0{}^2 \, H_0 \, \{q_0 z + (q_0 - 1) \, [(1 + 2q_0 z)^{\frac{1}{2}} - 1]\}.$$

At $z = 0.5$ the difference Δm in apparent magnitude between q_0 equal to 0.1 and 0.5 amounts to only 0.21 mag and increases to a more comfortable 0.40 mag at $z = 1$. Clearly Δm should be measured out to $z \sim 1$, for meaningful results.

2. TFR AT LOW REDSHIFT

Aaronson and Mould (1986) give the calibration for the HI velocity width-infrared luminosity relation as follows

$$H_{-0.5}^{c,abs} = 21.05 - 11.18 \, (\log \Delta V_{20}^c - 2.5) + 7.5 \, (\log \Delta V_{20}^c - 2.5)^2.$$

The spread around this relation is about 0.2 mag. From this study there appears to be little, if any dependance on morphological type. It shows the basic validity of the relation, although the calibration is still uncertain due to local calibration problems while the aperture correction of the H-magnitude above still depends on the diameters in the B-band as given in the RC2. Clearly improvements are still possible, especially by use of IR array-detectors for the photometry.

Rubin (1983) has derived the TFR from observed rotation curves and gives the results in terms of the maximum measured rotation velocity V_{max}. Using B-magnitudes she finds in her Fig. 7 different relations as a function of Hubble type:

Sa: $M_B = 3.65 - 10.1 \log V_{max}$ ($M_B = -19.6$ for $V_{max} = 200$)

Sb: $M_B = 2.11 - 10.0 \log V_{max}$ ($M_B = -20.9$ for $V_{max} = 200$)

Sc: $M_B = 3.31 - 11.0 \log V_{max}$ ($M_B = -22.0$ for $V_{max} = 200$)

From the above it follows that one should aim at infrared magnitudes in order to circumvent the dependance on morphological type.

Before turning to higher redshifts we need to address first the question of the definition of the linewidth ΔV. In the radio the integrated profile usually displays the well-known "double horn" structure and a common definition of ΔV is to measure the velocity width at 20% of the peak intensity in the profile. Can this be done also in optical lines?

The shape of the integrated profile depends both on the rotation curve and the radial density distribution. HI is usually distributed in an annular structure with a central depression and the rotation velocity is constant over the range in galactocentric radius where the HI is abundant. HII regions on the other hand often peak towards the centre (at least in Sc galaxies) and optical profiles could be different due to this. However, qualitatively we note the following: (1) Sb galaxies like M31 and our Galaxy also have central area's devoid of HII regions. (2) Long slit spectra (e.g. Rubin, 1983; Fig. 1) show essentially flat rotation curves over almost the whole extent where optical emission lines are recorded. Noting then that rotation curves usually reach V_{max} within a fraction of an optical disk scalelength h (e.g. Fall and Efstathiou, 1980) and that at least in Sc-galaxies the exponential scalelength of HII region number density is similar to that of the total light, it follows that in the optical we also expect double horned integrated profiles.

We can make this more quantitative with a simple analytical excersize: Assume a surface brightness of line emission $\sigma(R) = \sigma_0 \exp(-R/h)$. Now ignore first effects of velocity dispersion and take two extreme cases: (A) Solid body rotation $V_{rot} = (V_1/R_1)R$ out to R_{max} the "edge" of the HII-region disk. The integrated profile (edge-on) for $R_{max} = R_1 = ah$ then is

$$I(V)dV = 2(R_1/V_1)\sigma_0 h dV \int_{aV/V_1}^{a} y e^{-y} \{y^2 - (aV/V_1)^2\}^{\frac{1}{2}} dy,$$

which for $a \to \infty$ converges to

$$I(V)dV = 2\sigma_0 (R_1/V_1)^2 V K_1(R_1 V/hV_1)dV.$$

It has a sharp peak at the systemic velocity $V = 0$ and is illustrated for the case $R_1 = 3h$ in Fig. 1 with the dashed curve labelled SB.
(B) Flat rotation $V_{rot} = V_1 =$ constant and the HII disk extending from R_1 to R_{max}:

$$I(V)dV = 2h\sigma_0(V_1^2 - V^2)^{\frac{1}{2}} dV\{e^{-R_1/h}(R_1 + h) - e^{R_{max}/h}(R_{max} + h)\}.$$

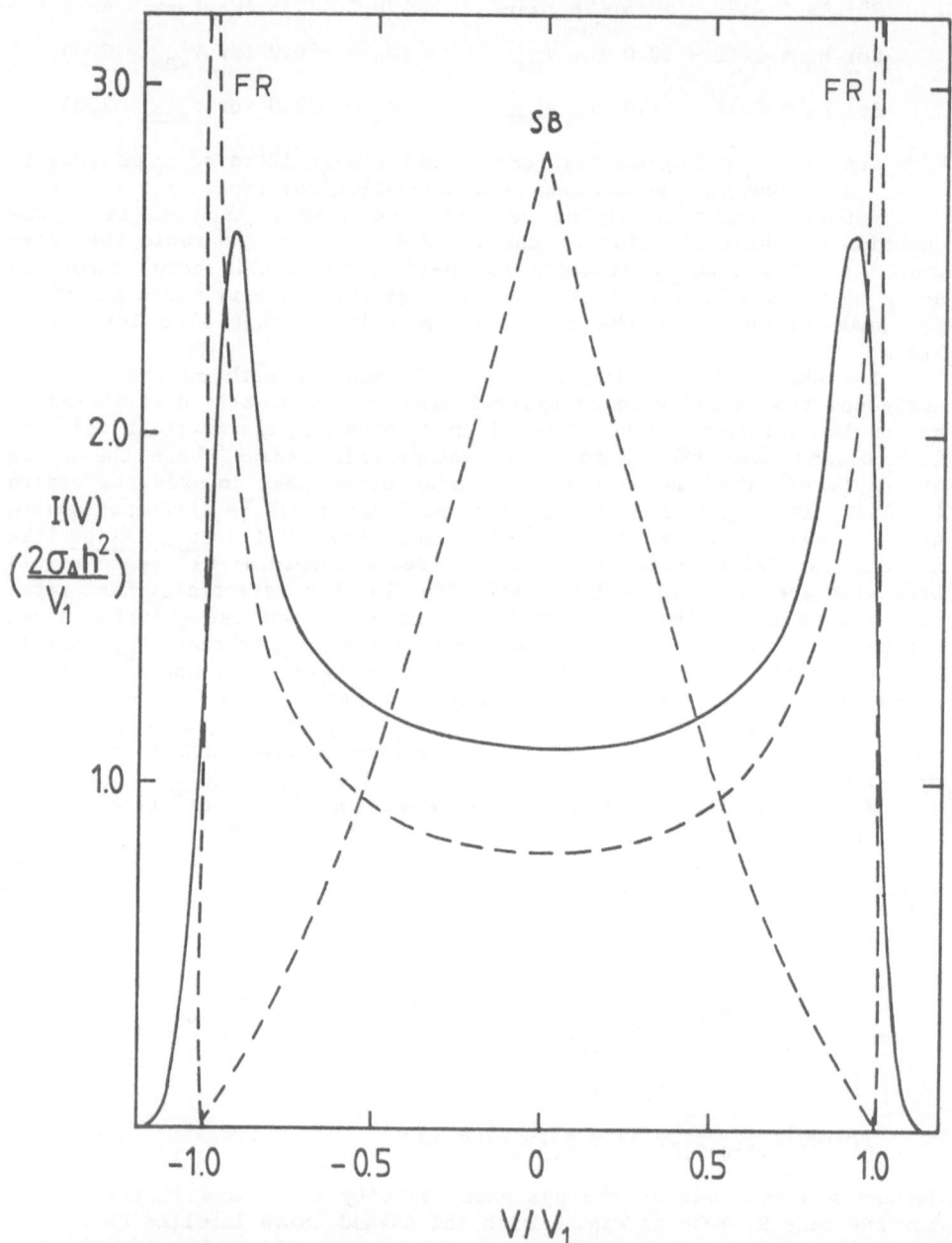

Fig. 1. Integrated, global Hα profiles of spiral galaxies. See text for explanations.

This curve has two "horns" and is shown - labelled FR - in Fig. 1 for the case $R_1 = 0$ and $R_{max} = 3h$.
It is clear that even for the exponential HII distribution, as long as the rotation cuve is flat over most of the observable extent, the profile will be similar to that at 21 cm. The full-drawn curve is for a rotation curve that rises as solid body from 0 to V_1 for R from 0 to $R_1 = 0.5h$ and is flat at V_1 beyond R_1 out to infinity. It has been calculated for a velocity dispersion of $0.1V_1$. This curve should approximate the actual expected observations.

3. PROSPECTS FOR HIGH REDSHIFTS

In this section we will make estimates of the expected flux rates of optical line photons and the required telescope-exposure time combinations. Of course Hβ is possible for $z < 1$ in the wavelength range shorter than 1μ. However flux considerations probably call for Hα which then is to be observed over the range 0.66 to about 1.3μ.

As our basic number we take the typical integrated flux of Hα + [NII] for Sbc and Sc galaxies from Kennicutt and Kent (1983) as 10^{-11} to 10^{-12} erg cm^{-2} sec^{-1}. Assume that 75% is Hα and we get

$$F(H\alpha) \sim 4 \times 10^{52} \qquad \text{photons per second.}$$

This flux has to be observed in velocity channels of about 20 km/s or so wide in order to reliably define the velocity width. The arriving flux in number of photons depends on the angular distance

$$D_{diam} = zc(1+z)^{-2}q_0^{-2}H_0^{-1} \{q_0 z + (q_0-1) [(1+2q_0 z)^{\frac{1}{2}} - 1]\}.$$

For definitness we will use $H_0 = 100$ km s^{-1} Mpc^{-1} and a typical observed velocity width of 250 km s^{-1}. The following flux rate estimates in photons m^{-2} sec^{-1} are then found for the total galaxy

z	λ_{obs} (μ)	Δλ (A)	$q_0 = 0.1$ D_{diam} (Mpc)	f	$q_0 = 0.1$ D_{diam} (Mpc)	f
0.01	0.65	5.5	60	960	60	960
0.2	0.79	6.5	910	4.0	870	4.4
0.5	0.98	8.5	1620	1.2	1470	1.4
0.1	1.31	11	2110	0.8	1760	1.0

The dependance of f on q_0 is minor and will be ignored in the following. The next thing is to compare this to fluxes from the sky background (f_{sky} in photons m^{-2} A^{-1} sec^{-1} arcsec^{-2}):

z	f_{gal}	μ_{sky} (mag arcsec^{-2})	f_{sky}
0.01	10^3	R = 20	0.03
0.2	4	I = 19	0.1
0.5	1.2	I = 19	0.1
1.0	0.8	J = 18	0.5

In order to calculate signal-to-noise ratios S/N we will assume a 4 arcsec^2 aperture, 20 km/s velocity channels (Res = 15000) and a continuum contrast of 10%. With ε the efficiency of the system and L, C and S the line, continuum and sky fluxes and R the readout noise, we have

$$S/N = \varepsilon Lt/(\varepsilon Ct + \varepsilon Lt + \varepsilon ST + R^2)^{\frac{1}{2}},$$

and with $\varepsilon \sim 0.1$ and assuming $R^2 \sim \varepsilon St$ typically this reduces to

$$S/N \sim \varepsilon^{\frac{1}{2}} t^{\frac{1}{2}} L /(11L + 2S)^{\frac{1}{2}}.$$

From this we now estimate the time T in hours needed to arrive at a S/N of 20 necessary for an adequate definition of the profile. We do this for a 4-m telescope and a 16-m equivalent telescope such as ESO's VLT:

z	line	cont	sky	T(4-m)	T(16-m)	$\lambda_{obs}(\mu)$
0.2	6	60	3	2.5	0.3	0.79
0.5	1.8	18	3	11	10.4	0.98
1.0	1.2	12	15	37	5	1.31

The conclusion is clearly that one needs a telescope like the VLT for a cosmological project involving the TFR to become feasible.

4. PROCEDURE AND CORRECTIONS

We will discuss in this section how to proceed in a project that aims at the determination of q_0 from an application of the TFR at $z \sim 1$. First we discuss the selection of objects, then the acquisition of the multi-object, spatially unresolved spectra and finally the correction procedures.

We start from the assumption that a sample of galaxy clusters out to a redshift of about unity is available. The first task is to find the galaxies that are (late-type) spirals and the diagnosis for this is the presence of Hα emission. For this purpose we need to survey the cluster at a spectral resolution of about 500 either with multi-object spectrograph, objective prism or scanning Fabry-Perot techniques. This will also give the approximate redshift of each spiral and foreground objects may be deleted at this stage. Then all spirals will need to be observed at a spectral resolution R > 10,000 in order to obtain the integrated line profile. We estimated above that even on a 16-m class telescope the required S/N prescribes an exposure time of order 5 hours and it is clear that some procedure – presumably the use of fibres – will be necessary to observe a number of galaxies simultaneously and make the project feasible in terms of telescope time.

Also one needs to do galaxies at lower redshift and local calibrations and for such systems it is not in general possible to observe the whole galaxy in a single resolution element. For this, scanning Fabry-Perot or scanning slit techniques are required to synthesize the integrated Hα profile. Of course a large fraction of this work can be performed on smaller telescopes, although we showed above that still 2.5 hours is necessary for an instantaneous measurement (no slit or wavelength scanning) on a 4-m telescope at z = 0.2. Also for these spatially resolved galaxies fibre techniques are less efficient in observing a number of galaxies at the same time.

Then photometry has to be done at near IR wavelengths to provide H-magnitudes. This involves some spectral coverage in order to correct the observed fluxes for redshift (K-correction) and relate all data to a standard band at zero redshift.

We believe that it will be necessary to observe as a minimum of order a dozen galaxies in each of a dozen clusters at various redshifts. For a solid and definitive result an even larger number of galaxies and redshifts is desirable, but clearly even a modest program requires an important allocation of time on next-generation very large telescopes.

We now turn to the various corrections that need to be made to the data in order to derive the absolute H-magnitudes and velocity widths.

- Aperture corrections to the observed magnitudes are required, especially at intermediate redshifts where the spirals may have angular extents of order 10 arcsec. This is of course most easily solved by observing with infrared imaging arrays which are now becoming available (e.g. IRCAM on UKIRT) and should soon acquire photometric stability and uniformity at low light levels.

- In order to avoid problems of internal absorption and morphology (see below) it is necessary to work in the IR. As stated above intermediate band colours or coarse spectral synthesis are required in order to apply K-corrections. In principle the photometric work can all or almost all be done using existing 4-m class telescopes.

- Inclination corrections are vital to correct the observed velocity width to edge-on view (full rotation in the line of sight). One should note that 20 kpc corresponds to about 2 arcsec at z = 1. There are four possible approaches: (1) Good or very good seeing measurements, which of course can be done in principle also in the optical. However this might not give the required accuracy at z = 1, also because one needs to check the influence of a significant bulge contribution to the light distribution, which can seriously affect the disk isophotes for large inclinations (near to edge-on). (ii) There should not be any problem if imaging with the Hubble Space Telescope were available. It is unlikely that this will be the case for the whole sample. (iii) Techniques for diffraction limited IR imaging may be developed to become practical at the required level. (iv) Finally, one may simply take a statistical approach by increasing the sample and assume random inclinations.

- Morphology has - as we have seen above - an effect on the TFR in the optical, but apparently gives little dependance in the infrared.
- Evolution clearly is the last certain effect. This is of course also the case for ellipticals in classical approaches, but there we certainly have a complicated star formation history (e.g. Pickles, 1985 and Ellis, O'Connell and Gunn in this volume) and more seriously, this history is important also for the question of whether these galaxies are indeed standard candles. For spirals the luminosity evolution and star formation history depends on (present) morphological type. In models with constant star formation a typical galaxy increases its L_V by a factor 3 between ages 1 and 2 Ggr, while its $(B-V)$ changes from about 0.25 to 0.55 (see e.g. Larson and Tinsley, 1978). These effects are clearly less in the IR. By using observed colours and $H\alpha$ fluxes per unit luminosity one should be able to derive a luminosity correction to the required accuracy of 0.1 mag. Since this requires a value for q_0 (and H_0) it must be done in iterative fashion.

5. CONCLUSION

We believe we have shown above that interesting cosmological questions can be addressed by application of the TFR at high redshift. We emphasize that this is a dynamical approach which is in principle better understood than the mechanisms behind the production of first-ranked elliptical galaxies in clusters as standard candles. It is true that a modest program described here requires a fairly large sample of galaxies, high-resolution imaging and probably prohibitively large quantities of time on 4-m telescopes. Definitive work with a very large sample of galaxies is feasible only on next-generation telescopes, but even then still needs much time, probably as a dedicated allocation of a major fraction of the time in a few years period. With very large samples it will however be possible to map systematic motions in the Universe over large volumes out to redshifts up to unity and that is material of fundamental nature.

PCK thanks the Institute of Astronomy, Cambridge for hospitality during the writing of this chapter.

REFERENCES

Aaronson, M., Bothun, G., Mould, J., Huchra, J., Schommer, R.A., Cornell, M.E. 1986, Ap. J. 302, 536
Aarsonson, M., Mould, J. 1986, Ap. J. 303,
Fall, S.M., Efstathiou, G. 1980, Mont. Not. R.A.S. 193, 189
Kennicutt, R.G., Kent, S.M. 1983, Astron. J. 88, 1094
Larson, R.B., Tinsley, B.M. 1978, Ap. J. 219, 46
Pickles, A.J. 1985, Ap. J. 296, 340

Rubin, V.C. 1983, IAU Symp. 100: Internal Kinematics and Dynamics
    of Galaxies, ed. L. Athanasoula (Dordrecht: Reidel), p. 3
Sandage, A.R. 1961, Ap. J. 133, 355
Sandage, A. 1972, Ap. J. 178, 1
Tully, R.B., Fisher, J.R. 1977, Astron. Astrophys. 54, 661